A HISTORY OF LACTIC ACID MAKING

CHEMISTS AND CHEMISTRY

A series of books devoted to the examination of the history and development of chemistry from its early emergence as a separate discipline to the present day. The series will describe the personalities, processes, theoretical and technical advances which have shaped our current understanding of chemical science.

A HISTORY OF LACTIC ACID MAKING

A Chapter in the History of Biotechnology

by

H. Benninga

KLUWER ACADEMIC PUBLISHERS

DORDRECHT / BOSTON / LONDON

Library of Congress Cataloging-in-Publication Data

Benninga, H. (Harm)
A history of lactic acid making : a chapter in the history of biotechnology / by H. Benninga.
p. cm. -- (Chemists and chemistry)
ISBN 0-7923-0625-2 (alk. paper)
1. Lactic acid industry--History. 2. Lactic acid--Biotechology--History. I. Title. II. Series.
TP248.L3B46 1990
660'.63--dc20 89-71683

ISBN 0-7923-0625-2

Published by Kluwer Academic Publishers,
P.O. Box 17, 3300 AA Dordrecht, The Netherlands.

Kluwer Academic Publishers incorporates
the publishing programmes of
D. Reidel, Martinus Nijhoff, Dr W. Junk and MTP Press.

Sold and distributed in the U.S.A. and Canada
by Kluwer Academic Publishers,
101 Philip Drive, Norwell, MA 02061, U.S.A.

In all other countries, sold and distributed
by Kluwer Academic Publishers,
P.O. Box 322, 3300 AH Dordrecht, The Netherlands.

Printed on acid-free paper

Printed in The Netherlands

TABLE OF CONTENTS

FOREWORD

It has been a pleasure for the CSM biochemical division and its working company CCA biochem to support the realization of this book on the history of lactic acid making by dr. Harm Benninga.

Dr. Benninga's partial early retirement created a unique opportunity for this historical review of the production of lactic acid. The combination of his professional experience in the lactic acid industry and his profound personal interest in history made him the most suitable person to cover this subject.

We congratulate dr. Benninga with the final result.

Gorinchem, October 23, 1989

CSM biochemical division
dr. J.A.J. VINK
managing director

PREFACE

The most striking feature of our age seems to be the impact of technology on almost every aspect of present-day life. Anyone who wants to acquire even a basic understanding of the events surrounding him (or her) is continually faced with the effects of technology and technological change.

The discipline which professes to help us best in understanding the time we live in is history. It cannot be said, however, that general historians, or for that matter, business historians pay more than casual attention to the impact of technology on the evolution of modern society. The interaction between technological "progress" and historical developments is a subject that is rarely studied.

The lack of knowledge of science and technology explains the minor attention given to this interaction by professional historians, but does not provide a justification. In a time when further progress of technology is dreaded by many, the influence of technology on the evolution of modern society deserves greater study and analysis than is usually given[1].

The impact of technology on society takes place through the intermediacy of the industry; so far as chemical engineering is concerned, the chemical industry is the intermediate body.

The history of technology has many overlaps with economic history: many (new) technologies have, in the past, had tremendous impact on industry and society. The other side of the picture is that changes in economic and political conditions have often spelled success or failure for a particular business or branch of industry.

What industry does and achieves is, however, not solely influenced by events taking place in society. Quite frequently, the activities of individual chemists and other workers of the industry who successfully applied current scientific ideas and new or adapted technologies to further their business, had unexpected and long-lasting effects on society.

It is, in particular, this interaction between science, technology and the personal approach and perseverance of the individual chemist and entrepreneur which is seldom treated in a satisfactory manner. Company historians tend to overemphasize the achievements of their particular enterprise. Interaction with society is sometimes not treated at all[2]. For business historians, it is often difficult to assess the role of technology in the interaction between a single company and society when the history of the technology at issue has not been accounted for in a readily accessible manner.

It is here that the individual chemist and worker in industry who believes that the interaction between industry and society warrants more interest could consider a contribution. This certainly applies when he has been at a crossroads of interaction between his particular branch of industry and society as a whole.

The belief that the impact of chemistry, in my case, that of biochemistry and biotechnology on modern society deserves much more attention than usually given, formed one reason why I embarked on this study. Such a study of the history of the lactic acid fermentation industry might serve as a small contribution to the general study of the historic interactions between (bio)technology and modern western society[3].

Lactic acid itself is a fascinating product, since prehistoric times it has been unwittingly used for preserving food products and, even for instance in the case of buttermilk, for industrial purposes. Its chemistry seems simple, but has, historically, posed large problems. The study of these problems implied the elaboration of large-scale laboratory preparation of the acid.

Furthermore, the manufacture of lactic acid is, in terms of fermentation technology and down-stream processing, a complicated matter involving many subdisciplines of the science of chemistry. The applications of lactic acid and a number of its derivatives encompass many fields of industrial activity and important aspects of food manufacture and use.

After my retirement from the position of managing director of a leading producer of lactic acid, I was reasonably confident that my experience in the manufacture and use of lactic acid made me sufficiently qualified to put the history of lactic acid making into a general historic context. This aim turned out to be much more difficult to realize than I envisaged when I started this study. Many hours of research were required to collect the historic material, many more hours to find relations between the events in this branch of industry and what happened elsewhere in science and technology, in politics, and in business.

The topics which I endeavored to cover encompass:

- the Industrial Revolution and its aftermath;
- the role of lactic acid in the history of chemistry and chemical engineering;
- the development of biotechnology and microbiology;

- lactic acid as a mordanting aid in the history of textile chemistry;
- the development of education in chemistry during the nineteenth century;
- the influence of governments on farm economy, industrial developments, and education;
- fermentation industries in the two World Wars;
- the rise of petrochemistry.

A basic guideline in my study has been the assumption that very few ideas, philosophies, inventions, and scientific theories emerge from nowhere, that there has, as a rule, always been somebody who prepared the way for a new development. I have searched as systematically as possible to discover which events led to new developments in the history of lactic acid making and in the applications of lactic acid.

Many times I felt that I was working on a three-dimensional jigsaw puzzle of which too many pieces were missing. Today, writing this preface, and after having finished the manuscript, I am aware of the fact that large gaps still exist in my account of events. Due to these gaps, the story is far from complete. Space and readability also dictated a presentation from different angles – different cuts through the three-dimensional jigsaw puzzle.

The first three chapters of the book treat the early history of the chemical substance lactic acid, how it was discovered, and made on a laboratory scale as well as in the small works of nineteenth-century pharmacists. Attention has also been paid to the development of the biotechnology required to make lactic acid on a commercial scale by fermentation.

The bulk of the story is concerned with finding markets for lactic acid and its salts and esters, and to the development of suitable production technologies to offer the most economical product with the right specifications. A certain amount of chemistry and technology is therefore an essential part of this book. I have tried to avoid the more difficult matters while treating these subjects, or otherwise relegated them have to special intermezzos which may be skipped by the non-initiated or less interested reader.

A separate intermezzo is devoted to the history of raw material supply, an essential topic in any study of the history of a fermentation industry.

Many libraries I have visited, from the east coast of the United States to the south of Germany and the Lorraine region of France. I owe very much to a great number of unknown library assistants who did not become tired or bored of helping me in collecting missing pieces of the great jigsaw puzzle. Without the aid of Jerome E. Anderson of the New England Historic Genealogical Society of Boston and Nancy Bradbury of the Littleton

Historical Society, I would not have been able to write a coherent, though still incomplete story of the Avery Lactate Company, the first manufacturer of lactic acid. The invaluable help I received from the second manufacturer, C.H. Boehringer Sohn of Ingelheim, Germany, is attested by a large number of references to internal documents of the company. The story of the use of lactylic esters in bread has been mainly provided by Larry F. Marnett who directed the research efforts of the C.J. Patterson Company in the development of dough conditioners. I am also grateful to many colleagues of the trade, to coworkers of mine, as well as to (former) personnel of a number of companies – C.H. Boehringer Sohn, CCA biochem, Clinton Corn Processing and (Croda) Bowmans – who have freely given information and advice, as well as putting archival and photographic material at my disposal. The information and assistance given by the Firmenarchiv of E. Merck, Darmstadt, and the Hagley Museum and Library (Du Pont) has also been of much help.

Scientific advice – on some of the chapters – has been provided by H.A.M. Snelders of the University of Utrecht, by E. Homburg of the Catholic University of Nijmegen and by R. Otto of CCA biochem, Gorinchem. W. Hornix, also of the Catholic University of Nijmegen, was so kind as to read the whole manuscript. His comments have resulted in essential improvements in the text.

Further acknowledgements go to M.H.M. Arnold, Mrs. L.N. Beattie, P. Bennecke, Mrs. M.E. Bowden, Mrs. L.T. Brueck, X. Casas, J.D.F. Companus, W. Deckers, W. Dehio, U. Dobler, M.J. Eerelman (for the preparation of the illustrations), Mrs. C. Ekdahl, C. Geitenbeek, Mrs. C. Gabrielli, J.B. van Gelder, E.C.N. Glimsholt, W. Goehtz, A.A. Hintze, Mrs. E. Jensen, L. van der Hoest, E. Keck, A. Ketterer, P.K. Kuipers, Mrs. E.S.K. Lim, W. Ludwig, F.J. van Maurik, R. Meindertsma, H. Metz, P.J.T. Morris, G. Nanninga, Mrs. M.L. Neighbour, J.M. Newton, K.G. de Noord, B.W. Peckham, W. Pleser, T.F. Reid, L. Osés, C. Parrott, Mrs. I. Possehl, Y. Shigematsu, Mrs. S.C. Swanson, C. den Uyl, J.A. van Velthuysen, J.A.J. Vink, W.F. Vogel, G. Vreeman, Mrs. C.K. van der Water, K. Yasumatsu, G.M. van Zanten, S. van der Zon, and the Institut für Gärungsgewerbe (Berlin).

Special thanks go to the managements of CSM biochemical division and CCA biochem b.v. for their generous financial and secretarial support as well as for allowing the free use of their files. Mrs. M. Schaap-Vermeer deserves much gratitude for preparing the final manuscript from my drafts which I kept "reprocessing" almost until the last minute.

Notes

[1] A recent and very knowledgeable presentation of the way in which scientific ideas and technological inventions have influenced and, in many cases, completely overturned our society is given by James Burke in "Connections" (MacMillan: London, 1979).
[2] Ernst Bäumler's book about the history of the Hoechst Company ("A Century of Chemistry"; Econ Verlag: Düsseldorf, 1968) is an illustrative example of the contrary: How the interaction between one company and world events can be put into an objective historic setting.
[3] Williams Haynes has, in "American Chemical Industry" (Six Volumes; D. Van Nostrand: New York, 1954, reprinted by Garland Publishing: New York, 1983), presented a masterly account of the history of the chemical industry in the United States from colonial times until the outbreak of World War II. This work, in which the interactions between the development of the chemical industry and other events in (American) society systematically treated, were has served the author as an invaluable example for this study.

Another book which I have consulted quite frequently, in particular on the subject of cartels, is Lutz. F. Haber's "The Chemical Industry, 1900–1930" (Clarendon Press: Oxford, 1971). This book is also an admirable example of the manner in which interactions between the chemical industry and society can be presented.

H. BENNINGA

CHAPTER 1

LACTIC ACID IN EARLY ORGANIC CHEMISTRY (1780–1840)

Lactic acid, in modern chemical terminology 2-hydroxypropionic acid ($C_3H_6O_3$), is a comparatively simple molecule. Today the study of the properties of lactic acid belongs to the realm of organic chemistry. It occurs as the acid ingredient of sour dairy products, fermented fruits and vegetables, and sausages, and as such mankind has known and used lactic acid since prehistoric times.

Lactic acid as a product of nature is even older than that. It occurs in nearly every form of organized life and supposedly was already present in the first forms of primitive life that existed on earth (Brin, 1965). The most important function in animals and men is related to the supply of energy to muscle tissue. Metabolic turnover of lactic acid in adult men has been estimated to be of the order of 120 to 150 g per day (FDA, 1978).

The discovery of lactic acid as a chemical substance was the achievement of a famous chemist, Carl Wilhelm Scheele, who lived in the middle of the eighteenth century. He could have been called the world's first organic chemist, were it not for lack of definition of the term organic chemistry.

Chemistry itself, that is the study of the properties of substances and their interactions, was on the verge of becoming a science in the modern sense of the word when Scheele made his investigations. Of course, practical chemistry like metallurgy, ceramics, the preparation of plant extracts, fermentation of plant and animal products and the tanning of hides, to name a few, had existed for ages, but none of these activities was placed in what could be called a scientific frame.

Then there was alchemy, a weird concoction of mysticism, scientific principles and secretiveness. Alchemy first of all also meant experimentation, laboratory practice and inventions ranging from sulfuric acid, mercury and saltpeter to plant elixirs and the like. It is this laboratory practice which attracted young men like Scheele and many others. It is from this laboratory practice that the science of chemistry gradually evolved in the 18th century.

The first scholar who seriously questioned the methods and theories of the alchemists was the Anglo-Irish chemist and philosopher Robert Boyle (1627–1691). Boyle exerted a profound influence on the future of chemistry through his experimental work (e.g. Boyle's Law), and in particular through his book "The Sceptical Chymist" published in 1661. This book already proclaims the device to which all chemists (and scientists) of good standing have adhered since:

Conclusions can only be drawn from experiments, never from unproven statements.

Boyle was a cofounder of The Royal Society of London for Improving Natural Knowledge, soon to become known as the Royal Society (1660). Another early member of the Royal Society was Isaac Newton (1643–1727), who was even more influential in establishing the foundations of the sciences of physics, mathematics and astronomy, than Boyle was in the science of chemistry. The Royal Society exerted much influence in Europe: Foreign inventors and scientists reported their inventions and observations to The Society and eagerly awaited the comments from the learned people in London. One of these men was Antonie van Leeuwenhoek, who reported in 1676 his observation of very tiny animals ("animalcules") he had seen under one of his simple self-made microscopes. Four years later Leeuwenhoek, who can be considered the first microscopist ever, was made a (corresponding) member of the Royal Society.

Production of chemicals at that time was very limited and primarily intended for use in metallurgical enterprises. Some special products were made for medical purposes in apothecaries, some of which also supplied "chemicals" to other users like alchemists. In fact it was sometimes difficult to distinguish between alchemists and apothecaries. The names of the various preparations, elixirs and tinctures were very exotic, the methods of preparation and the recommendations for use were described in the terms and symbols of the alchemists.

An early producer of chemicals and pharmaceutical preparations was Johann Rudolph Glauber (1604–1670), a German apothecary who established a manufacturing "plant" for chemicals in Amsterdam. He offered 700 different products for delivery throughout Europe; one of his products was "sal mirabile Glauber", crystallized sodium sulfate.

Starting with Boyle, chemists gradually stopped using Latin as their prime means of communication. The eighteenth century was the age of the Enlightenment, the age in which the ideals to liberate "the people" became a widespread movement in Europe: Liberation from feudalism, from superstition and from the restraints of the guilds system. One of the aspects of this liberation movement was the promotion of ordinary, "national" languages to vehicles of science and learning. The Enlightenment aimed at opening the intellectual, political and cultural life for everybody of good will, of good intellect and, not to forget, of appropriate means.

The impetus of the new movement was overwhelming, particularly in France. In 1751 the first volume of the "Encyclopédie ou Dictionnaire raisonné des sciences, des arts et des métiers", was published by Diderot and d'Alembert. This encyclopedia in 28 volumes gave a detailed and well illustrated account of the state of art, industry and technology of the time. It also fully endorsed and advocated the ideas of the Enlightenment, the idea of universal human progress and it treated traditional beliefs and doctrines with much scepticism. The views of the Encyclopedists in turn had been much influenced by the rather tolerant atmosphere in which new ideas in science and philosophy were generally received in England. Some of the Encyclopedists had experienced this tolerant atmosphere first hand when living there in exile.

The Industrial Revolution and Chemistry

In the beginning, the Industrial Revolution was an almost exclusively British affair. The long term impact of the Industrial Revolution would be even greater than that of the Enlightenment, if a comparison is allowed.

The ideas of the Enlightenment were very influential in the American uprising against the British (1776) and in the French Revolution of 1789. The armies of the Revolution and of Napoleon carried the Enlightenment ideas together with French legal and political institutions over most of continental Europe.

Napoleon's defeat came in the battles of Leipzig (1813) and, after his return from Elba, of Waterloo (1815). The economic strength and naval superiority of Great Britain were decisive factors in bringing down the French Empire. The ideas of the Enlightenment, although the members of the continental Holy Alliance tried to suppress them, were to stay, however. "Egalité et Liberté", equal rights and liberty would foster revolutionary spirit in Europe during the whole of the nineteenth century. Feudal opposition to these ideas formed one of the incentives for the large scale emigration of Europeans to the New World where freedom and liberty had become constitutionally enshrined rights.

In most books concerning the history of this period, the role of the new chemical industry is underestimated or completely neglected. The large scale production of what today are called heavy chemicals had its beginnings in the same period in which Richard Arkwright invented the spinning frame (1768),

James Watt the first workable steam engine (1769) and Edmund Cartwright, the first mechanical loom (1784).

The Industrial Revolution started in the textile industry and most of its major inventions were, in one way or another, related to the textile trade. Until the middle of the eighteenth century all common textile products were made of wool or linen; silk was a very exotic and expensive material and cotton was only available either rather cheap in the raw unbleached condition or in the bleached form, almost as expensive as silk. In the latter state it was put on the market as calico, cotton cloth, imported from Calicut, India.

Thanks to the new technologies the price of finished cotton fabrics dropped in less than 50 years to values lower than 10% of the price of say 1770. So cotton became in the nineteenth century the staple product for the clothing of mankind, and the country that produced by far the biggest quantities was Great Britain.

An important factor in the realization of the mass production of cheap cotton was played by the emerging chemical industry.

Sulfuric Acid Replaces Buttermilk in Textile Bleaching

Until well into the eighteenth century bleaching of linen and cotton was done in the open air, on bleaching meadows. The tissues to be bleached were first boiled in a dilute solution of potash (obtained by large scale burning of wood followed by leaching of the wood ashes!), then expanded on the meadow exposed to wet air and sunshine. Optimal conditions for this process were difficult to find and certainly did not exist in Great Britain, where sufficient wet air was usually present, but sunshine only occasionally. After the bleaching the cloth was soaked for two weeks or more in fermented sour milk to neutralize the potash. The first industrial application of lactic acid?

Nobody knew about lactic acid, few people were aware of the fact that an acid was needed to remove the potash. A Scottish physician by the name of Home, however, did know and by logical reasoning assumed that another acid might be more effective than sour milk. He tried sulfuric acid and found that with a dilute solution of this acid the cleaning time could be reduced from over two weeks to 24 hours or less. This important discovery led to a large increase in the demand of sulfuric acid, a product which until that time had only been manufactured on a very limited scale.

Another British physician, John Roebuck (1718–1794), took up the challenge to produce sulfuric acid in a more economical way. Today it may sound strange that physicians showed so much interest in chemical problems, but this was a consequence of the educational system: the only place to receive a university training in chemistry or pharmacy was in the faculty of medicine! Until Roebuck, sulfuric was made by burning sulfur in the presence of saltpeter with air in spherical glass vessels. Roebuck had learned from one of the books of Glauber that lead was highly resistant to sulfuric acid and it occurred to him that a large lead-plated chamber might be well-suited to convert sulfur into sulfuric acid on a much larger scale. In 1746 he founded in Birmingham the first "modern" sulfuric acid plant in the history of the chemical industry.

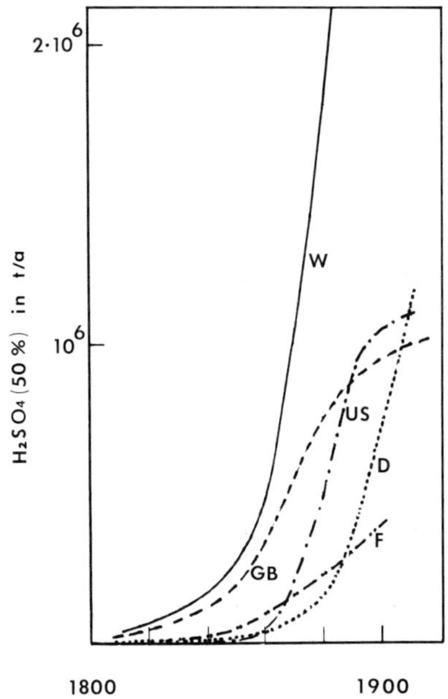

Fig. 1,1. Growth of sulfuric acid production in the nineteenth century.
GB = Great Britain; F = France; US = United States of America; D = Germany; W = world total.

It took Roebuck's company about 40 years to develop the lead chamber process to a reliable technology. In this period of time other companies in Great Britain and on the continent also began to produce sulfuric acid, the chemical workhorse of the new era.

Soon mass production of other chemicals followed, made possible by the availability of cheap sulfuric acid. In the first decades of the nineteenth century large scale production of soda, muriatic acid [= hydrochloric acid], bleaching powder and quite a few other chemicals was realized, which in turn helped to produce new or cheaper consumer goods. Of particular importance for the textile industry became the use of bleaching powder; thanks to this product chemical bleaching could be done much more cheaply in England than bleaching in the sun in India.

The new chemical industry had, due to its scale of production by necessity to master many technological problems in these initial years, problems not only related to the size of the operations, but also typical chemical problems like corrosion of production equipment and problems related to the disposal of obnoxious and poisonous by-products. The production of heavy chemicals like sulfuric acid is a good measure of the state of the chemical industry of a nation, at least during the nineteenth century. Figure 1,1 provides a comparison between the developments in the main industrialized nations of that century.

Carl Wilhelm Scheele (1742–1786)

At the time Roebuck experimented in the first sulfuric acid factories in England and Scotland, a young boy, who was to become the first organic chemist in history, went to school in Stralsund in Swedish Pomerania, today part of the G.D.R. (German Democratic Republic). In 1757 the boy, Carl Wilhelm Scheele, then 15 years of age, travelled to the Swedish mainland to be engaged as an apprentice to an apothecary. Scheele was not particularly interested in becoming an apothecary himself. He preferred to spend most of his time on chemical experimenting. He commenced with repeating the experiments he found in various books, but being a critical young man and a gifted experimenter he soon became aware of the errors and inadequacies in the older works. Knowingly or not of Boyle's device he learned to accept only the results of good experimentation as the basis for scientific considerations.

Until this time practical chemistry was more or less identical with the use of crude and drastic methods to change the properties of substances, methods like dry distillation, reduction with charcoal, etc., which were quite useful in metallurgy and in what we now call inorganic chemistry. Scheele developed more gentle methods to isolate and study substances from the vegetable and animal kingdoms. One of his first discoveries was tartaric acid, in 1769. The method for the preparation of this acid was a fairly simple one and cheap too. Because of this and because of its properties the new product soon became listed in recipe books for apothecaries, the pharmacopoeia. The recipe given in the first national Dutch pharmacopoeia is presented below.

Preparation of tartaric acid
(Bataafse Apotheek, 1807)

Dissolve 3 pounds of tartar in 15 pounds of boiling water, slowly add fine chalk, cool to room temperature. Collect the precipitate and add 1 pound of sulfuric acid, diluted with 10 pounds of rainwater, control for excess of sulfate. Separate the crystals [of gypsum] from the liquid, concentrate the liquid until crystallization [of tartaric acid] sets in. Cool, separate and wash the crystals, dry on blotting paper.

A few years later, in 1773, Scheele discovered the gas which was later named chlorine as a product of the reaction of pyrolusite [manganese dioxide] with muriatic acid [hydrochloric acid]. He also obtained chlorine by reacting pyrolusite with common salt and sulfuric acid, which became the accepted reaction for the industrial production of the gas. Scheele studied the properties of chlorine extensively and noticed as one of the properties the ability of chlorine to bleach plant leaves and flowers. This observation led to the development (by French and English chemists) of industrial bleaching processes, first for linen and later for cotton as well.

1780: Scheele Discovers Lactic Acid

The discovery of lactic acid was done in 1780; the exhibit below presents a translation of the original article (Dobbin, 1931). This article is difficult to read because of the unfamiliar names of the chemical compounds involved. A modern translation is therefore printed alongside.

I evaporated sour whey till about one eighth of it remained, when, meanwhile, all the curd separated from it: then I filtered the acid. In order next to obtain animal earth from it, in view of the fact that lime water precipitates this variety of earth … I saw no other way than to saturate the acid with lime. After this was done, I filtered the

"Translation": Sour whey is evaporated to one eighth of its volume, the curd [precipitated milk proteins] is then removed by filtration. The filtrate is saturated with milk of lime [calcium hydroxide], filtered again and diluted with three times its volume of water. [In this step a crude solution of calcium lactate, admixed with salts, amino acids

solution and diluted it with three times as much water. The acid of sugar was an excellent medium to obtain the lime now from its menstruum. I dissolved some in water and afterwards added it to the solution of lime until no more saccharated lime was precipitated, in doing which I specially took care that too much acid of sugar was not added, which could be easily ascertained by means of lime water. There now still remained the other substances which ought also to be separated from the acid of milk. I therefore evaporated the acid until it became as thick as honey. I then dissolved this inspissated acid again in the most highly rectified spirit of wine by which means the milk-sugar as well as the other foreign admixtures not pertaining to the acid were separated and the acid, consequently, was alone dissolved in the spirit of wine, which I then filtered. With this sour solution I mixed some pure water and again distilled off the spirit, when the acid of milk remained behind in the retort as pure as, in my opinion, it can be obtained by the aid of chemistry.

Scheele, quoted from Dobbin

and soluble peptides is obtained]. Oxalic acid (= acid of sugar) is slowly added until no further precipitation of insoluble calcium oxalate occurs. [The lactic acid is set free again in this reaction]. Scheele controls the endpoint of the reaction with an aqueous solution of calcium hydroxide. (After filtration) the lactic acid solution is concentrated to a viscous consistency. Separation from the other substances still contained in the solution is then effected by taking it up in concentrated alcohol, in which [most of] the impurities do not dissolve. After filtration some pure water is added and the alcohol removed by evaporation. The lactic acid produced is as pure as can be obtained by chemistry.

Scheele named the new acid after its origin Mjölksyra (acid of milk). In the eighteenth century every investigator had his own ideas about namegiving, no generally accepted rules existed. Sometimes this led to confusion and misunderstanding.

Scheele died in 1786, only 44 years old. In his short life he had discovered a surprisingly large number of substances, later recognized as elements or compounds. Besides the organic acids already mentioned he discovered malic acid, citric acid and a number of other acids. Citric acid he obtained as calcium citrate by adding powdered chalk to boiling lemon juice until frothing stopped. After separation by filtration from the precipitate it was converted with sulfuric acid into citric acid and gypsum. The acid finally was purified by crystallization. It typifies Scheele's experimental ingenuity that he immediately recognized an application for the new acid:

When one quintin [about 4 g?] of it is ground with six lod [about 15 g] of sugar in a glass mortar, a delicious lemonade powder is obtained, which does not become moist: two spoonfuls of this are dissolved in one quarter of spring water; and as lemon lozenges in apothecaries' shops are much disposed to deliquesce, this powder can also be used instead of the lozenges.

Dobbin

When Scheele started his experimental studies the prevailing theory of chemistry was the phlogiston theory. This theory presupposed that all combustible material contains phlogiston, an hypothetical substance congenial to fire. Carbon (charcoal) was considered to "contain" much phlogiston. Cavendish, who discovered inflammable air [= hydrogen] in 1766 thought that this gas was identical with phlogiston and the same view was held by Scheele. An example of the phlogiston theory is the explanation of the burning of lead:
When lead is burned it loses phlogiston and formes lead calx [= lead oxide]. By heating the calx with charcoal phlogiston is taken up again and lead is recuperated.

Antoine Laurent Lavoisier, the Father of Modern Chemistry

It was Lavoisier who demonstrated the fallacy of the phlogiston theory: when lead is burned it does not lose something like the phlogistonians believed, but it gains weight, which Lavoisier could prove in carefully designed experiments. He also showed that the increase in weight is due to the absorption of a part of the air during calcination. That part, originally named "vital air" was recognized by Lavoisier as an elementary substance, which he called "oxygen". The product of the calcination of lead should not be called a calx, but being a reaction product of lead with oxygen he proposed the name lead oxide, and the reaction he called an oxidation reaction. These and similar experiments of Lavoisier were based on the law of the conservation of matter, which he expressed explicitly:

The total mass of the substances taking part in a reaction is not changed by that reaction.

It was also Lavoisier (1743–1794) who proposed that a system should be worked out for the namegiving of all chemical compounds. He found a fellow countryman, the chemist Guyton de Morveau willing to join him in this task. In 1878 the new nomenclature was published, Lavoisier himself used it in his famous book “Traité élémentaire de Chimie” (1789). Thanks to the wide circulation of this book, in which Lavoisier’s new revolutionary ideas were comprised and used to explain the existing chemical knowledge, the new nomenclature rapidly became known and accepted by the chemical profession. Its principles have been applied ever since.

An example of the use of the new nomenclature: The “Traité” called Scheele’s acid of milk “acide lactique” whereas the salts are termed lactates.

Lavoisier borrows in his book Scheele’s recipe for the preparation of lactic acid. He also mentions a large number of lactates, many of which had already been prepared and described by Scheele (Dobbin, 1931). Scheele for instance had prepared a salt from vegetable alkali, which was deliquescent, an iron salt which was brown and did not crystallize, and crystalline zinc lactate (Brin, 1965).

Lavoisier, like Scheele, was an excellent experimenter. He was director of the French state gun powder works and a very wealthy man in the Ancien Régime, who could afford to have the best possible laboratory equipment built from his own resources. In the true spirit of the French Enlightenment and the intellectual atmosphere of the decades preceding the Great Revolution Lavoisier was an advocate of the idea that everything could and should be rationalized, be measured and reduced to quantitative formulas. He applied these ideas in his experimental studies and presented them in the “Traité”. Lavoisier’s work, borne by the spirit of the Encyclopédists, together with equally pioneering work by other French scientists in physics and mathematics, laid the basis for the dominant position French science achieved during the revolutionary period and the Napoleonic era.

With the “Traité” a beginning was made with the transformation of chemistry to a quantitative science, a science in which everything that could be measured also was measured. For this contribution to chemistry alone Lavoisier could already have earned the title “father of chemistry”; his other contributions strongly support the opinion that modern chemistry had its beginning with this outstanding French scientist. This, unfortunately, was not recognized by the revolutionary government of Robespierre. In 1794 Lavoisier was brought to trial, accused of conspiracy against the Revolution, although the main reason was probably jealousy: one of the leaders of the Terror Régime, Jean Paul Marat, had also written a book about chemistry, but this book was completely overshadowed by the “Traité”. Lavoisier was found guilty, his friends asked for mercy, but the judge said:

> France has no use for scientists.

The execution with the guillotine took place on the eighth of May, 1794.

Fortunately the "Terror" did not last very long and common sense came back to France. Not much later the ascent of Napoleon begun, who was an admirer of the intellectual ideas of the Revolution and a great promoter of science and technology.

In chemistry the lead was taken by a scientist worthy to be called the successor of Lavoisier: Joseph Louis Gay-Lussac (1778–1850), who already at the age of 30 was appointed member of the prestigious official body of French Science, the "Académie des Sciences". As professor at the École Polytechnique he attracted the best students from all over France and many from the rest of continental Europe as well. The importance of quantitative measurements was, of course, a main issue in the study of natural sciences at the École.

Berzelius About Organic Chemistry

Outside France the most influential advocate of Lavoisier's ideas was the Swedish chemist Jöns Jacob Berzelius (1779–1848). He made many contributions to the new science of chemistry:

- he introduced the present system of chemical symbols (H for hydrogen, C for carbon, O for oxygen and so on)
- he invented or improved the equipment used in the laboratory (from test-tubes and the use of filter paper to distillation vessels and fractionating columns)
- he wrote a new textbook of chemistry, based on the ideas of Lavoisier's Traité, but offering a wealth of additional information thanks to the rapid progress in chemistry. The first volume of the book appeared in 1808 in Sweden, a German translation, by Friedrich Wöhler, was published in eight volumes from 1820–1831.

Berzelius had repeated Scheele's recipe for the extraction of lactic acid from sour milk, he also found the acid in meat and other animal products (1808). At one time Berzelius believed lactic acid to be an individual compound and not a mixture of acetic acid with some impurities. But in some of his later writings he was more vague about the composition of lactic acid. Berzelius also prepared a number of salts of lactic acid and described their properties.

In his book Berzelius distinguishes the chemistry of plants and animals, which he called chemistry of organic compounds, from the chemistry of inorganic substances. Organic chemistry is that

> "part of physiology that describes the composition of living bodies, together with the chemical processes that occur in them".
>
> *- Berzelius, 1806, quoted from Fruton, 1972*

The idea that organic compounds could exist differing from those occurring in nature did not yet come up. As Berzelius wrote in the introduction to volume IV of his textbook which deals with organic chemistry:

> Every organic body distinguishes itself ... from an inorganic body in having a beginning, which we can observe. It develops itself, slows down, stops [in growth] and is [finally] destroyed. The inorganic body, on the contrary, existed already before we existed, and it keeps its existence, under whatever conditions.
>
> The inorganic elements of organic bodies also cannot be annihilated, but the real essence of these bodies will be irrevocably destroyed. The living individual which dies and returns his elements back to the inorganic nature, will never come back. The essence of a living body therefore does not rest on its inorganic elements, but is due to something else ...
>
> This *something,* which we call vital force, lies completely outside the realm of the inorganic elements and is not one of the original properties of these elements, such as (the properties of) weight, imperviousness, electrical polarity, etc. (are). We do not understand, however, what the vital force is, nor how it comes about, nor how it ceases to be[1].
>
> *- Berzelius, from Fierz-David, 1945, 218.*

Berzelius also was the man who introduced (in 1835) the term catalysis by which he meant the influence of a substance on a chemical reaction in which that substance itself does not take part. He recognized the similarity between the action of inorganic catalysts (e.g. of sulfuric acid on the forming of ether from alcohol) and that of ferments[2], like the newly discovered ferment diastase, in the hydrolysis of starch.

Berzelius' influence would become very great in later years, mainly through his many contributions to chemistry, his book and his papers, but also because of his strong and dominating personality.

Braconnot Finds Lactic Acid in Soured Plant Juices

Of about the same age as Berzelius and Gay-Lussac was the French chemist Henri Braconnot (1780 or 1781–1855), Professor of Natural History and

Director of the Botanical Garden at Nancy. He had studied chemistry and pharmacy in Strasbourg during the few years when he was enlisted in the revolutionary army ("la levée en masse") and served as a military pharmacist in the same city. After the Treaty of Amiens in 1802 Braconnot enrolled in the faculty of medicine at the University of Paris. In 1807 he was appointed professor in Nancy, not far from his birth place Commercy. He was involved in the first French efforts to produce beet sugar (As a consequence of Napoleon's Continental System Europe had been shut off from imports of colonial products. Large scale efforts were then made in Prussia and France to replace cane sugar by beet sugar and starch sugar).

At the University of Nancy Braconnot taught chemistry and pharmacy and performed many experimental studies in the field of chemical constituents of plants.

A number of these studies concerned lactic acid, a product he named nanceic acid, because he believed it to be different from the acid discovered by Scheele in sour milk. (The naming of a city, in this case Nancy, was forbidden in the new nomenclature rules, as the editors of the Annales de Chimie et de Physique pointed out in a footnote to Braconnot's paper!)

In this paper Braconnot (Braconnot, 1813) describes his studies on the acid constituents of souring media. He investigated rice water, beet juice, boiled beans and peas, and soured suspensions of baker's yeast and of vomit nuts *(Strychnos nux vomica)*.

In all of these cases he discovered the same acid ingredient, "acide nancéique". [A few years later, in 1817, the German chemist Vogel reported on experiments, which showed that lactic acid and nanceic acid were identical compounds (Holten, 1971)].

In his conclusion Braconnot carefully expresses the opinion that in rapid fermentation, in which first alcohol is produced, only vinegar will be formed (in the second phase). Products which undergo slow fermentation or souring form, in contrast, mainly nanceic acid. In his article Braconnot also describes the properties of seventeen salts of the acid.

Through the studies of Braconnot, Berzelius and others it now became generally accepted that lactic acid was a normal product of the fermentation of many vegetable materials and that it also was present in meat and animal tissue. Later we will return to the problem of what a fermentation process was considered to be.

Many years later Braconnot again reported on lactic acid, which he persistently kept calling nançeic acid (Braconnot, 1832). A tanner had asked him to investigate the composition of the sour liquid used in the treatment of hides to effect their swelling and to remove the lime remaining in the hide after the dehairing operation. (The role of lactic acid in tanning became an important subject of investigations in the last decades of the century. These studies would lead to the first large scale use of lactic acid. See Ch. IV, 4, 103).

Joseph Louis Gay-Lussac (1778–1850)

In the years around 1800 many chemists whose names have survived up until today through the reactions or laws which bear their names, lived and worked more or less solitarily in England, Prussia, other German countries, Sweden or elsewhere.

The main center of European science, however, was France and that meant Paris. In Paris the right intellectual atmosphere existed to foster collaboration, exchange of ideas and the publication of scientific discoveries. The old Sorbonne University played some role in this, but the real centre of science was at the *École Polytechnique* (founded in 1794, with the main purpose of training men for the engineer and artillery corps of the army).

In the center of the science of chemistry stood Gay-Lussac, the scientific heir of Lavoisier.

Gay-Lussac had studied at the École and from 1802 onwards taught chemistry and physics there. He was appointed Professor of Practical Chemistry at the École in 1807 and in 1809 also became Professor of Physics at the Sorbonne.

Gay-Lussac was an outstanding scientists with great experimental skills. A few of his achievements were:

Two balloon expeditions in 1804, the first one together with the physicist Biot to study the magnetism of the earth; the second one alone to make meteorological observations. In this last ascent a height of 7,016 m was reached!

An important discovery of Gay-Lussac was his law of combining volumes, stating that the volumes of the gases involved in a chemical reaction are in the ratio of small whole numbers (1809).

In later years Gay-Lussac became more and more involved in the problems of industrial chemistry. Best known is his contribution, the Gay-Lussac tower, to the sulfuric acid industry. In this (counter current extraction) tower dilute sulfuric acid trickles down over acid resistant stones with which the column is filled and in doing so adsorbs the nitrous oxides contained in the exhaust fumes from the lead chambers. Before this invention the nitrous fumes escaped freely, heavily polluting the atmosphere. Also in the field of organic chemistry many important discoveries came from the hands of Gay-Lussac and his collaborators.

Before turning our attention to Gay-Lussac's involvement in the chemistry of lactic acid his colleague, the physicist Biot, should be introduced. For the history of fermentation Jean Baptiste Biot (1774–1862) is of particular interest, because he was the man who put Pasteur on the trail that led to the foundation of the science of microbiology (Ch. 3).

Jean Baptiste Biot was appointed Professor of Mathematical Physics at the College de France in 1800. He was an excellent physicist, theoretical as well as experimental.

One of Biot's studies concerned the polarization of light, discovered by Malus in 1808. In 1835 Biot found that some liquid substances and solutions had the property to rotate the plane of polarized light. (The physical background of this phenomenon is given in Intermezzo 1, 70). He studied the rotatory power of various solutions and proved in one of his experiments that solutions of cane sugar and beet sugar behaved similarly. They must therefore be identical substances, Biot concluded. Based on these studies he developed the polarimeter for the determination of the concentration of sugar in solutions. This in turn led him to the discovery of the phenomenon of inversion of sugar when it is hydrolyzed (the angle of rotation changes from right to left, is "inverted" during the hydrolysis).

Gay-Lussac Addresses Himself to the Lactic Acid Controversy

In 1833 Gay-Lussac and Pelouze[3] published the results of extensive studies of lactic acid and lactate salts conducted in Gay-Lussac's private laboratory at the École Polytechnique (Gay-Lussac and Pelouze, 1833). In the introduction to their paper the authors refer to the studies of Scheele and Braconnot and the description given in Berzelius' book and the controversy still existing regarding the true nature of lactic acid:

> ... lactic acid salts are not yet known in the pure state. It will be the prime task of future chemists ... to determine whether [the substance] indicated by the name lactic acid is not composed of a mixture of two each other resembling acids, but yielding different salts. - *Berzelius, quoted from Gay-Lussac and Pelouze, 1833*

The next paragraph of the paper of Gay-Lussac and Pelouze, also in translation, reads:

In the conviction that one of the largest problems in the purification of certain organic bodies [substances] usually is caused by the small amounts of the substance to be investigated, we decided to produce a quantity of lactic acid sufficient for the elaboration of different methods of purification and analysis. We consequently processed several hundreds of liters of sugarbeet juice ....

- Gay-Lussac and Pelouze, 1833

The preparation adopted was long and tedious, the most essential part of it consisted in subjecting the beet juice to fermentation for a period of about two months at a temperature between 25 and 30°C. (conform to one of Braconnot's trials). The solution then contained lactic acid, mannitol and many other substances which were not investigated. The lactic acid was purified and concentrated in a number of steps involving concentrating, filtering off mannitol and other insolubles, extracting with alcohol, converting the acid into its zinc salt, treating the hot zinc lactate solution with animal charcoal and recrystallizing the zinc lactate until very pure white crystals were obtained. The acid was released again with baryte [barium hydroxide] and sulfuric acid and received a final purification through extraction with ether.

Gay-Lussac and Pelouze thereupon investigated the properties of the acid to determine whether it was an individual compound and not a mixture of two acids. This they proved by making and investigating 15 different salts, by subjecting the acid to heat treatment and by careful analysis of the acid. Most remarkably they find the behavior of pure lactic acid when exposed to heat: part of the acid sublimes (solidifies in the receiver without an intermittent liquid state being noticed); this product has almost completely lost its acid taste, but regains the acidic properties after prolonged boiling in water. [What they had obtained was the cyclic dimer of lactic acid, dilactide].

In 1845, reporting on further investigations regarding the chemical properties of lactic acid, Pelouze called the sublimate lactide, a name already proposed by Gerhardt a few years earlier (Pelouze, 1845). He also observed that the distillation residue after cooling became solid. This is the first recorded mentioning of a polymeric derivative of lactic acid (polylactide). Pelouze, of course, could not know this; he thought he had found the dehydrated form of lactic acid. [Until about 1840 chemists almost universally believed that all acids were derived from oxides. Often the – postulated – oxide was considered to be the real acid. The formula for lactic acid would then be written as $C_3H_4O_2 = C_3H_6O_3 - H_2O$. More about the formulation of chemical formulas in Intermezzo 1].

The First Quantitative Analysis of Lactic Acid

The first reliable method for the quantitative determination of the elementary composition of organic substances had been developed in 1810 by Gay-Lussac and Thénard (Snelders, 1979). The method required much experimental skill and this invited some of the best chemists of the time to search for improvements. One of these chemists was Justus Liebig (1803–1873).

Justus Liebig had studied chemistry and pharmacy at the universities of Bonn and Erlangen, but was so dissatisfied with the quality of chemical education in Germany that he applied for a grant to pursue his studies in Paris. The Hessian government complied with his request and so Liebig was able to become acquainted with the newest ideas in chemistry and the methods of quantitative analysis in the private laboratory of Gay-Lussac (1822–1823).

Soon after his return to Germany in 1823 Liebig was appointed Professor at the University of Giessen in Hesse. Here he established his own private laboratory – the university had neither room nor funds for something as extravagant as a laboratory. One of the main tasks Liebig set himself upon his return was to develop an improved and easier to use method for the analysis of the elementary composition of organic compounds. Six years later such a method was finally found and from then on the laboratory in Giessen had many years in which several hundred elementary analyses were executed. The results were usually very accurate, even by today's standards. They were of immeasurable help in unravelling many of the problems faced by the students of organic chemistry.

Liebig's new method was published in 1831 and for the next sixty years remained the best method for the analysis of non-nitrogen containing organic compounds. (A method to determine nitrogen was developed by Jean-Baptiste Dumas (1800–1884), another student of Gay-Lussac).

The quantitative elementary analysis of lactic acid reported by Gay-Lussac and Pelouze in their 1833 paper was based on the copper oxide method, they said. This was probably not yet the method of Liebig, but an improvement from Gay-Lussac's own laboratory over the original method of 1810. Liebig, who published a translation of the paper written by his French friends in the "Annalen der Chemie und Pharmacie"[4], added a short note from himself and Mitscherlich (1794–1863), a student of Berzelius and at the time Professor of Chemistry at the University of Berlin. In this note Mitscherlich and Liebig (Mitscherlich and Liebig, 1833) pronounced the opinion that the paper of Gay-Lussac and Pelouze offered a good opportunity "to test the [new] equipment for organic analysis".

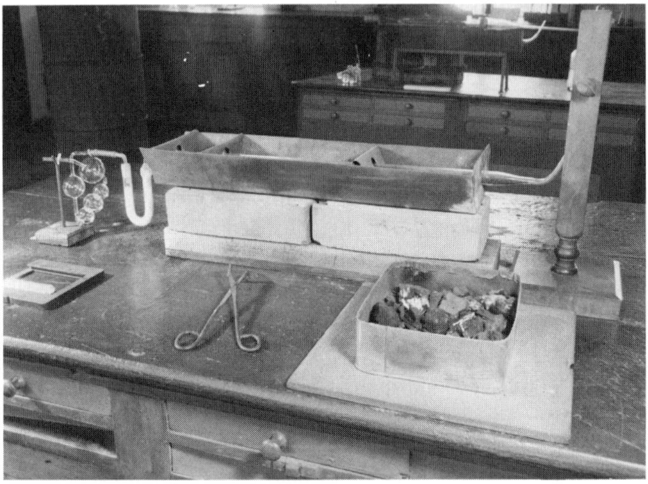

Fig. 1,2. Liebig's apparatus for elementary analysis
Courtesy: Liebig-Museum, Giessen, F.R.G.

The results of Mitscherlich and Liebig on zinc lactate were almost identical with those of Gay-Lussac and Pelouze, the conclusion being that the formula of lactic acid was $C_6H_{12}O_6$. (In fact this is the formula of the double molecule, lactic acid today is presented by the formula $C_3H_6O_3$).

Justus Liebig Makes Germany the Center of Organic Chemistry

Justus Liebig, in later years von Liebig when he was made a baron, was the most famous student of Gay-Lussac. He is the man who played a decisive role in making Germany the center of the science of chemistry in the world, taking over this position from France. His laboratory in Giessen became the main training ground for students of chemistry from all over Europe as well as from the New World. Liebig's influence, directly and through his many students, has been of crucial importance in the development of higher education in the German countries.

After the Napoleonic war the German countries (Länder) had formed a loose confederation, which exerted practically no power. In 1834 this confederation was superseded by a customs union – the "Zollverein" – in which Prussia was the leading country. Much rivalry between the separate countries remained, in particular on cultural and intellectual levels: they tried to outdo each other for instance in the funding of existing universities and in the creation of engineering schools (Technische Hochschulen). The latter often molded after the famous example of the École Polytechnique in Paris.

These institutes of higher learning began to pour out large numbers of well trained engineers and scientists in the second half of the century, in the period in which the large German chemical industries were founded. The coincidence was not really accidental; without a large supply of well trained chemists and engineers these industries could not have developed so rapidly.

Liebig became best known through his work on agriculture: he advocated the use of mineral fertilizers and through this work he can be considered the pioneer of the first "green revolution", the enormous upsurge in agricultural production achieved by the application of inorganic fertilizers.

His book on agricultural chemistry, published in 1840, gave sound scientific arguments for the replenishment of the soil with the mineral constituents taken up by the plants. In the twenty years following the publication of this book it was much applauded, and also severely criticized, but around 1860 the breakthrough came and mineral supplements found markets in all important agricultural regions. The first products accepted on a grand scale were phosphates (Liebig himself had already elaborated in 1841 a method to convert insoluble tricalcium phosphate with the aid of sulfuric acid to soluble monocalcium phosphate), Chili saltpeter, ammonium sulfate from the gasworks, and, somewhat later, potassium salts (discovered as secondary deposits in salt mines at Stassfurt, Prussia, now G.D.R.).

The increase in productivity was almost unbelievable, some agricultural soils produced up to ten times as much grain as previously and fertile fallow lands were taken into production all over Europe. Unexpectedly large and rapidly increasing amounts of agricultural products came available to feed the strongly increasing workforce of the industry and to supply raw material for the industrial production of e.g. beet sugar and potato starch. From 1800 to 1900 the population of Europe more than doubled, from about 200 million to 420 million.

Preceding his work on agricultural chemistry and following the footsteps of Berzelius, Liebig had devoted much time to the study of the chemistry of plants and animals. Like Gay-Lussac he had a particular interest in fermentation processes. From Berzelius Liebig accepted the view that something like a vital force exists, a force responsible for keeping living organisms intact.

When substances were extracted from living organisms they were relieved from the influence of this vital force and would only keep their properties by inertia (Florkin, 1972). "Dead" substances, like dead plants and animals are therefore prone to break down into simple molecules or their original elements. This can be the result of the application of heat, of air (oxygen) or of chemical actions. Such processes are called fermentation, putrefaction or slow combustion, between which in Liebig's view it is often difficult to distinguish.

Before turning to the controversy with Pasteur over the nature of fermentation processes we will, in Chapter 2, first pay attention to the history of fermentation technology until about the 1850's when Pasteur began his investigations into the nature of fermentation processes.

Notes

[1] This statement of Berzelius is seen as the beginning of the science of organic chemistry, the chemistry of living bodies. Gradually, in the next two or three decades, it became evident that under laboratory conditions organic substances could be formed and react, in ways very different from those occurring in nature. As a consequence the meaning of the term organic chemistry changed from "the chemistry of organic life" to "the chemistry of carbon compounds", because of the fact that all major chemicals extracted from living matter (except salts and water) as well as their laboratory derivatives contain the element carbon.
[2] The latin word for yeast is ferment. The original meaning is related to boiling or to the evolution of gas – carbon dioxide – in alcoholic fermentations. In the times of Berzelius a distinction was usually made between living ferments (now: microorganisms), of which yeast was the best known example, and dead ferments, obtained by extraction from living matter (example: diastase, today called amylase).
[3] Théophile Jules Pelouze (1807–1869), student of Gay-Lussac at the École Polytechnique. Held a professorship in Lille (1830), later at the École Polytechnique in Paris. Coeditor (with Thénard and Dumas) of a "Dictionnaire de technologie".
[4] Liebig had accepted the position of editor of the "Annalen" to earn the money he needed to run his private laboratory! He turned this journal into one of the leading chemical journals of the century. It is still usually referred to as "Liebig's Annalen".

Books Consulted

The New Encyclopaedia Britannica; 15th ed.; Encyclopaedia Britannica: Chicago, 1987.
Nouvelle Biographie Générale; Fimin Didot Frères: Paris, 1863 (on Braconnot).
Fierz-David, H.E. *Die Entwicklungsgeschichte der Chemie*; Birkhäuser; Basel, 1945.

Fruton, J.S. *Molecules and Life*; Wiley-Interscience: New York, 1972.
Gillespie, C.C., Ed. *Dictionary of Scientific Biography*; Charles Scribner's Sons: New York, 1974.
Greiling, W. *Chemie erobert die Welt*; Wilhelm Limpert: Berlin, 1938.
Snelders, H.A.M. *Hoofdstukken uit de geschiedenis van de scheikunde*; 2nd ed.; Rodopi: Amsterdam, 1979.

References

Braconnot, H., **1813**. *Ann. Chim. Phys.*, **86**, 84.
Braconnot, H., **1832**. *Ann. Chim. Phys.*, [2], **50**, 376.
Brin, M., **1965**. *Ann. N.Y. Acad. Sci.*, **107**, 1084.
Dobbin, L., **1931**. *The Collected Papers of Carl Wilhelm Scheele*; G. Bell & Sons: London.
FDA, **1978**. *Evaluation of the Health Aspects of Lactic Acid and Calcium Lactate as Food Ingredients*, U.S. Department of Commerce, National Technical Information Service, PB-283 713; National Technical Administration: Washington, DC.
Florkin, M., **1972**. In *Comprehensive Biochemistry*; Florkin, M., and Stotz, E.H., Eds.; Vol. 30; *A History of Biochemistry*; Elsevier: Amsterdam.
Fruton, J.S., **1976**. *Science (Washington D.C.)*, April 23, **1976**, 328.
Gay-Lussac, J.L., and Pelouze, T.J., **1833**. *Ann. Chim. Phys.*,[2], **52**, 410.
Holten, C.H., **1971**. *Lactic Acid*; Verlag Chemie: Weinheim.
Mitscherlich, E., and Liebig, J., **1833**. *Ann. Chem.*, **7**, 47.
Pelouze, T.J., **1845**. *Ann. Chim. Phys.*, [3] **53**, 113.

CHAPTER 2

THE PRESCIENTIFIC ERA OF BIOTECHNOLOGY

Fermentation processes have been used by mankind for many thousands of years in the preparation of a variety of food products like bread, cheese, wine and beer, but no knowledge of the nature of these processes existed, the manufacture was solely a matter of experience.

The processes applied in those old fermentations had, of course, undergone many improvements in the course of time, but as long as the manufacture remained in the domain of the household or small craft, changes were few and small and seldom of a technological nature. The new era starting with the Industrial Revolution in Britain and the spread of French revolutionary ideas all over Europe would soon also exert a large influence on fermentation technology.

Lager Beer

The first fermentation process to which more scientific methods were applied was the brewing process. In the middle of the eighteenth century it became known that a special method for making beer had been developed in Bavaria, Germany. The method was described in detail by the Bavarian brewer Benno Scharl (1741–1812); it is believed that Scharl himself contributed much to the success of the new beer which he called winterbeer, but today is known as lager beer or Pilsner beer. (The story goes that one day the process was stolen together with a good sample of the yeast and sold to a brewer in the Bohemian town of Plzen, Czechoslovakia, or Pilsen as the Germans call it).

The important invention the Bavarian brewers had made consisted of the use of bottom yeast instead of top yeast for the fermentation of the wort. In the older process most of the yeast formed during the fermentation rose to the upper layers, some remained in suspension and another portion settled at the bottom. How much settles depends on the temperature of the fermenting wort, the density, alcohol content and the development of gas (carbon dioxide). In the winter the temperature of the broth was, quite naturally, lower, the fermentation slower, the gas production less vehement and there appeared more sediment (yeast) at the bottom. Somebody then had the idea to use only the bottom sediment for the inoculation of the next fermentation. By doing this the brewing process apparently was better kept under control.

The method had a drawback, however: the fermentation was so much slower that the conversion was not completed. Then the second part of the "invention" was made: the intermediate product was stored in a cold cellar (in German Lager) for a number of months. In this additional fermentation period full conversion of the remaining sugars was achieved and, surprisingly enough, a final product with much improved keeping qualities obtained.

Scharl's recipe is a fine example of the importance of good manufacturing practice in biotechnology (the original recipe is given in Delbrück and Schrohe, 1904):

> The production should take place between St Michaelis [29 September] and St Georgi [23 April], in the rest of the year only summerbeer can be made. Good and wholesome beers do not need anything else for their preparation than water, wort, hop and good sediment [bottom yeast] from the brewery. When in September the production of the winterbeer has to be started bottom yeast from the production of summerbeer can be taken, but only from good productions. Otherwise supply from another brewery, which made a good start with winterbeer earlier, is advised. In case such a yeast is also not attainable, dried yeast from the winter production of the preceding year can also be tried. In general this dried yeast is not very suitable.
>
> The fermentation should be executed in a cellar 4 to 5 feet deep in the earth to protect against winterfrost. The wort, to which hops are added, is boiled first, cooled to 10° R [= 12.5°C] and then provided with the bottom yeast. After the first fermentation the beer is drawn off to a second cellar, preferable laying below the fermentation cellar, in which the temperature naturally stays above the freezing point.
>
> *- Scharl, quoted from Delbrück and Schrohe, 1904*

Schrohe mentions that Scharl devoted much attention to the quality of the yeast used for the inoculation: Scharl even advised to remove the upper layer of the yeast sediment because it might contain impurities. These early brewers of lager beer, however, had no notion that yeast was composed of living matter, of animalcules to use the expression of Leeuwenhoek. By experience they must have known that hygiene was important, but they did not know why, they did not even know the word.

Spoilage and Spontaneous Generation

But something was known about hygiene and spoilage, partly by experience, partly from microscopic studies of Leeuwenhoek and his eighteenth century successors.

An example of the first aspect can be found in the works of Scheele: one day Scheele was confronted with the problem of the spoilage of vinegar. He found four methods to prevent this and the one he recommends most is to boil the vinegar for a short time prior to bottling!

Much more attention was given to the problem of infection by the Italian priest and scientist Lazzaro Spallanzani (1729– 1799). Spallanzani first took up the study of the animalcules of Leeuwenhoek: from where did these little creatures come? Some people said they originated just from the air, from nothing so to speak, by a process called *spontaneous generation*. Spallanzani could not believe this theory of a kind of ongoing creation. In a large series of experiments, and only after having mastered with great difficulty the use of a still very primitive microscope, he succeeded in demonstrating that heating will prevent the appearance of animalcules even in "soups" in which they normally soon show up in large numbers. He also showed that air had to be excluded permanently if one wanted to protect organic matter against the incursion of microorganisms. Turning the argument around he proclaimed that spontaneous generation did not exist; wherever it was found some infection from the air must have taken place.

The ideas of Spallanzani found a practical application through the work of Nicholas Appert (1760–1840), son of an innkeeper and champagne bottler in Chalons-sur-Marne. For him the problem was a military one: as a result of the conscription introduced during the French Revolution – the "Levée en masse" – the armies had increased to hundreds of thousands of men, and the supply of these armies had become an almost insurmountable problem, particularly when Napoleon set out to try to dominate the whole of Europe. His armies could no longer be fed from the territory they were occupying or passing through. The Imperial Government decided to offer a reward to the man who could solve this supply problem. Appert had already been experimenting for some time with champagne bottles filled with meat stew, soup, milk, peas, beans and various fruits, placing the bottles in boiling water for different periods of time and sealing them with corks held in place with wire cages, like he was accustomed to do with champagne bottles. After a few years Appert believed he had mastered the new technique and in 1806 he established a bottling factory near Paris. Here he employed fifty workers in bottling a large variety of foods. The French navy did the first thorough investigation, shipping a whole load to the Caribbean and back. In 1809 Appert received a large prize of 12,000 francs from the Imperial Government for his new conservation method.

In 1811 Appert's idea was introduced in England, where instead of champagne bottles, metal cans were preferred as containers. Initial production was mainly sold to the Royal Navy, but in 1830 the cans filled with food products like peas, tomatoes and sardines, also appeared in the shops. The price was extremely high and the cans could only be opened with hammer and chisel! But a new method of food preservation was found, in the beginning of limited use only in military expeditions and in peaceful explorations of exotic parts of the earth. But in the course of the century, with the development of the technology of canning, canned foods also became available to the ordinary citizen. The invention of Appert and the development of the canning industry again provide an interesting example of the use of technology based on scanty knowledge, in this case knowledge about putrefaction which was, in general, not shared by the scientists of the era.

The Chemistry of Life and the Causes of Fermentation

The problem of what caused fermentation remained in the background. Sure, Lavoisier had been aware of the importance of the problem, he induced the Académie des Sciences in Paris in 1792 to put up a prize for the elucidation of the problem of fermentation and putrefaction, phenomena considered to be the reverse of the forming and origin of living beings. The well known chemist Louis Jacques Thénard, Professor at Napoleons École Polytechnique suggested that the animalcules (of Leeuwenhoek) had something to do with fermentation, but the time was not yet ripe to prove or disprove this idea.

Gay-Lussac examined, in 1810, some of Appert's bottles and found that the enclosed air was devoid of oxygen. This led him to the view that oxygen plays a role in putrefaction and fermentation (Fruton, 1972, 42).

The leading chemists of the time realized themselves that the chemistry of organic life, a science they usually called physiology, was different from the chemistry of inorganic substances. A special problem for them was how reactions took place in living organisms. Berzelius, who had coined the word catalysis, suggested that in the organic world special types of catalysts are active. These catalysts, he assumed, could promote reactions under the mild conditions of life, which without their presence would not take place. As an example Berzelius gave the hydrolysis of starch by diastase, a substance obtained as an alcoholic precipitate from an aqueous extract of fresh barley sprouts (by Anselme Payen and Jean Persoz in 1833). Diastase, it was shown, could readily transform gelatinized starch at room temperature into a mixture of dextrins and a sugary substance[1].

More examples of the action of organic catalysts, acting outside the living material from which they had been extracted, were found in the following years. The idea that [nearly] all reactions in living organisms take place under the influence of similar organic catalysts was not yet widely accepted, however.

Liebig also believed that a special force, a vital force, must exist, which supervises in living bodies the attractions of the atoms, the movement of the molecules and controls the way in which the aggregates of living matter are formed (Florkin, 1972, p.141).

At death the action of the vital force stops and "normal" chemical forces take over. A slow process of decomposition then follows, leading in the end to (complete) decay of the organic material and to the formation of simple inorganic molecules.

Fermentation is the first step in this decomposition process. Fermentation and putrefaction require the presence of organic nitrogenous substances in a decomposing state. Such substances easily transfer the decomposition to other organic substances, which, like starch for instance, do not easily decompose by themselves. In Chapter 3 we will see how Liebig applied this theory to the laboratory preparation of lactic acid.

The Use of the Microscope in Fermentation Studies

Professional brewers had since long appreciated the importance of a good yeast, but what was the role the yeast actually played, how did it control the chemical reaction in which alcohol and carbonic acid were formed, and how could this be investigated? In the same period, in which the study of organic chemistry began to depart from pure physiology, a few isolated scientists began to apply improved microscopes in the study of yeast particles (globules) and other "animalcules", which always (or often?) appeared to be present in fermentation and putrefaction processes[2].

One of these men was Charles Cagniard-Latour (1777–1859). He picked up the problem of the cause of fermentation and putrefaction first formulated by Lavoisier in 1792. Although the prize (one kg of gold) offered for the solution of this problem was withdrawn by Napoleon in 1803, the young engineer Cagniard-Latour had become so fascinated with the problems related to fermentation, that he took up the search nevertheless.

When he started his studies, around 1810, he could not avail himself of a good microscope, which he believed to be indispensable for a study of fermentation processes. Only in 1835 was he able to procure a good instrument, magnifying 300–400 times and provided with a micrometer-eyepiece. Now he could observe the yeast cells and measure their dimensions[2]. A new world was opened to him, in a similar way as it had been to Leeuwenhoek and Spallanzani before him. But he saw more:

Cagniard-Latour saw the cells – which he called globules – changing form and appearance, forming pairs, and dividing themselves. He concluded that yeast globules are organized living creatures belonging to the kingdom of the plants (Cagniard-Latour, 1838). These little plants, he told the Académie des Sciences in a preliminary report on June 18, 1836, exert a certain action on sugar, which seems to support their vegetation and is accompanied by the development of carbonic acid and the formation of alcohol. These actions go on only as long as the globules are alive.

In the same years of 1836 and 1837 the German physician and biologist Theodor Schwann performed studies on spontaneous generation. He showed that meat will not spoil when it has been exposed to boiling during a certain period of time and thereafter is exposed for a number of weeks to a continuous stream of air, which had been disinfected by passing it through red hot glass tubings. The fermentation of sugar he investigated in a similar manner:

> Four bottles with solutions of cane sugar were kept at 100° C for about 10 minutes, placed upside down in a mercury bath and cooled to ambient temperature. Thereupon about one third to one quarter of the solution was replaced by air; in two of the bottles by "normal" untreated air, in the other two by heat treated [sterilized] air. The two bottles with "fresh" air started to ferment, to develop gas after about 4–6 weeks, the other two did not show any change over a much longer period of time.
>
> - *Schwann, quoted from Delbrück and Schrohe, 1904*

In retrospect, one must conclude that Schwann had been rather lucky in his experiments, it being just a matter of chance what kind of microorganism he would have added with this "fresh" air.

Schwann also did microscopic studies. He noted that beer yeast consisted of globules, usually connected in strings, sometimes branched strings as if a new string had grown out of the first one. He concluded from these observations that the (beer)ferment is a plant which in many ways is similar to certain fungi. Schwann repeatedly noticed that when (alcoholic) fermentation takes place yeast cells – he says sugar fungi – are always also present. He concludes that most probably the growth of the yeast globules initiates the fermentation of the sugar.

In the same years a third investigator, F.T. Kützing reached the conclusion that vinegar also was produced by the action of a living organism, this time on alcohol.

Liebig Ridicules the Use of the Microscope

The observations and theories of Cagniard-Latour, Schwann and Kützing were met with scorn and satire by the leading chemists of the time, in particular by Liebig and Wöhler. They even saw fit to publish a pseudo-scientific article in Liebig's Annalen der Pharmazie (**29**, 100, 1839) in which they depict the animalcules of Schwann and Cagniard-Latour as a kind of living distillation retorts equipped with sucking trunks and most of the organs other animals have, except eyes and teeth. "Under the microscope", the anonymous writers tell,

> you can closely observe how these animals devour the sugar and almost instantaneously convert it into wine spirit and carbonic acid. It can actually be seen how, probably under the influence of magnetic and electric forces, the bladder enlarges tenfold and then in bursts excretes the carbonic gas through the enormous genitals. The wine spirit, however, is excreted more continuously from the anus, it is detected as an ascending stream of fluid visible through the difference in gravity with the solution.
>
> *- NN, 1839*

So much from this curious article, more a blame for the "unknown" authors than for the proponents of the idea that living beings – microorganisms – played a decisive role in fermentation processes. One has to recognize, however, the basic truth of the objection of Liebig and his partisans, that "chemical reactions cannot be seen under the microscope, and will in the future also never be seen under this instrument, irrespective of the degree of magnification".

The opposition of the chemical establishment was so strong and influential that the ideas of Schwann, Kützing and Cagniard-Latour almost disappeared from the scientific scene until the early 1850's when a young French chemist by the name of Louis Pasteur approached anew the problem of what causes fermentation (Ch. 3).

Lactic Acid, a New Fermentation Product

Although around 1840 the causes of fermentation were still almost completely unknown, a new product based on fermentation processes was about to enter the growing market of new drugs and chemicals. This was the result of the initiative taken by Gay-Lussac and Pelouze (Ch. 1, 15) and followed up by Pelouze and his student Frémy. True, the investigations conducted by Gay-Lussac and Pelouze did not have the goal to prepare a new chemical, their objective was to produce lactic acid in sufficient quantity to enable a good analysis of the product and the determination of its chemical properties, in short to ascertain the individuality of the acid.

But as good scientists they also had doubts about the existing views on fermentation. In the summer of 1837 Pelouze visited his friend Liebig in Giessen, who had a lively interest in natural processes occurring in animals and plants. Very probably the two exchanged their views on the subject of fermentation. A short time later Pelouze instructed one of his students, Edmond Frémy, to take up the study of (the causes of) the fermentation of sugar to lactic acid[3].

A first report of Frémy's work was presented to the "Académie des Sciences" by Pelouze on June 19, 1839 (Frémy, 1839):

> Thoroughly washed calfs rennet[4] has been found to convert at a temperature of 40°C sugar into lactic acid. This reaction could be repeated indefinitely with the same membrane.
>
> *- Frémy, 1839*

Gay-Lussac, who was present at the meeting, remarked that he had noted – some years earlier – that sugar and other vegetable products not only produced lactic acid when mixed with animal substances, but that this (reaction) also had resulted in a conserving effect upon those animal substances. (One wonders to what experience Gay-Lussac is referring here; maybe to the preparation of fermented sausages?)

Frémy tried to explain the observed facts:

> ... [the existence of] a decomposing force which can act on all organic material, could explain [many] phenomena in the physiology of animals and plants. Before arriving at a more general result it is necessary to investigate the conditions needed to achieve these decompositions, the influence of animal matter in such reactions and the kind of substances which are formed.
>
> *- Frémy, 1839*

Liebig[5], commenting on the paper by Frémy, also believed that much additional study was required:

> The transformations which organic materials alone and on contact with other materials undergo, are a separate new field of investigation. A thorough understanding of these transformations can hardly be expected before the products of decomposition of separate materials have been investigated. As an example I mention [the investigation of] the change which the animal membrane experiences when exposed to water at elevated temperatures.
>
> *- Liebig, 1839.*

The fermentation studies were continued by Frémy and F. Boutron, his father-in-law[3]. The results of their extensive studies on the fermentation of sugary substances were published in 1841 (Boutron and Frémy, 1841a, 1841b).

Boutron and Frémy reported on experiments with quite a number of different organic substances thought suitable of carrying out the fermentation: calfs rennet, dog stomach, bladders, casein and malt sprouts. They found two recipes useful for making larger quantities of lactic acid:

> 1. Moistened malt sprouts are exposed to air for two to three days, mashed and mixed with water of 20–25°C. The liquid warms up (by itself) and becomes very acid within a few days. After filtration it is saturated with lime; with alcohol calcium lactate precipitates.
> 2. Fermentation of milk.
>
> To three or four liters of milk a solution of 200 to 300 g of sugar of milk [lactose] is added. This liquid, placed in an open vessel at a temperature between 15 and 20°C, is exposed to air until it becomes very acid, which takes a number of days. Thereupon it is saturated with sodium bicarbonate. Within 24 to 36 hours the liquid turns acid again and is saturated anew with sodium bicarbonate. This procedure is repeated until all of the lactose has been transformed into lactic acid.
>
> The casein is then precipitated by boiling and filtering off; the filtrate concentrated to a syrupy consistency. Alcohol is added to dissolve the sodium lactate, then sulfuric acid is added to precipitate sodium sulfate. The sodium sulfate is subsequently filtered off, the filtrate concentrated by evaporation of the water and alcohol, which yields lactic acid in a rather pure form. (Through crystallization of the calcium salts a completely pure lactic acid can be obtained, the authors remark finally.)
>
> *- Boutron and Frémy, 1841*

The reason behind the use of sodium bicarbonate in the fermentation recipe was a theoretical one. It was assumed that when milk curdles a casein-lactic acid complex is formed. Such a complex could be opened up again with sodium bicarbonate, the casein would then be liberated and go into solution again. Being back in its original state the casein would thereupon react again as a decomposing agent on the sugar of milk. [The explanation is wrong; what really happened was that the fermentation stopped when the broth became too acid; through the addition of bicarbonate sufficient acidity was neutralized to make conditions favorable for further fermentation.]

Liebig's Theory of Fermentation

A theoretical explanation of the experimental results of Boutron and Frémy (Boutron and Frémy, 1841b) was given in a footnote to their paper by Liebig when he published a translation in his Annalen. After stating that in his book on "Organic Chemistry applied to Agriculture and Physiology" he had explained the action of various ferments, Liebig continued:

> Ferments decompose under the influence of peculiar processes during which they exert a certain action on the other organic substances present. These other organic substances then also start to decompose in a similar fashion.
>
> *- Liebig, 1841*

Liebig uses the word ferment here to define organic substances capable of inducing fermentation, which means transforming organic material into another form in a way resembling the processes by which cheese, wine and beer are made. Almost triumphantly Liebig continues in his footnote:

> Boutron and Frémy have taken my views as the basis for their experimental work and have obtained results that provide a more decisive and beautiful proof of my ideas than could be expected.
>
> *- Liebig, 1841*

Boutron and Frémy indeed open their paper with an indirect reference to the new and helpful ideas of Liebig, which they formulate as follows:

> All experiments on fermentation we have performed deliver proof of the existence of animal substances which are subject to peculiar decomposing forces, which substances on contact with other (organic) bodies also induce decomposition in such bodies. It has been established that many organic substances can be converted by the aid of fermentation and that for the conversion of different substances different ferments are required.
>
> *- Boutron and Frémy, 1841a*

The actions of diastase (amylases) on starch, of emulsin on amygdalin and the formation of mustard oil and of pectinic acid are cited as examples.

Boutron and Frémy compare the action – in the German translation the word "Kraft" = "force" is used – of decomposing ferments on other organic substances with the phenomena occurring when organic substances are decomposed under the action of gradually increasing temperatures. In that case – that of dry distillation – one observes a successive change in the composition of the distillate and something similar, they believe, happens when decomposing ferments exert their action. In general, they say that in fermentation the following sequence will be observed:

> First lactic acid (is formed), then mannitol, thereupon a viscous material and finally alcohol and carbonic acid [carbon dioxide].
>
> - *Boutron and Frémy, 1841*[a]

Today this theory seems far-fetched, but at the time it was a step forward, it provided a basis for further systematic research. The two French investigators also established some of the conditions under which no decomposition (= fermentation) would occur:

> When the material is too dry, or is excluded from contact with air, or first has been heated to 100°C.
>
> - *Boutron and Frémy, 1841*[a]

One can only wonder what they would have concluded if they had followed their experiments with the microscope, like Schwann, Cagniard-Latour and Kützing had done in the fermentation of alcohol and vinegar. They would probably have observed that in the three above cases, in which no fermentation occurred, no or only very few living microorganisms were present; too few to support lactic acid fermentation. But they would certainly also have earned Liebig's scorn instead of his praise.

Thanks to the investigations of Pelouze, Frémy and Boutron methods had now been developed to produce lactic acid by fermentation in fairly large quantities from sucrose (ordinary sugar) as well as from lactose (milk sugar).

Soon after the publication of the article by Boutron and Frémy an iron salt of lactic acid [ferrous lactate] prepared by apothecaries was introduced into medicine.

The Role of Apothecaries in the Preparation of Chemicals

In the first decades of the nineteenth century Great Britain was the most important producer of heavy chemicals, followed at some distance by France. Small scale production of drugs and fine chemicals was carried out by apothecaries (pharmacists, or in England: chemists). Particularly in Germany, where the use of plant extracts and elixirs for medicinal purposes had long been popular, apothecaries took to larger scale production, but the same trend was found in France, Great Britain and the United States. In various countries small chemical works were established in those years, which produced a variety of chemical substances, like for instance morphine and quinine, which were sold to colleague apothecaries. (A very early enterprise of this kind was Glauber's in Amsterdam in the seventeenth century).

One of the most successful enterprises of this era was Heinrich Emanuel Merck's at Darmstadt. Heinrich Emanuel Merck (1794–1855) was the descendent of a family which since 1668 had owned an apothecary shop in the city of Darmstadt (Hesse, Germany). From 1810 until 1812 he studied pharmacy with Johann Bartholmä Trommsdorff, a well-known Professor of Chemistry and Pharmacy in Erfurt. He concluded his studies in Berlin and took over the direction of the family apothecary in 1816. In 1827 Emanuel Merck (he did not use the given name Heinrich) published a booklet describing "large scale" production methods for the preparation of plant extracts, in particular of alkaloids. This publication marks the beginning of the manufacturing activities of the firm E. Merck. Merck was very well regarded for this scientific work, he corresponded with many of the leading chemists and pharmacists of the time and became close friends with Justus Liebig.

Some of the producers of "heavy" chemicals also diversified into the production of small scale chemicals. An example is the firm set up in 1808 by Philippe-Charles Kestner (1776– 1846) at Thann, in the Alsace region of France. This enterprise, originally called Charles Kestner, later (1872) Fabrique de produits chimiques de Thann, and still later (1884) Fabrique de produits chimiques de Thann et Mulhouse, specialized in the manufacture of chemicals for the textile industry. In 1809 for instance the company made: sulfuric acid, nitric acid, tartaric acid and tin chloride (Soc. Ind. Mulhouse, 1902).

Similar developments, albeit on a smaller scale, took place in other European countries and in North America:

In America, the center of pharmaceutical production was Philadelphia, where to give an example the firm of Wetherill & Company, as early as 1826 carried tartaric acid as one of its products (Haynes, 1954).

Iron Lactate, the First "Pharmaceutical" Based on Lactic Acid

France was probably the first country where apothecaries tried to introduce salts of lactic acid and later lactic acid itself for medical uses. This was not very surprising since Gay-Lussac and his students had elaborated a number of methods to prepare lactic acid in larger quantities. In 1839 iron lactate (ferrous lactate) was introduced in France for the treatment of anemia and chlorosis (green sickness). It was recommended over other products containing iron, because it was a combination with lactic acid, an acid the body could easily assimilate because it had [mistakenly] been found to be normally present in the human stomach.

As we have seen before, tartaric acid was already used by apothecaries in the beginning of the century, and citric acid also found its way into the pharmacopoeia's before lactic acid (for instance in the Pharmacopoeia of Baden 1841). Both of these earlier accepted acids were extracted from vegetable raw material, which was comparatively easy because the calcium salts could be readily precipitated from the extracts as had first been shown by Scheele. After washing the precipitates were converted into the corresponding acids with sulfuric acid, and the acids prepared in a pure solid form by crystallization.

A similar method which tried to extract lactic acid from sour milk did not work so well because of the high solubility of the calcium salt and because the acid itself did not crystallize. With the new fermentation method of Boutron and Frémy much higher concentrations of lactic acid (in the form of sodium lactate) could be obtained than present in souring milk or other souring liquids, thanks to the intermediate use of sodium bicarbonate. This opened the way to a more economical method of preparing lactic acid in large quantities and with less by-products. It should be noted here that E. Merck as early as in 1842 mentions lactic acid as one of his products (Löw, 1952)! Could it be that Merck, thanks to his connections with Liebig, had put Frémy's method almost immediately into practice?

The book by Schneider (1975), from which the information about the use of iron lactate has been taken, also mentions the use of zinc lactate (Pharmacopoeia Saxonia, 1867), a product recommended for the treatment of epilepsy.

The first pharmacopoeia of the new German Empire of 1871, the "Deutsches Arznei Buch" (DAB) of 1872 includes lactic acid itself, iron lactate and the magnesium salt. The properties are given in the Hager-Kommentar of 1876 (Hager, 1876):

– acidum lacticum helps digestion; it is used in the treatment of diphtheria, for mouth waters and tooth cleaning;
– magnesia lactica is a mild laxative.

The French pharmacopoeia (Dorvault, 1872) has additionally:

– bismuth lactate, an antispasmodic and thirst causing agent;
– calcium lactophosphate against rickets and osseous affections that accompany disorders of the digestive system.

Improvements in the Preparation of Lactic Acid

The method of Boutron and Frémy for the first time allowed the preparation of lactic acid on a kg scale, even a 100 kg scale. The preparation, however, was long and tedious, leading to a high price for the acid and its salts. The Parisian apothecary Gobley[6] found in particular the price of iron lactate (ferrous lactate) of 40 francs per kg much too high and set himself the task of finding a cheaper way to prepare lactic acid (Gobley, 1844). [Merck offered ounce quantities for sale in 1841 at a price of about 150 francs per kg]. Commenting on the "beautiful experiments" of Boutron and Frémy he noted that the transformation of lactose into lactic acid is very simple indeed:

> one equivalent of anhydrous milk sugar is converted into four equivalents of anhydrous lactic acid under the assimilation of the elements of one equivalent of water [lactose ($C_{12}H_{22}O_{11}$) + water (H_2O) gives four lactic acid ($C_3H_6O_3$)].
>
> - *Gobley, 1844*

Gobley executed a large number of experiments, from which he derived the following improvements over Boutron and Frémy:

1. sodium bicarbonate was replaced by finely ground chalk (calcium carbonate),
2. the quantity of milk was diminished to one third,
3. instead of whole milk skimmed milk was taken.

The idea to use chalk Gobley had learned – as he states in his article – from Pelouze and Gélis (Pelouze and Gélis, 1844) who investigated the butyric fermentation and who had found it inconvenient to use a (neutralizing) ingredient like sodium bicarbonate that had to be added step-by-step and under supervision. All the calcium carbonate needed for the conversion to butyric acid could be supplied at the beginning of the fermentation reaction (a reaction which took several weeks to complete).

Gobley tried to ferment different raw materials like ordinary sugar, potato starch and gum [gum arabic?]; the best results, however, he obtained with lactose. With other raw materials the conversion usually was not so complete, which in Gobley's view could be a consequence of the gradual coagulation of the casein. When the fermentation lasted too long butyric acid was formed (a confirmation of Liebig's theory of stepwise fermentation). (See also Intermezzo 5). Instead of casein Gobley also tried brewers yeast as the decomposing agent, with favorable results.

Gobley gives as his best recipe:

> Take a number of small tureens of earthenware, each of three liter capacity; fill these vessels each with 250 g of sugar of milk, 200 g of pulverized chalk, one liter of skimmed milk and fill up with water; hold the pots at temperatures between 25 and 30 °C, stir occasionally and replenish evaporated water. After 24 hours fermentation begins with the generation of carbonic acid, which goes on until the eleventh or twelfth day. The mixture smells like sour milk, toward the end more like cheese. The end of the fermentation will also be noted by the appearance of many floating lumps of solid material, sometimes the whole mass more or less solidifies [separation of crystalline calcium lactate].
>
> \- *Gobley, 1844*

To prepare pure calcium lactate the fermentation broth is heated to coagulate the casein, filtrated, crystallized again, separated from the mother liquor and dried.

Lactic acid itself is made by reacting calcium lactate with oxalic acid, followed by separation of lactic acid from insoluble calcium oxalate.

The economics of the process were very favorable, Gobley writes:

> 1 l of skimmed milk costs 20 centimes, 250 g of lactose 70 centimes, the cost of heating is next to nothing in the summertime, the yield of the process is 340 g of calcium lactate [this equals 83% of the theoretically possible yield, assuming that Gobley obtained the pentahydrate of calcium lactate].
>
> \- *Gobley, 1844*

From these data Gobley concludes that the cost price of calcium lactate decreases to a value of not more than 6 francs per kg. How he proceeded to make iron lactate he did not mention. Information on the prices of pharmaceutical lactates of the time is difficult to obtain (see Table 2, I).

Table 2.1. Prices of pharmaceutical lactates (for small amounts of about 100 g).

Year	Product	Price/kg	Converted into US$/kg	Source
1843	lactic acid	87 M	22	Vershoven (1958)
1860	lactic acid	44,64 Fl	19,50	Merck
	zinc lactate	13,39 Fl	5,85	Merck
1871	lactic acid	19,64 Fl	8,60	Merck
	zinc lactate	6,79 Fl	3,00	Merck
	calcium lactate	3,12 Fl	1,36	Merck
1872	zinc lactate	15 Ffr	2,84	Dorvault (1872)
	ferrous lactate	17,50 Ffr	3,32	Dorvault
1880	lactic acid	20 M	5,10	Vershofen (1958)

M = mark of the German Empire of 1871
Fl = gulden of the German Zollverein
Ffr = French franc

The next improvement came from a German pharmacist, Dr. A. Bensch, who formulated two objections against Gobley's recipe (Bensch, 1847):

the forming of lactic acid will be followed by that of butyric acid as it is not always possible to stop the fermentation at the right moment.

The second (and related) objection is that one does not notice the end of the lactic fermentation because the fermentation mass usually does **not** solidify. Moreover, he found that ordinary sugar could very well be used instead of the sugar of milk, provided it is first transferred with the aid of tartaric acid into grape sugar by a process we now call hydrolysis or inversion (See Ch. 10, 395).

Reading Bensch's recipe is more difficult than those of his French colleagues, because he did not use the metric system. The writer has converted his weights into metric units, assuming that a Pfund (pound) is a Cologne pound of 467.7 g and an Unze (ounce) is a sixteenth of a pound, which then is 29.23 g. The "translated" recipe is presented in the exhibits on the next two pages, the first giving the essential elements of the fermentation process, the second those of the purification method. (The exhibits are intended to provide for easy comparison with later preparative and manufacturing processes, to allow the technically oriented reader to follow the historical developments in lactic acid production technology more closely.)

Bensch's recipe was presented in Dorvault's "Officine" as well as in most pharmacopoeia's of the second part of the century as the (standard) method to prepare lactic acid.

The papers by Gobley and Bensch show that the application of an incorrect theory of fermentation (viz. that of Liebig and Pelouze) nevertheless could lead to the development of a useful process for the fermentative preparation of lactic acid and lactates. It is also evident that the way in which inoculation occurred (by rotten cheese) was an inferior practice compared to the state of art achieved in, for instance, the brewing of lager beer.

EXHIBIT 2,I
BENSCH'S FERMENTATION PROCEDURE
lit. Bensch, 1847

raw material	cane sugar
concentration	about 160 g/l
sterilization	sugar is dissolved in boiling water to which 0.12% of tartaric acid has been added (to bring about the inversion of the sugar) and then left standing for a number of days. The sugar concentration is 23% (w/w).
organic nutrients	about 3% of old, stinking farmhouse cheese is divided in skimmed milk; from this mixture about one part is mixed with three parts of the sugar solution
inorganic nutrients	supplied by the skimmed milk
neutralization by	chalk powder, added at the beginning, 50% on sugar [which is about 15% short of what is required for complete conversion]
fermentation temperature	30/35°C
fermentation vessels	probably of earthenware, contents about 15 l
sterility of operation	??
mixing	a number of times each day
acidity	slightly acidic
fermentation time	8–10 days
control of end	the broth turns into a stiff porridge of calcium lactate
yield	?? (the yield of the fermentation process cannot have been more than 42% [7]
type of bacterium	Bensch did not know that bacteria were involved in his fermentation process!

EXHIBIT 2,II
BENSCH'S PURIFICATION PROCEDURE
lit. Bensch, 1847

treatment of broth	a boiling solution of 15% quick lime in water is added (60% on the broth), the whole kept boiling for half an hour
filtration	following the alkaline treatment
concentration	to a syrupy consistency
crystallization	slowly, should take two days
purification	the gritty calcium lactate is pressed free from adhering liquid, 10% water is kneaded in and liquid is pressed out again. This operation is repeated 2 or 3 times
conversion to acid	with the equivalent amount of sulfuric acid at boiling temperature
filtration	to obtain crude lactic acid
purification	the filtrate is boiled for half an hour with zinc carbonate (30% excess)[8]
filtration	hot
crystallization	after some time completely colorless zinc lactate crusts appear; mother liquors are worked up
conversion of zinc lactate	the zinc lactate is dissolved in 7.5 parts of boiling water, hydrogen sulfide is passed through until no further reaction can be noticed
separation of lactic acid	filtration, followed by heating to remove residual hydrogen sulfide, concentrating at 100°C to a syrupy consistency

Notes

[1] In 1847 Augustin Pierre Dubrunfaut discovered that the sugary substance was maltose, which, upon treatment with acid, could be converted into dextrose (glucose). At the end of the century diastase was renamed in amylase because of its ability to hydrolyze starch (Latin: amylum). About the discovery of these and other (dead) ferments obtained by extraction, see Fruton, 1972.

[2] Since the days of Leeuwenhoek, who applied in his famous work a microscope with only one lens, compound microscopes had come into use. In these microscopes at least two lenses, the objective lens and the eyepiece, were mounted. The early lenses, however, all showed the phenomenon of chromatic aberration, that is of coloring and blurring of the image as a consequence of the change of the refractive index of the light with the wavelength (color). This problem was solved, both theoretically and practically, by the Englishman Joseph Jackson Lister in 1830. It then became possible to construct lenses in which the chromatic aberration was automatically corrected.

[3] Edmond Frémy (1814–1894) studied chemistry with Gay-Lussac and Pelouze. At

the age of 17 he became Pelouze's assistant and in 1850 he succeeded Gay-Lussac as Professor at the Muséum d'Histoire Naturelle. Frémy was primarily an experimentor and disliked theoretical studies. He worked in almost every field of chemistry, was at one time consultant to the Compagnie de Saint-Gobin. He also was editor of a 93 volume chemical encyclopedia. Frémy married (in 1842?) the daughter of a well-known Parisian pharmacist, Antoine François Boutron-Charlard (1796–1878).

[4] Rennet, the lining membrane of the fourth stomach of a calf, is used in cheese manufacture for the curdling of milk.

[5] In the Annalen Liebig regularly published translations of the work of French scientists, sometimes with annotations.

[6] Nicolas Théodore Gobley (1811–1876) was a student of Pierre Robiquet, a well-known French pharmacist and chemist. In 1837 Gobley inherited a large fortune from an uncle, which made him financially independant. He bought a pharmacy annex chemical shop (officine), which he exploited until 1861. He was (honorary) Professor of Pharmacy at the École de Pharmacie de Paris in the 1840's. Known as an excellent chemist, he published articles about phosphatides, plant extracts and fat chemistry. He discovered lecithin in 1850 (Dictionnaire de Biographie Française, Prevost and d'Amat, Paris, 1956).

[7] The solubility of calcium lactate pentahydrate at 30–35°C is in the range of 80–110 g/l; starting with 160 g/l of sugar and complete conversion a final concentration of about 260 g/l of calcium lactate can be calculated. As soon as solidification occurs the fermentation stops, however. Hence the maximum conversion obtained can hardly have been more than about 110/260, equaling 42% of what is theoretically possible.

[8] The purification through crystallization of zinc lactate goes back to Mitscherlich, who discovered "the first method by which lactic acid could be prepared in a fully pure state" (Berzelius, 1837).

Books consulted

The New Encyclopaedia Britannica; 15th ed.; Encyclopedia Britannica: Chicago, 1987.

Burke, J. *Connections*; Book Club Associates/MacMillan: London.

Delbrück, M. and Schrohe, A. *Gärung und Fäulnis*; Paul Parey: Berlin, 1904.

Fierz-David, H.E. *Die Entwicklungsgeschichte der Chemie*; Birkhäuser: Basel, 1945.

Florkin, M.A. *History of Biochemistry*; Vol. 30 of Comprehensive Biochemistry; Florkin, M. and Stotz, E.H., Eds.; Elsevier: Amsterdam, 1972.

Fruton, J.S. *Molecules and Life*; Wiley-Interscience: New York, 1972.

Gillespie, C.C., Ed. *Dictionary of Scientific Biography*;Charles Scribner's Sons: New York, 1974.

Greiling, W. *Chemie erobert die Welt*, Wilhelm Limpert: Berlin, 1938.

Lechevalier, H.A., and Solotorovsky, M. *Three Centuries of Microbiology*; Dover: New York, 1974.

Schelenz, H. *Geschichte der Pharmazie*; Govi-Verlag: Berlin, 1904.

Schneider, W. *Lexicon zur Arzneimittelgeschichte*; Govi-Verlag: Frankfurt, 1975.

Snelders, H.A.M. *Hoofdstukken uit de geschiedenis van de scheikunde*; 2nd ed.; Rodopi: Amsterdam, 1979.

References

Soc. Ind. Mulhouse, **1902**. *Histoire Documentaire de l'Industrie de Mulhouse*; Tome II, Veuve Bader & Cie: Mulhouse.
Bensch, A., **1847**. *Ann. Chem.*, **61**, 174.
Berzelius, J.J., **1839**. *Lehrbuch der Chemie*; 3rd Ed.; German translation by Wöhler, F.; Arnoldische Buchhandlung: Dresden.
Boutron, F., and Frémy, E., **1841a**. *Ann. Chim. Phys.,[3]*, **2**, 257.
Boutron, F., and Frémy, E., **1841b**. *Ann. Chem.*, **39**, 181.
Cagniard-Latour, C., **1838**. *Ann. Chim. Phys.*, [2], **68**, 26.
Dorvault, **1872**. *L'Officine de Pharmacie Practique*; 8th ed.; Dorvault: Paris.
Frémy, E., **1839**. *Ann. Chem.*, **31**, 188.
Gobley, N.T., **1844**. *J. Pharm. Chim.* [3], **6**, 54.
Hager, H., Ed. *Handbuch der pharmazeutischen Praxis*; Vol.I; Springer: Berlin, **1876**.
Haynes, Williams, **1954**. *American Chemical Industry; Vol. 1; Background and Beginnings*; D. Van Nostrand: Toronto; reprinted by Garland: New York, 1983.
Löw, C., **1952**. *Die Chemische Fabrik E. Merck, Darmstadt*; E. Merck-Darmstadt: Darmstadt.
Merck. *Preisliste*, **1860** and **1871**.
Pelouze, T.J. and Gélis, **1844**. *Ann. Chim. Phys.* [3], **10**, 434.
Vershofen, W., **1958**. *Die Anfänge der Chemisch-Pharmazeutischen Industrie*, Vol. III; Aulendorf: Stuttgart.

CHAPTER 3

MICROBIOLOGY BECOMES A SCIENCE

Louis Pasteur's[1] interest in the chemistry of natural substances was first aroused when he investigated the differences in optical behavior between tartrates and paratartrates, a subject to which Professor Jean Baptiste Biot had called attention in 1847. Because of Pasteur's success in unraveling this problem Biot asked him, in 1849, to study the optical activity of amyl alcohol, a by-product of alcoholic fermentation.

The studies of the optical properties of tartrates and, later, of amyl alcohol were, almost logically, followed by the study of fermentation processes and microbiological phenomena, and, still much later, by that of infectious diseases. In all these fields, Pasteur's work would become of paramount importance:

Pasteur's ideas about the cause of the differences in optical activity between otherwise identical substances would eventually lead to the formulation of stereoisomerism by van 't Hoff and Le Bel in 1874.

Pasteur's work on fermentation and the causes of infectious diseases forms the beginning of the science of microbiology. They also would become of prime importance in the manufacture of less perishable goods (through pasteurization). The scope of Pasteur's ideas about the prevention and the combatting of infectious diseases also can hardly be overestimated.

Pasteur on Optical Activity

Pasteur had observed that the crystals of the sodium ammonium salt of racemic acid (the inactive form of tartaric acid), which he had obtained by slowly crystallizing, consisted of two kinds of crystals. The fact that he noticed this proves his keenness of observation, which was helped by his near-sightedness. Most of the crystals were very small and minor differences between them had escaped earlier investigators.

Pasteur also noticed that the two kinds of crystals were mirror images of each other (Figure 3, 1). He separated them by hand according to their symmetry, dissolved the two different forms and found their solutions to be optically active. The one form was dextrorotatory to the same degree as the corresponding salt of normal tartaric acid; the other form showed exactly the opposite activity.

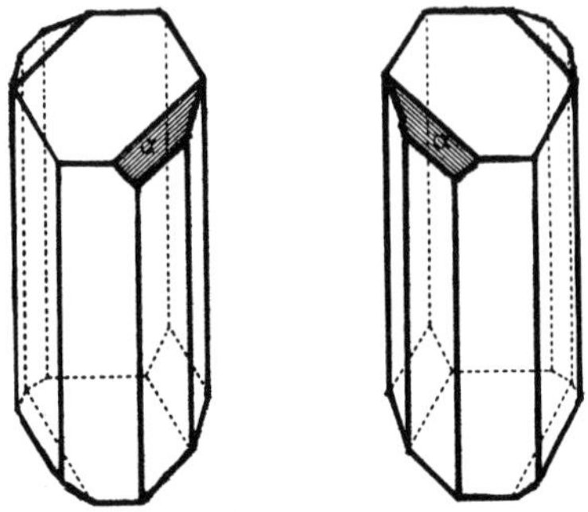

Fig. 3,1. Crystals of the two optical isomers of sodium ammonium tartrate. Reproduced with permission from Holleman, A.F., Leerboek der Organische Chemie, 13th ed.; by J.P. Wibaut Ed.; J.B. Wolters: Groningen, 1941.

In this way it was clearly demonstrated that racemic acid was composed of an equal mixture of two optically active forms of tartaric acid: a dextrorotatory form indicated as d-tartaric acid and a levorotatory form called l-tartaric acid or paratartaric acid.

The manner in which Pasteur had tackled the problem of the tartaric acids was very typical of him and showed the kind of thoroughness and versatility in approach that would become the hallmark of his scientific work. The inactive form of tartaric acid, called racemic acid, had been discovered in the mid-1820's by Charles Kestner, owner of a chemical works in Thann in the Alsace. Kestner had brought his discovery to the attention of professor Biot, *the* authority on the phenomena of optical activity.

Pasteur began his research with a visit to the tartaric acid plant in nearby Thann to find out whether the forming of racemic acid was related to the manufacturing process or to the origin of the wine from which the tartar was collected. Feeling that he needed more information on both possible influences, he made a journey (largely by train, the new means of travelling, which had supplanted the age old coach in the preceding ten years) to Germany, Austria and Bohemia to discuss the details of local manufacturing processes.

Back in his laboratory in Strasbourg, he started studies aimed at finding methods to separate the optical isomers and in fact in the course of the next few years he found two more methods to achieve this. (In the first method he uses an optically active alkaloid, which forms a precipitate with only one of the isomers. In the other method the natural d-tartaric acid is digested by microorganisms, which leave the paratartaric acid intact.)

The studies of the isomers of amyl alcohol and of tartaric acid led Pasteur to the very general conclusion that

> optical activity and life are somehow intimately associated, and the production of a single optically active substance unaccompanied by its mirror image is indeed nature's prerogative except under highly exceptional and "basically" asymmetrical conditions.
>
> *Pasteur, quoted from Geison (Gillespie)*

Up until today this postulate has remained one of the basic axioms of the chemistry of organic life.

The optically active properties of lactic acid and the developments that lead to the explanation of these properties are presented in Intermezzo I.

Pasteur on Fermentation

In 1854 Pasteur was appointed Professor of Chemistry and dean of the new Faculty of Science at the University of Lille. Here the young professor and teacher of chemistry devoted much time to the practical aspects of chemistry. In his courses he taught the students the principles of the processes used in the industries in the Lille region, at the time one of the most flourishing industrial regions of France.

In the fall of 1856 a local distillery asked Pasteur's advice on a problem it was experiencing in the fermentation of beet juice. Investigating with very primitive means the action of the brewer's yeast used in the distilleries on sugar (the university had no budget for equipment), Pasteur nevertheless was able to discover that two fermentations were taking place:

First a *lactic fermentation*; under the influence of a peculiar grey substance he called "lactic yeast". This special "yeast", composed of little globules or very short segmented filaments, he found to be responsible for the forming of lactic acid, which in turn played a critical role in the growth of the yeast cells and in the production of alcohol[2].

Secondly an *alcoholic fermentation* took place under the influence of another living organism, the yeast, which Pasteur described as a plant capable of reproduction and able to live and grow without free oxygen being present. Pasteur also established that without yeast, sugar will not undergo alcoholic fermentation, thus confirming the investigations of Schwann and Cagniard-Latour.

These studies of Pasteur, begun in Lille in 1856 and continued in Paris after his appointment as Director of Scientific Studies at the famous École Normale Superieure in 1857, demonstrated the role of microorganisms in the souring of beet juice (by forming lactic acid) and in the alcoholic fermentation itself. It also helped the distillery industry in solving some of their fermentation problems.

The first paper of Pasteur on lactic fermentation appeared in 1857 (Pasteur, 1857). The author paid tribute to the views of Liebig on fermentation and referred to the experiments of Frémy and Boutron, which seemed to support those views. But, he continued

> I have developed an entirely different point of view. I propose to provide proof in the first part of my paper, that just as always an alcoholic ferment, the *yeast of beer* [italics added], is found when sugar is converted into alcohol and carbonic acid [carbon dioxide], so always a special ferment, a *lactic yeast* [italics added], is found when sugar is transformed into lactic acid. And, furthermore, when any plastic nitrogenated substance is able to transform sugar into that acid, the reason is that it is a suitable nutrient for the growth of the [lactic] ferment.
>
> \- *Pasteur, 1857*

In a second article (Pasteur, 1858) on the same subject Pasteur recognized the fact that

> a large number of different yeasts does exist, which all have different actions. It presents great difficulties to isolate these different yeasts [microorganisms] and to find the appropriate conditions for their mutually exclusive developments.
>
> \- *Pasteur, 1858*

Pasteur also gave a description of the microorganism:

> The lactic yeast, consisting of globules or of other short corpuscles swollen at the extremities with diameters of about 1/400 mm, has much of the general properties of the yeast of beer, but it never forms alcohol and carbonic acid. The principle product of its action [on sugar] is lactic acid, a product never formed by fermentation under the influence of the yeast of beer.
>
> \- *Pasteur, 1858*

Pasteur's recipe for lactic fermentation was quite simple:

> I mix a solution of sugar with a small quantity of an ammonium salt, phosphates and precipitated chalk. After 24 hours the liquid turns turbid and some gas starts to develop. The fermentation continues the next days, the ammonium disappears, the chalk dissolves and calcium lactate is formed. Correlate to these phenomena one observes the formation of a sediment of lactic yeast, almost always associated with its infusoria [animalcules = microorganisms]. Often also calcium butyrate is formed. In

one word all of the characteristics so well defined earlier by monsieur Frémy do apply.

- Pasteur, 1858

On the subject of the origin of the lactic yeast present in these experiments, Pasteur remarked that its provenience could only have been the atmospheric air. This he admitted could possibly lend support to the view that *spontaneous generation* was a real phenomenon. He rejected this "evidence" implicitly by reporting further experiments, which showed that no lactic acid will be formed when air admitted to previously boiled solutions of sugar is passed through a red-hot platinum tube [that is when the experiments are performed under sterile conditions].

In sharp contrast with his nice title – Director of Scientific Studies – were the experimental facilities Pasteur had at the École Normale in Paris. He did not have a laboratory of his own, nor did he receive an allowance for research expenses. After many requests he first secured two rooms in the attic of the École, and in 1859 a small pavilion was put at his disposal to conduct his experimental work.

His next studies concerned butyric fermentations, which presented new problems. He had noticed that the butyric ferment was very sensitive to air, in fact, on exposure to air the ferment would die within about two hours. Carbonic acid [CO_2], however, would not harm the ferment at all. From these observations Pasteur concluded that fermentation processes do not always need air: the energy the little "animals" need to live and grow was apparently supplied by the substrate molecules only, in Pasteur's experiments cane sugar. The postulate *"fermentation is a form of life without air"* was another very important contribution of Pasteur to the emerging science of microbiology.

Pasteur on Spontaneous Generation

The scientific community had severe objections to the ideas of Pasteur. Liebig and his followers did not accept the idea that living creatures were able to perform chemical reactions, which could not be performed in vitro. They were convinced that "chemical reactions could not be seen under the microscope" and believed that at best catalytic processes were responsible for the reactions that occurred in fermentations.

The proof Pasteur gave of ways to prevent fermentation and putrefaction by sterilization of perishable substances and of the air to which they are exposed, dealt a decisive blow to the doctrine of spontaneous generation.

The problem of the origin of Leeuwenhoek's "animalcules" had of course been a difficult problem, no account of their "creation" was really satisfactory. Hence the idea that they could form spontaneously out of non-living matter was not rejected easily: everybody could observe that most natural substances left to themselves would soon start to ferment or putrefy, processes apparently always accompanied by the formation of microscopic living creatures.

This was, however, doubted by others, particularly by those who took the Biblical account of a completed Creation in seven days literally. Pasteur was one of those, in a lecture at the Sorbonne University in 1864 he even

> emphasized that the doctrine of spontaneous generation (like materialism in general) threatened the very concept of God the Creator... [But] he had approached the issue without preconceived ideas and would willingly have announced in favour of spontaneous generation had 'experiment imposed the view on me'.
>
> *- Geison on Pasteur in Gillespie, 1974*

In one of his experiments on spontaneous generation Pasteur used guncotton as adsorbing material and showed that air passed through the guncotton wad did not induce fermentation (nor putrefaction). A piece of the germ-laden cotton when placed in a sterile substrate solution, however, would always invoke bacterial growth. In a memoir on the subject, published in 1861, Pasteur concluded that spontaneous generation never occurs (Geison, 1974).

Some of his opponents still resisted his views: heat-treated air would have lost the capacity of spontaneous generation was an objection to some of his experiments; another more serious objection was that solutions considered to be sterile sometimes appeared to be putrefact[3].

Lister Prepares a Pure Culture of a Lactic Acid Bacterium

Pasteur's discovery that normal air, the air everybody breathes, is laden with innumerable microorganisms, was of immense importance for the new science of microbiology and was going to have many practical applications. The importance of this discovery for the treatment and prevention of diseases, for the health of mankind, would become even greater. Pasteur himself soon became aware of this, as is evidenced by his famous work on the causes and treatment of anthrax and hydrophobia. Also of immeasurable importance for the health of injured men was the application of Pasteur's findings in surgery. This was largely the work of the celebrated English scientist and surgeon Joseph Lister[4].

Around 1860 Lister became aware of the possible influence of bacteria on the inflammation of surgical wounds. At the time, putrefactive processes were a common phenomenon following even minor operations and often caused mortality rates of over 50%. These were ascribed to the influence of oxygen. Chemists, the most famous and influential of whom was Liebig, taught that putrefaction was a kind of combustion which most organic (= living) substances underwent in the presence of air. Explanations of fermentative processes in terms of actions of microorganisms, as those put forward by Cagniard-Latour and Schwann, had been ridiculized by Liebig cum suis in such an outrageous manner that the proponents of such theories were ignored by the scientific community at large (Ch. 2, 29). For a few isolated exceptions preceding Lister, see note[5].

In 1865 Lister begins to experiment with disinfecting chemicals: potassium sulfite and carbolic acid [phenol]. At about the same time his attention is called to the studies of Pasteur on fermentation and putrefaction. Later he reports on this period of his life:

> it occurred to me that decomposition in the injured part might be avoided without excluding air, by applying as a dressing some material capable of destroying the life of floating particles.
>
> *- Lister, quoted by Gillespie, 1974*

The introduction of aseptic methods by Lister, first in Glasgow, and after his appointment there also in Edinburgh, were a tremendous success; the death rate from major amputations dropped from 45% to 12%! His appointment of Professor of Surgery at King's College was the official recognition of these achievements.

The new teacher at King's College delivered his first lecture in October, 1877, on the subject of fermentation, a remarkable and uncommon subject for a professor of clinical surgery. In his Scottish years Lister had not only made a thorough study of the works of Pasteur, he had also read the earlier articles of Schwann and Cagniard-Latour. Through these studies and his own microscopic examinations he had become convinced of the correctness of the views of these investigators about the influence of microorganisms on fermentation and putrefaction and about the destructive influence microorganisms could exert in the healing of wounds.

Although the practical importance of Lister's methods could not be denied (the most important of which was the use of carbolic acid for the treatment of wounds as well as for the disinfection of surgical instruments), the view that infections could be transmitted by air his colleagues still found difficult to believe. He had, therefore, as the main subject of his inaugural lecture in London in 1877 chosen the influence of air-borne microorganisms on fermentation and putrefaction (Lister, 1878). In the introduction he explained the importance of the study of fermentation processes, very much in line with the ideas set forth by Pasteur (a remarkable fact because Pasteur was a chemist by education and most medical men did not believe that a chemist could contribute much of importance to their science.)

Alcoholic fermentation, he went on, only starts when a little of a yeast plant, for instance from fermenting grape juice, is added to a sugary solution, provided it contains other ingredients requisite for the nutrition of the yeast plant. Then he demonstrated that (oxen) blood collected under sterile conditions will not putrefy even when stored for a period of six weeks. Which proves that the blood has no inherent tendency to putrefaction. "Putrefaction, then, is a fermentation, a true fermentation, characterized by the power of self-multiplication of the ferment". Liebig's theory that all fermentations are caused by the action of so-called chemical ferments was rejected by the speaker. This purely chemical explanation is certainly the correct one for the decomposition amygdalin undergoes through the action of emulsin, he said, because emulsin is without doubt a chemical ferment destitute of life.

> But the fermentation which I have been especially investigating, has not been the putrefactive, but one which seemed to me more convenient for the purpose, the lactic fermentation, by means of which milk sours and curdles, through conversion of milk into lactic acid.
>
> *- Lister, 1877*

Boiled milk – Lister continued – will not ferment by itself (he showed the audience a flask with milk, 6 months old, which did not show any decomposition). On the other hand, in and on the surface of milk, exposed to dust from the air fungi and bacteria of various sorts will develop. But, referring to Liebig's theory, "may it not be that, by boiling the milk, you have destroyed certain chemical ferments", which otherwise would have provoked fermentation?

The problem Lister then faced was how to obtain sterile milk from a cow. He conducted a large series of experiments, but found that all the samples of milk he had collected underwent fermentation of many different kinds, but souring did not occur. How to explain this:

Simply this. If the *Bacterium lactis* had been here, it would have taken the precedence of all other organisms in its development and the changes which it would have induced would have made the milk an unfit soil for these other numerous species. And the novelty of the appearances depended not on the presence of an unusual variety of organisms, but merely on their having enjoyed an unprecedented opportunity for coming forward. Under ordinary circumstances they would have been smothered –killed – by the effects of the *Bacterium lactis* and the other ferments that commonly develop in its wake.

- Lister, 1877

In this paragraph Lister gives a formulation of how one kind of bacterium can overgrow other microorganisms, a principle which would later achieve great practical value in the development of industrial fermentations.

After an unsuccessful experiment in the cow-house where too many germs were present in the air, and another in the open air that failed because of too much wind, Lister performed a third experiment:

The same cow was taken out again into the open air, and this day the elements were in my favour. It had been a drizzly morning, and I might fairly hope that some of the multitudes of organisms existing in the little orchard might have been washed down and that the air might thus have been somewhat purified. I was also more careful in this respect. I got the dairywoman to milk the cow without drawing the hand over the teat, performing the operation by an action of the fingers in succession, so that the end of the teat should always be exposed. Her hands were washed with water, and the cow's udder also, and she having squirted out a little milk to wash away any organisms from the orifice of the duct, the glass cap which protected the larger tube from dust was removed and the end of the tube was held in the immediate vicinity of the teat; a few drachms [1 drachm = 3.55 ml] were introduced, then the cap was readjusted, and then these little glasses were filled by the simple expedient of alternately relaxing and compressing with the finger and thumb on the caoutchouc, so that there was as little disturbance as possible of the organisms that might be supposed to be introduced in spite of my care.

- Lister, 1877

Of the six samples so taken four had grown bacteria or fungi within weeks, two however, remained perfectly fluid and kept a perfectly sweet taste, and, under the microscope Lister "could not discover any organisms of any kind whatsoever". Therefore the conclusion was justified that in unboiled milk from a healthy cow no constituent is present "having the power of giving rise to organisms or producing the lactic or any other fermentative change".

Lister then decided to investigate if one bacterium was sufficient to start (lactic) fermentation. For this experiment he needed to know the number of bacteria in a certain volume of liquid, that is he had to make what was later called a bacterial count.

> This [bacterial count] being done with a particular specimen of souring milk, I found that it was needful to add no less than one million parts of boiled water to the milk to ensure that there should be rather less than one bacterium, on the average, to every drop. Then with drops of that size I inoculated five glasses of boiled milk, and the result was that out of the five only one curdled; but one did curdle and soured, and that one had the *Bacterium lactis* in abundance; the others did not curdle, underwent no fermentation whatsoever, and had no bacteria in them.
>
> *- Lister, 1877*

The experiment was repeated with boiled milk to which so many droplets of sour milk were added as contained, by calculation, two, respectively four bacteria. This time all of the glasses hatched, in all of them curdling did occur, and on microscopic investigation it appeared that in all of them *Bacterium lactis* was present.

Lister then concluded that the lactic fermentation of milk was caused by a living bacterium and that without such a bacterium milk would not sour. He also expressed the view that *Bacterium lactis* does not form spores or germs, in that case he would have observed rather different results in his counting and dilution experiments.

The experiments of Lister are a landmark in the development of microbiology:

It had now been established without doubt that one, and only one bacterium would suffice to bring about fermentation (or putrefaction); Lister had also shown that his *dilution method* made it feasible to prepare pure cultures, cultures containing only one kind of microorganism.

The Preparation of Pure Cultures

Joseph Lister had not been the first person who tried to make pure cultures of microorganisms. The first studies to this end had been performed by the mycologist O. Brefeld in 1872. He worked with media "solidified" with gelatin. With fungi this worked satisfactorily, but with bacteria the results were poor. The idea that the isolated spots of discoloration shown by so many food products when left standing were composed of colonies of only one type of microorganism had become quite natural in the 1870's as a result of the growing acceptance of the views of the pioneers of microbiology during the preceding forty years. The celebrated physician, turned scientist Robert Koch also picked up the idea and set to work out a practical method to isolate pure cultures. First he tried sterile slices of potatoes, which he inoculated with bacteria. The results were, however, not very satisfactory. He then gave gelatin another try; he used gelatin solutions containing nutrients required for bacterial growth. From these solutions plates were poured and these were, after solidification, infected by the streak method: the gelatin plates were streaked in thin lines with a previously sterilized platinum needle infected with a minute quantity of bacterial cells from the material to be investigated. The streaks caused local pointlike infections of isolated bacteria. Koch also developed a method for multiplying the pure colonies of microorganisms by transferring them to tubes filled with nutrient gelatin and placing them in a slanted position (slant cultures). In 1876 Koch used the new methods to prepare pure cultures of the anthrax bacterium. A couple of years later Koch's cleaning help asked the great doctor why he used gelatin as a hardener, she found it too temperature sensitive for use in the kitchen and preferred agar. Since that time agar plates have replaced gelatin plates in microbiology.

Pasteur and the Brewing Process

In 1876 Pasteur published the results of investigations he had made on beer brewing. This research was a strange consequence of the Franco-Prussian war of 1870–1871, which had turned Pasteur, all his life a very patriotic Frenchman, into an enemy of the Germans. In the aftermath of the war Pasteur decided to help the French brewing industry to attain or surpass the level of quality achieved by the German [lager] beer producers, in particular with respect to the shelf life of the beer. Ironically enough one of the causes of the better shelf life of German beers was due to the introduction of a heat treatment of bottled beers at 55°C, a "pasteurization" step advocated by Pasteur in 1866 for the improvement of the stability of wine.

In a 1873 patent Pasteur had already laid down the most important principles to be applied in the production of beer:

- the preparation and use of pure yeast cultures,
- the use of clean equipment,
- prevention of infection of the wort through the use of closed containers, and by covering the cooled wort with germ free air and carbon dioxide.

Hansen Introduces Pure Cultures in Beer Brewing

One of the brewers who read Pasteur's treatise on the production of beer with great interest was J.C. Jacobsen (1811–1887), founder of the Carlsberg Brewery in Copenhagen. After reading it, he decided almost immediately to have a modern and well-equipped laboratory built to introduce the new scientific methods in his brewery. In 1878 the laboratory was opened, it had cost 1.5 million francs. As first head of the Physiological Department the botanist Emil Christian Hansen (1842–1909) was appointed. Pasteur in his treatise on beer had described (hygienic) methods which would favor the growth of the desired yeast culture, he even claimed to be able to ascertain whether he had prepared an absolutely pure culture. Hansen was not convinced that Pasteur really had obtained pure cultures and set about to prepare such cultures himself (Klöckner, 1913; Wettstein, 1983). He first tried the dilution method applied by Lister to obtain cultures of *Bacterium lactis* and he succeeded in isolating colonies of yeast from one single yeast cell.

In the fall of 1882 Hansen visited the laboratory of the now famous Robert Koch in Berlin and learned there the use of nutrient gelatin plates and slant cultures for the isolation of pure cultures of microorganisms. He rushed back to Copenhagen to apply the new methods for the isolation of brewery yeasts, and of the "wild" yeasts Hansen believed to be responsible for the occurrence of unsaleable beers once in a while.

In November 1883 Hansen had cultured enough of a strain of pure yeast to run a trial at the Gamle Carlsberg brewery (Holter and Møller, 1976):

> The result was a most convincing success. Jacobsen's last doubts disappeared, and he used his great authority to speedily communicate the reforms as widely as possible, among others by sending notices to a number of famous breweries, offering them the pure yeast for experimentation. Hansen's aim had also been to provide Gamle Carlsberg with additional income through his inventions, and he therefore suggested that Jacobsen might establish a factory to produce pure strains to supply users all over the world with pure yeast. But Jacobsen did not agree; whatever was discovered at the Carlsberg laboratory belonged to the entire world. So, Jacobsen ... in fact gave away large sums of money.
>
> *Klöckner (1910), quoted by Holter and Møller*

The problems Jacobsen had experienced in his brewery in Copenhagen which resulted in beers having bad flavours and off-tastes, naturally existed almost everywhere in the brewing industry. Similar problems were experienced in the distillery branch. German distillers had taken notice of studies of Schwann and Pasteur on alcoholic fermentation even earlier than Jacobsen, and like Jacobsen they became receptive to the application of scientific methods in their trade.

In 1857 the German distillers founded the “Verein der Spiritusfabrikanten in Deutschland” (The Association of Distilleries in Germany) to support the technical development of their industry. In the beginning the members of the association sought to increase their knowledge by exchanging ideas, by reporting technical progress in the meetings and by inviting investigators to publish the results of their research in the columns of the journal of the “Verein”, the “Zeitschrift für Spiritusindustrie”.

A Research Institute for the German Fermentation Industries

Since the founding of the “Verein” the members had on many occasions discussed the establishment of their own research institute. This idea received a strong impetus from the publication of Pasteur’s work on beer technology (1873). The unification of the German countries in one Empire under the leadership of Prussia’s Bismarck induced the right climate for a joint effort to introduce also in this branch of industry the best scientifically based technology. Not a merely coincidental factor favoring the establishment of a research institute was the publicity aroused by Liebig who in a few lectures given in the years just preceding the Franco-Prussian war of 1870–1871 again had denounced Pasteur’s theories on fermentation, theories which then had achieved much interest and appreciation amongst brewers. (It was not so easy to take a stand against Liebig, the giant of German chemistry! The best thing, it was felt, was to make a really scientific and independent study in a German institute)[6].

The foundation of the “Versuchsanstalt für die Brennereigewerbe” (Research Institute for the Distillery Industries) took place in September 1874, the seat of the Institute would be Berlin, the capital of the new Empire. The first director was also appointed: Dr. Max Delbrück, (1850– 1919), a promising young scientist of 24. He would direct the research at the Institute for the next 45 years with impressive scientific and technological achievements[7].

Pure Cultures and “Natural Pure” Cultures

The new institute under Delbrück’s energetic direction did much to promote the use of pure cultures prepared according to Hansen’s methods in the

German distillery and brewing industry. The introduction of pure cultures in the industry proved to be more complicated than initially thought. One very important discovery was the observation that laboratory cultures when transferred to the brewery sometimes changed properties notwithstanding the fact that all fermentation conditions had been copied carefully. Scaling up factors like the rate of mixing, the oxygen concentration and the uniformity of temperature in the large vessels of the brewing industry apparently could have much influence on the behavior and the growth of yeast. Delbrück and his collaborators at the Institute investigated these technological problems in detail and were able to give the industry much valuable advice. One of Delbrück's important contributions to the fermentation industry was the formulation of the principle of *"natural pure culturing"* (natürliche Reinzucht), a principle which would also become of great importance in industrial lactic acid production:

In natural culturing use is made of certain characteristic properties of the (industrially) preferred type of microorganism to promote the growth of that organism while slowing down or suppressing the growth of other, undesirable microorganisms. (The basics of this principle can be found in Pasteur's early work on alcoholic fermentation, and also in the study of Lister on lactic fermentation.)

The Study of Lactic Acid Bacteria at Delbrück's Institute

Pasteur's first experience with lactic acid bacteria (he spoke of lactic yeast), was when he studied the problems of the distilleries in Lille in the 1850's. In Delbrück's Institute the same problems were tackled again in a very systematic manner:

It was established that alcoholic fermentation could best be performed under acidic conditions, microbial infections would then not have much chance against the fast growing yeast cells, they would be "overgrown" so to speak. A prime requirement to achieve such a favorable condition for the alcoholic fermentation was the use of pure strains of yeast as Hansen had already proved earlier at the Carlsberg Brewery in Copenhagen, and which the Institute had began to deliver to the industry in 1892.

In practice the souring of the mash (on brewing technology see Intermezzo 3) was usually achieved by lactic fermentation. Delbrück and collaborators found a temperature of 50°C optimal for this fermentation. At that temperature (many) lactic acid bacteria normally present in the mash still showed good activity, while (most) infecting bacteria were suppressed in their growth. As a matter of fact this was the first example Delbrück gave of his principle of "natural pure culturing". The selection in this fermentation was related to differences in temperature sensitivity, whereas in the ensuing alcoholic fermentation acidity was the selective principle.

Wilhelm Henneberg Introduces Pure Cultures in Other Industries

In the fall of 1897 the young scientist Wilhelm Henneberg joined Delbrück's renowned Institute as unsalaried collaborator[8]. He became assistant to Dr. Paul Lindner, head of the Botanical Department, sometimes referred to as the Botanical Institute, at other times called the Hygienic Institute.

Dr. Lindner had spent some years in Robert Koch's research institute before he came to the institute of Delbrück. Being very well versed in microscopy and microbiology, he was an ideal teacher and instructor for Henneberg. Lindner was also the originator of a very extended collection of microorganisms, the culture collection of the Institute. (The collection was lost completely in the turmoil following the capitulation of Hitler's Reich in 1945.)

Lindner charged his young assistant with the classification of microorganisms and the determination of their physiological properties. Henneberg showed great proficiency in this work and gradually more responsibilities were entrusted to him. In 1904 he was made head of the Technological Department, in 1906 of the newly organized Technical and Scientific Department, a position he kept until 1922 when he was called to head the Bacteriological Department of the Dairy Institute in Kiel.

On the occasion of the fiftieth anniversary of the Institute an overview of its scientific contributions was written by Glaubitz (1925), who said of the importance of Henneberg's work for the fermentation industry:

> The biological research [of Lindner's department] was not confined to yeast, on the contrary all other organisms of interest for the fermentation industry were part of the field of research. Thorough investigations by Henneberg on the use of fungi in the distillery industry were concluded in 1903 [these researches led to an improved method for the production of alcohol from potatoes, called "Amylobrennerei"]. The same researcher then studied the action of lactic acid bacteria in the distillery industry, in beer, in milk, in baker's yeast, in molasses, in sauerkraut, in gherkins and in leaven. And the use of *Bac. Delbrücki* [systematic name *Lactobacillus delbrueckii*] in lactic acid factories.

In 1906 the use of pure cultures of lactic acid bacteria was introduced [by Henneberg] in German distilleries. The important question of the cause of acid production in a fermenting yeast mash was solved. Infections can always be traced to "wild" lactic acid bacteria. These infections could be fully prevented by 1° of lactic acid[9].

The introduction of pure cultures of yeast in the brewery, the distillery and in the production of baker's yeast gradually increased to an impressive number of deliveries to these industries. It was a logical next step to also deliver the most applicable pure cultures of microorganisms to other fermentation industries. [Other sources report much earlier deliveries of pure cultures to the lactic acid industry; see Ch. 5, 144.]

- *Glaubitz, (1925) on Henneberg*

Henneberg was famous for his excellent microscopic studies, he is said to have worked every day for about four hours with the microscope. He was endowed with a very sharp eye, a critical and scientific mind, and the gift to make good drawings of what he had observed under the microscope. He also developed practical methods to control industrial fermentations, and taught practical microbiology to trainees from the industry.

Henneberg laid down his experiences in a textbook on fermentation bacteriology (Handbuch der Gärungstechnologie, 1st Ed., 1909; 2nd Ed., 1926; Paul Parey: Berlin). The book became a standard textbook for many generations of German and foreign microbiologists.

The End of the Liebig – Pasteur Controversy

It is appropriate to conclude this chapter about the beginnings of scientific microbiology (as far as related to the history of lactic acid making) with a short remark about the insight scientists at the end of the nineteenth century had acquired of the chemistry of fermentation processes.

The very basic discovery that the fermentative conversion of sugar into alcohol (and carbon dioxide) was caused by dead ferments, was made in 1896 by Eduard Büchner (1860–1917). In his now classic experiment, Büchner destroyed the cells of brewery yeast by ball milling with finely ground silica pellets, and produced a yeast extract completely free from living cells. This dead extract, added to a sugar solution, was capable of converting sugar into alcohol and carbon dioxide, albeit only in limited amounts and for a short period of time.

This experiment did on the one hand seem to confirm Liebig's view that chemical transformations taking place in fermentation were not caused by microorganisms:

No one will ever be able to see a chemical reaction under the microscope.

On the other hand it also gave reasons to believe that for a sustained fermentation process living (yeast) cells are a prerequisite to deliver fresh enzymes for the continuation of the process.

Tragically neither of the two adversaries lived long enough to witness this remarkable outcome of their acrimonious debates, Liebig (von Liebig) died in 1873, still fully convinced that Pasteur was wrong. Pasteur died in 1895, having lived long enough to become aware of the shakiness of some of his concepts and of the important role of enzymes (biocatalysts) in fermentation processes.

For his outstanding work, Büchner received in 1907 the Nobel prize for chemistry. In the first World War the German Empire needed his skills for higher duties: he died in action in Romania in 1917.

Biochemistry and Vitalism

Büchner's experiments mark the beginning of the science of biochemistry, the study of chemical processes taking place in the cells of microorganisms and other living matter. It became gradually accepted that the chemistry of "life" had to be understood in terms of enzymatic processes. The notion of vitalism, however, did not completely disappear, because the problem of the origin of the enzymes (and other complicated substituents of living cells) themselves had not been solved. Living cells were apparently involved in the production of enzymes, which could lead to a "corrected" hypothesis of vitalism:

Life is an essential condition for the genesis of enzymes, of biocatalysts.

The synthesis of enzymes only came within the grasps of science more than fifty years later when Crick and Watson discovered the role of DNA and RNA in the synthesis of proteins (1953).

Even after this discovery the arguments still went on:

... efforts to explain the phenomena of life in terms of the specific chemical properties of such things as enzymes or nucleic acids are doomed to fail because biochemical analysis cannot discern the complex integration necessary for specific biological organization and for the purposeful behavior of living things.

- Fruton, 1976

On this subject see also: Quastel, 1984.

Notes

[1] Louis Pasteur (1822–1895) studied chemistry at the École Normale in Paris. Among his teachers was the famous Professor Jean-Baptise Dumas, who in turn had received his education from Gay-Lussac at the École Polytechnique at about the same time as Pelouze and Liebig. Dumas was the first university teacher of chemistry who gave experimental courses in chemistry as an integral part of the curriculum. These courses became very popular and attracted many students, of French as well as of foreign nationality. Dumas became an ardent supporter of Pasteur, whom he gave much help in his scientific and academic career.

[2] The importance of performing the alcoholic fermentation under acidic conditions had already been pointed out in 1841 in a German textbook for the distillery industry by Lüdersdorff. The book also gives a (the first) titrimetric method to measure acidity quantitatively (Delbrück and Schrohe, 1904, 149).

[3] Later, in 1879, it was indeed shown by Chamberland in Pasteur's laboratory that certain microorganisms can withstand (for a short period of time) temperatures up to 115°C. In the case of spores, intermittent heating is required to achieve complete sterilization. Since Chamberland's thesis, the use of autoclaves and the flaming of glassware have become standard procedures in microbiological work (Gillespie, 1974).

[4] Joseph Lister (1827–1912) of Quaker descent, born in London, even as a child showed a lively interest in microscopy. This he had not learned from a stranger: his father, Joseph Jackson Lister, a wine merchant and amateur opticist was the inventor of the achromatic microscope (1830). This invention had paved the way for the scientific studies of microorganisms reported in Chapter 2, as well as for the researches of Pasteur.

After a general medical training in London he went, in 1853, to Edinburgh to become a surgeon. In 1860 he was appointed Professor of Surgery at the University of Glasgow, in 1869 he returned to Edinburgh. In 1877 Lister was appointed Professor of Surgery at the prestigious King's College in London, where he taught until 1893. Lister invented many new valuable methods and instruments for the surgical practitioner and is most credited for the introduction of aseptical surgery in the operating room.

[5] The American physician Oliver Wendell Holmes (1809– 1894) in 1835, and, independently, his Hungarian colleague Ignáz Fülöp Semmelweis (1818–1865) in 1847, both intuitively sought the cause of large mortality rates in hospitals in the transmission of disease by the hands and the instruments of surgeons. Both advised to wash hands (and instruments) with a solution of chlorinated lime.

[6] On the Pasteur – Liebig controversy of those years see also Lechevalier and Solotorowsky, 1974, 361 ff.

[7] The Institute would gradually expand its activities to other fermentation industries; the brewing industry and the starch manufacturers joined in 1883, other fermentation industries followed later (Dellweg, 1974).

[8] Wilhelm Henneberg (1871–1936) was the son of a physician in Magdeburg, he studied chemistry, botany and microbiology at the universities of Brunswick and Göttingen and obtained his doctor's degree at the last university in 1896. His

dedication to the study of microbes was strongly fostered by the early death of his mother from tuberculosis. Before coming to the Institute in Berlin he spent about one year at Halle University investigating acetic acid bacteria. Biographies: Windisch (1971 and 1974); Henneberg (1971).

[9] The term one degree of acid (1°) means 50 milli equivalent of acid per liter, in the case of lactic acid 1° = 0.45%.

Books Consulted

The New Encyclopaedia Britannica; 13th ed.; Encyclopaedia Britannica: Chicago, 1987.

Delbrück, M. and Schrohe, A. *Gärung und Fäulnis*; Paul Parey: Berlin, 1904.

Dellweg, H., *Die Geschichte der Fermentation*; in "100 Jahre Institut für Gärungsgewerbe und Biotechnologie zu Berlin", 1874–1974; Festschrift, Berlin, 1974.

Geison, G.L.; Louis Pasteur, in *Dictionary of Scientific Biography*; Vol. 10, Gillespie, C.E., Ed.; Charles Scribner's Sons: New York, 1974.

Glaubitz, M., *Die Biologie der Gärungsgewerbe*; in *"Das Institut für Gärungsgewerbe und Stärkefabrikation zu Berlin"; Festschrift zur Feier des 50-jährigen Bestehens am 29. September, 1924*; Parey: Berlin, 1925.

Holter, H. and Møller, K.M., Eds. *The Carlsberg Laboratory, 1876–1976*; Carlsberg Foundation; Rhodos: Copenhagen, 1976, 178.

Lechevalier, H.A. and Solotorovsky, M. *Three Centuries of Microbiology*; Dover: New York, 1974.

Nicolle, J. Louis Pasteur, *A Master of Scientific Enquiry*; Hutchinson: London, 1961.

Stanier, R.Y., Adelberg, E.A., and Ingraham, J.L. *General Microbiology*; 4th ed.; MacMillan: London, 1977.

Windisch, S., *100 Jahre Institut für Gärungsgewerbe und Biotechnologie zu Berlin, 1874–1974*; Festschrift Berlin, 1974.

References

Fruton, J.S., **1976**. *Science*, April 33, 327.

Henneberg, G., **1971**. *Dtsch. Molk.Z.* (Kempten-Algäu), *92*, 47.

Klöckner, A., **1913**. *Medd. Carlsberg Lab.*, *9*,, I.

Lister, J., **1878**. *Q.J. Microsc. Sci.*, *18*, 177.

Pasteur, L., **1857**. *C.R. Hebd. Seances Acad. Sci.*, *45*, 913.

Pasteur, L., **1858**. *C.R. Hebd. Seances Acad. Sci.*, *48*, 337.

Quastel, J.H., **1984**. *Can. J. Biochem. Cell Biol.*, *62*, 110.

Wettstein, D. von, **1983**. *Emil Christian Hansen Centennial Lecture;* EBC Congress 1983.

Windisch, S., **1971**. *Branntweinwirtschaft*, *111*, nr. 2.

MOLECULAR STRUCTURE AND OPTICAL ACTIVITY OF LACTIC ACID

The development of a comprehensive formulation system of organic substances during the nineteenth century is a very fascinating story in the history of chemistry. Some aspects of this story deal with lactic acid and should therefore be brought to the forefront.

As pointed out in Chapter 1, Berzelius helped the systematization of chemistry greatly by proposing the use of capital letters as symbols of the elements (H for hydrogen, C for carbon, and so on).

His views on what elementary particles were, would cause much confusion, however. With John Dalton (1766–1844) he considered an elementary atom as the smallest particle of an elementary substance, that is of an element, and considered elementary gases as composed of separate atoms. The existence of elementary molecules consisting of several atoms, for example H_2 and O_2, as proposed by Amadeo Avogadro (1776–1856) was rejected by Berzelius. (In 1911 Avogadro introduced the world molecule, Berzelius spoke about "composed" atoms.)

From the volume ratio 2:1 in which hydrogen gas and oxygen gas reacted to form water, Berzelius concluded that the formula for water is H_2O. Because of the weight ratio of eight of oxygen and hydrogen in H_2O the ratio of the atomic weights then is 16 to 1. He also used a formula with the ratio 8 to 1, in which the weight of two H atoms, written as Ħ, is chosen as unity.

Because of problems related to the unambiguous definition of atomic weight, many chemists used until about 1860 equivalent weights to indicate the weight ratios in which elements react with each other: the equivalent weight of oxygen in the reaction with hydrogen is eight. The formulas in use for chemical compounds therefore either denote the number of atoms of the different elements in the [molecule] or the number of equivalents. Much confusion arose as a consequence of this double system of formulation, as is evidenced for instance in the case of lactic acid.

In the paper of Gay-Lussac and Pelouze which presents the first analysis of lactic acid (Gay-Lussac and Pelouze, 1833a), lactic acid bears the formula $C_6H_6O_6$ (C=6, H=1, and O=8), whereas in Liebig's translation (Gay-Lussac and Pelouze, 1833b) the formula reads $C_6H_{12}O_6$ (C=12, H=1, and O=16).

A further confusion originated from the introduction of so-called rational formulas which pretend to define how the [molecule] of a compound is composed of [molecules] of other compounds. Gay-Lussac and Pelouze for instance write lactic acid also as $C_6H_4O_4$ + 2HO, meaning that lactic acid is considered to be the hydrate of anhydric lactic acid – $C_6H_4O_4$.

The rationale for the definition of lactic acid as a hydrate is also related to the fact that the analysis was carried out via the zinc, copper or calcium salt. These salts could be obtained in crystal form well amenable to analysis which was not possible with the syrupy concentrated lactic acid. The composition of the zinc salt for instance was written as $C_6H_4O_4$ + zinc oxide (C=6, H=1 and O=8). In this definition (which follows Berzelius) a salt is defined as a combination of an acid, that is an anhydride, with a base, that is a metal oxide.

When syrupy lactic acid was exposed to dry distillation Gay-Lussac and Pelouze obtained a sublimate with the composition $C_6H_4O_4$ (C=6, H=1 and O=8), which they believed to be (anhydric) lactic acid. In their view the "rational" formulas are therefore:

syrupy lactic acid $C_6H_4O_4 + 2HO$
zinc lactate $C_6H_4O_4 + HO + ZnO$
(anhydrous) lactic acid $C_6H_4O_4$

Berzelius doubted whether the foundations of these relationships were correct (Berzelius, 1837):

> It is by no means certain that the sublimate is anhydrous lactic acid. It could be a substance which reacts with water to [form] that acid.
>
> *- Berzelius, in Wöhler's translation, 1837*

[We know now that Gay-Lussac and Pelouze's solid lactic acid was dilactide, the cyclic interesterification product of two molecules of lactic acid.]

Liebig: Lactic Acid Is Dibasic

From about 1837 Liebig assumed that many acids could be considered as compositions of one molecule of acid anhydride with a number of molecules of water, which could be replaced by one or more molecules of metal oxide, forming acid salts, neutral salts or double salts. An acid which "contained" more than one replaceable water molecule he called polybasic.

A simple criterion for di- and polybasicity was formulated by the great

master in 1838:
acids are polybasic when they can form mixed salts.

In 1847 Engelhardt and Maddrell, working in Liebig's laboratory at Giessen, reported the existence of two double salts of lactic acid, viz.:

- acid barium lactate which they had prepared from the neutral barium salt and an amount of lactic acid equal to the amount present in the barium lactate, and
- acid calcium lactate which was preferably obtained by crystallization from a solution containing an excess of lactic acid and recrystallization from anhydrous alcohol (Engelhardt and Maddrell, 1847).

According to Liebig's criterion the existence of these two acid salts offered conclusive proof of the dibasic character of lactic acid.

The Discovery of Sarcolactic Acid

In those years Liebig himself was studying the properties of "the constituents of the liquids of meat". In a lengthy article on this subject (Liebig, 1847) he reports as one of the constituents a nitrogen free acid, identical with the acid (lactic acid) formed when lactose, starch, dextrose and cane sugar are brought into contact with "decomposing animal substances". In a note Liebig, however, makes a critical remark about the similarity of the two acids:

> Engelhardt and Maddrell have established that calcium lactate obtained by fermentation contains five molecules (29%) of crystal water, and zinc lactate four molecules (18%). This contradicts with my observation of four molecules (25.8%) of crystal water in the calcium salt, and two molecules (13.4%) in the zinc lactate. The difference may be due to the different manners in which the crystallizations have been effected.
>
> *- Liebig, 1847*

Not content with this unsatisfactory conclusion, Liebig asked Engelhardt to investigate the matter further. After a thorough study of the crystal forming habits, the solubilities and some other physical properties of a number of lactate salts, Engelhardt came to the conclusion that the salts of lactic acid from meat (to which he gave the name *sarcolactic acid*) have properties different from those of the salts of lactic acid extracted from souring substances – the fermentation product (Engelhardt, 1848). In the same year

Professor W. Heintz of the University of Halle (Prussia, now GDR) proposed to use the more systematic name of *paralactic acid* for the new substance.

The Structure of the Lactic Acid Molecule

In the 1850's lactic acid was the subject of quite a number of investigations of its chemical reactivity, and the nature of the products which could be prepared from it (for a summary see for instance Wurtz, 1870). The interpretation of the results caused a profound controversy between Hermann Kolbe (1818–1884) and Charles-Adolphe Wurtz (1817–1884). Wurtz was a proponent of the type theory of classification. In this theory organic compounds were regarded as substitution products of simple inorganic substances like H^2, H^2O^2 (O=8) and NH^3.

In Wurtz' view the formula of dibasic lactic acid can be derived from that of two molecules of water –H^4O^4– by replacing two H atoms by the divalent lactyl radical $C^6H^4O^2$:

$$\left.\begin{matrix} C^6H^4O^2 \\ H^2 \end{matrix}\right\} O^4 \quad (C=6,\ O=8)$$

According to Wurtz the two "typical" H atoms of lactic acid are responsible for its acidic character. Lactic acid is therefore dibasic.

Kolbe, who modernized the radical theory of Berzelius, considers lactic acid as a carbonic acid ($C^2O^4 = 2CO^2$) in which one O atom has been replaced by the ethyl radical C^4H^5 (C=6, H=1) yielding the formula of "propionic acid" $C^4H^5.C^2O^3$. The next (hypothetical) step is the replacement of one H atom in the ethyl radical by hydroxyl O^2H ($O={}^8$, $H={}^1$). Lactic acid is the hydrate (water is OH) of this compound, hence the formula of lactic acid becomes:

$$\left(\begin{matrix} C^4H^4 \\ O^2H \end{matrix}\right) C^2O^3 \cdot OH \quad (C=6,\ O=8)$$

In a similar way Kolbe writes propionic acid also as an hydrate:

$$\left(C^4H^5\right)\ C^2O^3 \cdot OH$$

Propionic acid and lactic acid are both monobasic because the basicity depends on the number of oxygen atoms in C^2O^4 that has been replaced: C^2O^4 is dibasic, $X.C^2O^3$ is monobasic and $Y.Y.C^2O^2$ has no acid function. The hydroxyl radical distinguishes lactic acid as an alcohol.

Also around this time, August Kékulé (1829–1896) and, independently, the Scottish chemist Archibald Scott Couper (1831–1892) introduced the concept of the quadrivalency of carbon and the notion to write (fixed) linkages between the atoms and radicals in the molecule and the carbon atom. This enabled the writing of formulas of organic molecules, in the formulas the relative positions of the various groups in the molecule now could be presented directly. This new theory, the structure theory, soon replaced the older type and radical theories.

The next important step in the systematization of organic chemistry came from Stanislao Cannizzaro (1826–1910), who proposed at an international congress of chemists in Karlsruhe in 1860 to base the definition of atoms and molecules on determinations of their molecular weight. In the course of the next ten years the old atomic (= equivalent) weights of 6 for carbon and 8 for oxygen were abandoned, and modern writing of organic formulas became fully accepted. In 1859 Kékulé compared lactic acid with glycolic acid of which he also had views regarding the structure which differed from those of Wurtz. Both compounds he said

> have the character of a monobasic acid, although they contain two typical hydrogen atoms:

$$\left.\begin{array}{r} C_2H_2''O \\ H_2 \end{array}\right\} O_2 \qquad\qquad \left.\begin{array}{r} C_3H_4''O \\ H_2 \end{array}\right\} O_2$$

glycollic acid lactic acid

> The two [typical] hydrogen atoms of glycollic acid are of unequal value, ... One behaves exactly like the typical hydrogen of alcohol, the other exactly like the typical hydrogen of acetic acid ..., the one lies, as in acetic acid, in the neighbourhood of *two* oxygen atoms, the other, as in the alcohols, in the neighbourhood of *one* oxygen atom.
>
> *- Kékulé, 1859, quoted from Partington, Vol IV, 1964, 525*

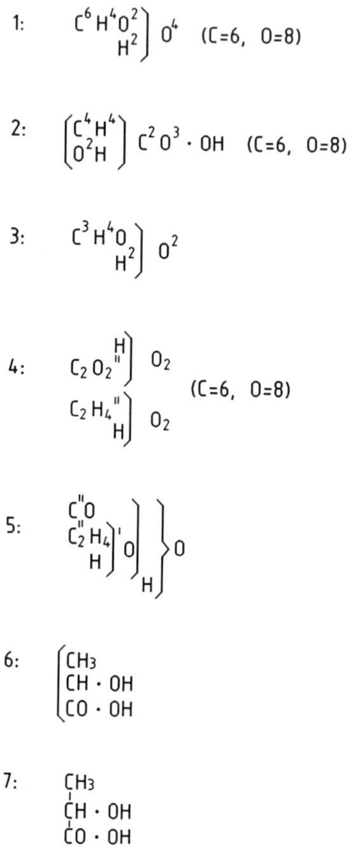

Fig. II,1. Formulas of lactic acid (1850–1873).
1. Wurtz[1] in 1857. 2. Kolbe in 1859. 3. Wurtz in 1864 (in response to Kolbe). 4. Perkin[2] in 1861. 5. Wislicenus in 1863. 6. Von der Brüggen in 1868. 7. Wislicenus in 1873.
Literature: Wislicenus, 1863b. Von der Brüggen, 1868. Wurtz, 1870. Wislicenus, 1873. Partington, 1964.

Graphic formulas, Kékulé says, make the unsymmetrical character (of glycolic acid and lactic acid) clear, but, unfortunately, he did not give these formulas. [In modern writing the formula of glycolic acid is

$$HCH(OH)–CO_2H.$$

Replacing one H by CH_3 gives the formula of lactic acid

$$CH_3–CH(OH)–CO_2H.]$$

Once the monobasic character of lactic acid was accepted, salts having two different metal atoms (or one metal and one hydrogen atom) could be explained as being double salts, that is salts formed by cocrystallization of two different singular salts, or of one salt plus the acid. Wurtz writes in his Dictionnaire de Chimie (l.c.) acid calcium lactate as:

$$(C_3H_5O_3)_2Ca + (C_3H_6O_3)_2 + 2\ H_2O.$$

To give an idea of the state of confusion which must have existed before the introduction of valence linkages a number of the formulas given for lactic acid in the years between 1850 and 1873 are presented in Figure I1,1.

Johannes Wislicenus (1835–1902)

Kékulé's structure theory was not able to explain the existence of "optical isomerism" of lactic acid nor the difference in properties of the salts of the isomers of lactic acid. The study of the structure of lactic acid was in 1858 also taken up by Johannes Wislicenus, then assistant of Professor Wilhelm Heinrich Heintz (1817–1880) of Halle University. Wislicenus' investigations of lactic acid lasted with some interruptions about 15 years and finally provided the clue for the elucidation of its structure.

Wislicenus, who was of Polish descent, was a young student of chemistry at Halle University, when his father, a Lutheran pastor with modernistic ideas, for religious reasons decided to emigrate with his family to America (1853). In 1853 and 1854 Wislicenus was assistant to Professor Horsford at Harvard University (Cambridge, Massachusetts)[3]. In 1855 Wislicenus operated an analytical laboratory in New York, where he also lectured at the Mechanics' Institute. A year later he went back to Europe to follow a scientific career.

From 1857 until 1859 Wislicenus worked with Heintz at Halle University, he went to Zürich in 1859 and later held professorships in Würzburg and Leipzig (Partington, 1964).

One of Wislicenus' first publications on the properties of lactic acid concerns the role of the two reactive hydrogen atoms in the molecule (Wislicenus, 1863a). To explain the different types of reactions of the two hydrogens Wislicenus uses Kolbe's theory. He concludes, however, that the only difference between the isomers is the order in which the radicals "are distributed in space, because our present formulae can do no more than represent a picture from the compound in one plane" (Wislicenus, 1863, quoted from Partington, 1964, Vol. IV, 760).

In the next ten years a number of chemists tried to use Kékulé's structure theory to explain the phenomenon of optical activity. Wislicenus would come closest to a solution of the problem which finally was solved by van 't Hoff and Le Bel (See for instance Brooke, 1985).

What is Optical Activity?

Light can be described as a wave phenomenon in which the vibration of the waves is perpendicular to the direction of propagation. This bears some similarity to ocean waves, which also vibrate perpendicularly to the direction in which the waves travel. Waves of ordinary light, however, vibrate around the axis of propagation in all possible perpendicular directions, there is not one specially favored direction.

The discovery Malus made in 1808 was that light which hits a reflecting surface at a rather narrow, well-defined angle is partly reflected and partly refracted. The angle at which this happens is called the polarization angle. The reflected light only has waves which vibrate parallel to the surface of the reflecting surface, the waves of the refracted light vibrate at an angle of 90 degrees of that direction. The original light beam has been split into two polarized beams of light.

Two decades later the Scottish physicist William Nicol devised a special prism in which this difference in reflecting properties was used to allow only one of the two perpendicular phases of light to pass through. This prism came to be called nicol or polarizing prism. When two nicols are placed behind each other the amount of (polarized) light coming through the second nicol depends on the angle the second nicol is rotated relative to the first one (Figure I1,2). Biot placed cuvettes filled with organic solutions between two crossed nicols and found that in some cases light passed through and in other cases this did not happen. When the organic substance had this influence, darkness could again be found when the second nicol was rotated over a certain angle. Such a substance was said to be optically active. When the nicol had to be turned to the left (counter clockwise) to obtain darkness it was called levorotatory, when the opposite was the case it was called dextrorotatory.

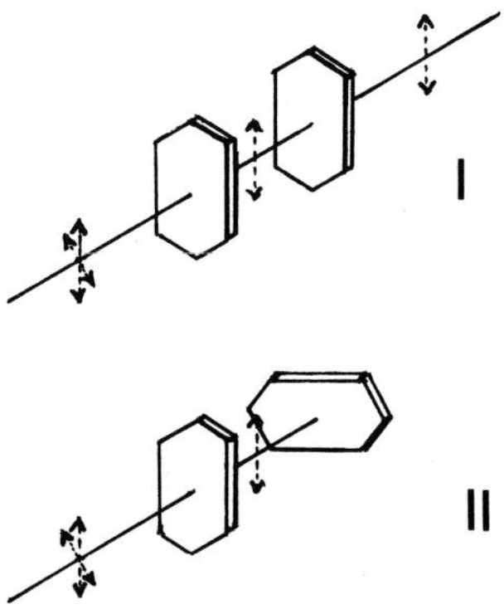

Fig. II,2. The effect of polarizing prisms on light.

I A beam of light passes two parallel nicols: the first transmits only one phase, it polarizes the light.
The second nicol transmits the polarized beam completely.

II The nicols are crossed: the second nicol extinguishes the polarized beam completely.

The Stereochemistry of Lactic Acid

The first person who had related optical activity to asymmetric properties of the molecule was Pasteur. From the studies of the optically active properties of amyl alcohols and tartaric acids he had conceived the idea that optical activity had something to do with molecular dissymmetry (Pasteur presented his ideas to the Paris Chemical Society in 1860. Nicolle, 1961).

This was quite a bold idea:
the notion of the existence of discrete elementary particles (atoms and molecules) had been gradually accepted in chemistry since the days Dalton

had formulated the atomic theory in 1808. The study of the structure of molecules, however, was considered to be the domain of the physicist, for chemists the existence of molecules was a useful hypothesis only and molecular formulas served primarily as a means to explain the multiple functions of the molecules in chemical reactions. A possible influence of the geometry of the molecule was not taken into consideration.

In the following decade a number of scientists picked up the idea that the geometry of the molecule could give an explanation of some properties of carbon compounds (see Partington, Vol. IV, 1964). It was again Wislicenus who made some of the decisive investigations. In 1869 he gave a lecture in Innsbruck on the various modifications of lactic acid:

It is very remarkable, Wislicenus says, that three isomers of [hydr]oxypropionic acid are known: ethylidene lactic acid [1-hydroxypropionic acid], ethylene lactic acid [2-hydroxypropionic acid] and sarcolactic acid. The last acid seems to be a mixture of ethylene lactic acid and a modification of ethylidene lactic acid. Sarcolactic acid is optically active, it rotates the plane of polarization about 3.3 degrees to the right. He concludes his lecture with the remark (Wislicenus, 1869):

> Such delicate isomeries can possibly be well explained by a geometric presentation of the arrangements of the atoms.
>
> *-Wislicenus, 1869*

In a lengthy paper on the properties of paralactic acid Wislicenus discussed in 1873 possible explanations of the fact that paralactic acid and fermentative lactic acid, although having identical chemical structures, nevertheless differ much in optical activity (Wislicenus, 1873). The salts also differ in properties, they show differences in solubility, crystal form and bound crystalwater[4]. The only reasonable explanation for these phenomena is to suppose a difference in the geometrical arrangement of the atoms in the molecular structure, Wislicenus concludes. He proposes the expression *geometrical isomerism* for this phenomena, which he still regards to be of a physical nature.

A year later the Dutch chemist Jacobus Henricus van 't Hoff[5] who was much influenced by Wislicenus' studies and thoughts on the isomerism of lactic acid, proposed a geometrical model to explain the phenomenon of optical activity and some other "irregularities" of the existing theories in explaining the structures of unsaturated compounds. Van 't Hoff's drawings of the two isomers of this acid are given in Figure I1,3:

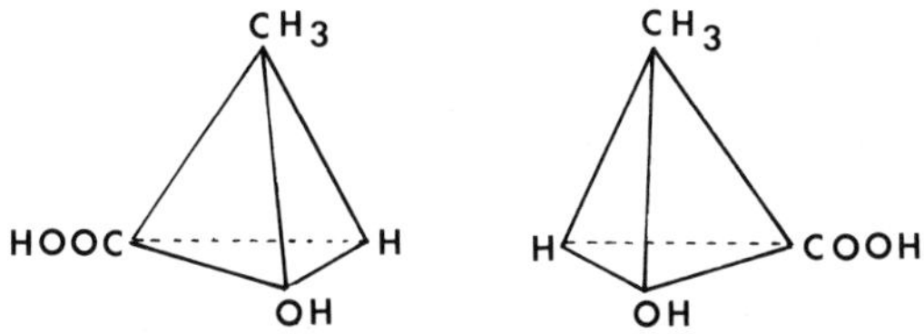

Fig. II,3. Van 't Hoff's presentation of the two isomers of lactic acid. The central asymmetric carbon atom is not depicted, it is situated in the center of the triangular pyramid.

In the same year as van 't Hoff published his ideas on the stereochemical presentation of organic formulas, the French scientist Joseph Achille Le Bel (1847–1930) independently arrived at almost the same explanation of the phenomenon of optical rotation. Le Bel in his paper explicitly names Louis Pasteur as the man who first laid the relation between optical rotation and molecular asymmetry.

Modern Nomenclature Relating to Stereoisomerism

The scientific name for optical isomers, stereoisomers, of which the molecules geometrically are mirror images, today is enantiomers. An example of true enantiomerism is given by the two optical isomers of lactic acid. Sarcolactic acid, the only lactic acid "natural" to humans and mammals is called L(+)-lactic acid or according to the latest convention S(+)-lactic acid. The other enantiomer is then the D(-) or R(-) form[6]. Their optical activities are exactly opposite, the precise value depends on concentration, temperature, the wave length of the light and the content of interesterification products (see Holten, 1971). The salts of lactic acid, in particular zinc lactate, are more amenable to the determination of a reproducible value of the optical activity and, hence, to the determination of enantiomeric purity. Salts of lactic acid have an optical activity of the sign which conforms with the letter indication of the enantiomerism: L = levorotatory = minus, and D= dextrorotatory = plus.

The use of the l-sign for the geometry of one of the isomers of lactic acid as well as for the negative optical rotation of the other isomer has led to much confusion. In the literature from the days of Wislicenus up to the 1940's l-lactic acid can mean a configuration related to that of the corresponding amino acid (alanine), or a negative rotatory power. The right configuration of sarcolactic acid was established by Walden in 1899. He then expressed the view that the acid should be named l-lactic acid, although its optical rotation is to the right (Walden, 1899).

In this book, confusing indications have in general been replaced by modern symbols, unless noted otherwise. The designations of the latest convention S = sinistra (Italian) and R = rechts (German) have not been used.

Notes

[1] Charles Adolphe Wurtz (1817–1884) studied medicine and chemistry in Strasbourg and Paris, spent a summer semester (1842) with Liebig in Giessen and became assistant of Dumas in 1845. In 1853 he succeeded Dumas as professor of chemistry at the "École de Médicine" in Paris. One of his many research subjects concerned the properties and structure of lactic acid.

[2] William Henry Perkin (1838–1907), famous for his accidental discovery of mauveine (1856), the first synthetic organic dyestuff, studied chemistry at the Royal College of Chemistry in London with professor August Wilhelm von Hofmann (1818–1892), also a student of Liebig.

[3] On Horsford, who, indirectly, played a role in the first industrial production of lactic acid, see Ch. 4, 80.

[4] In one of his numerous studies on the chemistry of lactic acid Wislicenus also reported the discovery of a new synthesis for [racemic] lactic acid via lactonitrile (Wislicenus, 1863b): acetaldehyde was reacted with hydrogen cyanide to lactonitrile, the nitrile thereupon hydrolyzed with hydrochloric acid. This was a very simple reaction with high yields. Almost a hundred years later when acetaldehyde and hydrogen cyanide had become cheap (petro-) chemicals this reaction would become the basis for the synthetic production of lactic acid.

[5] Jacobus Henricus van 't Hoff (1852–1911) studied chemistry in Delft and Leyden in the Netherlands, then with Kékulé in Bonn and with Wurz in Paris. He received his doctor's degree in Utrecht in 1874, the same year in which he published his ideas about the spatial arrangements of the atoms in organic molecules (Voorstel tot uitbreiding der structuurformules in de ruimte).

[6] D(-)-lactic acid was not discovered until 1890 when Schardinger obtained it by fermentation (Schardinger, 1890).

Books Consulted

Dictionnaire de Chimie; Wurtz, C.A., Ed.; Vol. II, 2; Librairie Hachette: Paris, about 1870.

Grote Winkler Prins (Encyclopedie), 7th Ed.,; Elsevier: Amsterdam, 1966–1975.

Alphen, J. van. *Overzicht van de Geschiedenis der Organische Chemie voor 1870*; Stenfert Kroese: Leiden, 1933.

Fierz-David, H.E., *Die Entwicklungsgeschichte der Chemie*; Birkhäuser: Basel, 1945.
Holten, C.H. *Lactic Acid*; Verlag Chemie: Weinberg, 1971.
Partington, J.R. *A History of Chemistry*, Vols. 3 and 4; MacMillan: London, 1964.
Snelders, H.A.M. *Hoofdstukken uit de Geschiedenis van de Scheikunde; Rodopi: Amsterdam 1979.*

References

Berzelius, J.J., **1837**. *Lehrbuch der Chemie*; translated from the third Swedish edition by F. Wöhler; Arnoldische Buchhandlung: Dresden, p. 135.
Brooke, J.H., **1985**. *Recent Developments of the History of Chemistry*; Russell, C.A., Ed.; The Royal Society of Chemistry: London.
Brüggen, N. von der, **1868.** *Ann. Chem.*, **148**, 224.
Engelhardt, H. and Maddrell, R., **1847**. *Ann. Chem.*, **63**, 83.
Gay-Lussac, J. and Pelouze, J., **1833a**. *Ann. Chim. Phys.*[2], **52**, 410.
Gay-Lussac, J., and Pelouze, J., **1833b**. *Ann. Chem.*, **7**, 40.
Liebig, J., **1847**. *Ann. Chem.*, **62**, 257.
Schardinger, F., **1890**. *Monatsh. Chem.*, **11**, 569.
Walden, P., **1899**. *Ber. Dtsch. Chem. Ges.*, **32**, 2861.
Engelhardt, H., **1848**. *Ann. Chem.*, **65**, 359.
Wislicenus, J., **1863a**. *Ann. Chem.*, **125**, 41.
Wislicenus, J., **1863b**. *Ann. Chem.*, **128**, 22.
Wislicenus, J., **1869**. *Ber. Dtsch. Chem. Ges.*, **2**, 550.
Wislicenus, J., **1873**. *Ann. Chem.*, **167**, 302.

CHAPTER 4

INDUSTRIAL PRODUCTION OF LACTIC ACID BEGINS IN AMERICA

Bensch's recipe (Ch. 2, Exh.2,I and II) for making lactic acid, published in 1847, found wide publicity, and was probably used quite frequently by apothecaries and others. It allowed for the preparation of the acid on kilogram scale and gave a good, albeit elaborate method for the purification of the acid. The recipe had been developed without any knowledge of the fermentation process involved, knowledge which would only gradually become available during the following 20 to 30 years, primarily thanks to the work of Pasteur and his disciples.

It is an amazing fact that, although all of these early developments in fermentation "technology" took place in Europe, the first commercial production of an industrial chemical by a fermentation process, viz., that of lactic acid, took place in the United States of America. The United States had a number of disadvantages in the evolution of chemistry and chemical technology in the nineteenth century, disadvantages which persisted up to World War I. The country was slow in picking up the new scientific thoughts coming from the other side of the Atlantic, and when they became known, was not always eager to accept the new ideas. Research was considered to be elitist, a thought held even by prominent men of science. Veysey (Veysey 1965) cites as examples Charles W. Eliot, a chemist who became president of Harvard in 1869, and Daniel Coit Gilman, a geographer and the first president of Johns Hopkins University in Baltimore in 1875. These and other leading men on the boards of American universities had considerable objections to allow the "German" doctrine of materialism to become part of American higher education. The philosophy of materialism in which "all being and processes can be explained as manifestations or results of matter" (Webster's Dictionary, 1974) had guided most European scholars in physics and chemistry since the days of Laplace and Lavoisier, and had proven to be a very useful hypothesis. For quite a few of these European scholars it was, however, also a kind of belief, almost a religion, and to this American educators objected strongly.

In the practical application of chemistry the United States also lagged behind, particularly in respect to Britain. During the Revolutionary War (1775–1783) and in the first decades thereafter, American chemical manufacturers had gradually learned to make the chemical products hitherto imported from England. Centers of chemical activities were in particular the Philadephia and New York areas and some regions of New England (for instance Massachusetts). As in Europe, the manufacture of chemicals was first carried out by apothecaries and by a few chemically trained or self-trained men, who saw practical applications for the products of the new, intriguing, and marvelous "science" of chemistry. At the universities chemistry was one of the courses given in the medical departments, a situation which gradually changed in the first half of the nineteenth century. Haynes (Haynes, 1954) writes that although by 1853 already more than 100 courses in chemistry (in departments of science) were offered by American institutes of higher education, only four of them had a chemical laboratory and provided facilities for experimentation and research. The interest in experimental chemistry had been much stimulated by the teachings of Benjamin Silliman Sr. (1779–1864), Professor of Chemistry at Yale University, New Haven, CT. Silliman had observed the importance of applied chemistry in the early days of the Industrial Revolution while studying in London and Edinburgh, and upon his return he had transferred Yale into a center of chemical education. Even more influence was exerted, from 1840 onwards, by the new schools of chemistry at the European universities, which attracted many young Americans who wanted to become acquainted with the latest theories and discoveries in the field of chemistry. They followed in the footsteps of Silliman, but more and more they went to Germany, the nation which, thanks to people like Liebig, was now becoming the Mecca of chemistry.

European chemistry was evolving into a systematic science; it had adopted a useful nomenclature, the elements had received comprehensible abbreviations and the formalization of the composition of organic molecules was nearly complete (see Intermezzo 1). Of great importance also was the conviction that chemistry had to be based on quantitative measurements, on analysis. Liebig, for instance, trained his students in the three aspects of chemistry that he considered of prime importance: analysis, preparative work and research. In his view research not only implied the synthesis of new products, but also, and even more, the search for the principles governing chemistry.

Chemistry in Massachusetts

The Eastern part of Massachusetts, with Boston as its center, was one of the areas where chemistry found a fertile soil. In colonial times Boston was one of the most important harbors the British owned on the American coasts. In the first half of the nineteenth century, when New England became a center of industry, Boston as a harbor and industrial town reached an even more prominent position. The textile industry of the region was one of the first in the world to learn and apply the new methods invented and developed in Britain during the first stages of the Industrial Revolution a few decades earlier. From a craftsmanlike type of operation around 1800 the New England textile industry was transformed in the course of some thirty years into an industry of integrated textile mills, which took in raw cotton at the one end and poured out finished and colored fabrics at the other end. As in Britain the American textile producers required many chemicals for their operations, some of them in large quantities – sulfuric acid, caustic soda, bleaching powder, starches, dyes, etc. Small chemical factories were set up to supply these chemicals, either by small entrepreneurs or by the textile works themselves. Other chemicals were imported from England, and raw materials came from many parts of the world, mainly through Boston harbor.

As early as 1803 a canal, the Middlesex canal, was dug to connect Chelmsford on the Merrimack River with Boston to facilitate commerce. By 1840 quite a few small chemical enterprises had been established near this canal and other waterways in the region. Then railroad building began in the State of Massachusetts. By 1855 fifteen lines were in operation, greatly improving the transportation of men and goods of all kinds from and toward Boston (See Map 4,1). The city of Chelmsford, renamed Lowell, by the middle of the century had acquired so many textile mills that it became known as the American "Manchester". But the Boston-Lowell area also had many other flourishing industries: paper mills, tanneries, paint shops, brick and tile works, etc. All of these industries needed chemicals in larger or smaller amounts.

Chemical Education in Boston

Today we do not believe that chemicals can be produced without a good knowledge of the science of chemistry, without the involvement of professionally trained chemists. This view, however, was not held during the first half of the nineteenth century. Chemists usually acquired their skills through apprenticeships with other chemists and apothecaries, in a few cases through studies and practical training in England and in later years in France and Germany.

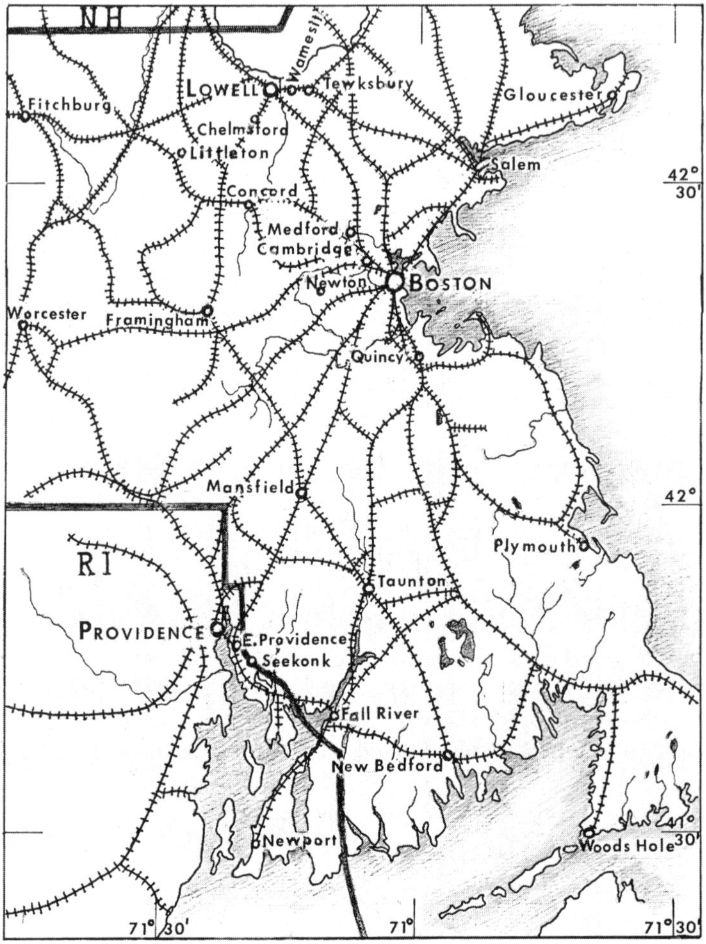

Map 4,1. The State of Massachusetts around 1890.
I-I-I-I railroads; NH = New Hampshire; RI = Rhode Island.

The oldest university of the United States, Harvard University, founded in 1636 at Cambridge within the bells of Boston, had no course in chemistry until the 1840's, when the Lawrence Scientific School was established. One of the first professors of chemistry at this school was Eben Norton Horsford

(1818–1893). Horsford's career provides an excellent illustration of the way in which European science came to the United States and was put to good use[1].

At the age of 26 Horsford went to Europe for a period of two years to receive additional training in chemistry in the already world-famous laboratory of Justus Liebig at Giessen, Germany. Here he worked together with many European chemists, here he absorbed the newest ideas in chemistry and learned the newest methods. Soon after his return to the United States in 1845 Horsford was appointed to the Rumford Chair of the Application of Science to the Useful Arts, a professorship in chemistry at the Lawrence Scientific School. In this position the new Professor organized and equipped the Chemistry Department in accordance with the principles he had learned from Liebig. Thanks to Horsford the study of chemistry as an experimental science, with strong emphasis on analysis, became firmly entrenched at Harvard. Under Horsford the Lawrence Scientific School became one of the first schools in America in which analytical chemistry was taught systematically to individual students.

Horsford was a man of broad interests. One of his favorite subjects was the chemistry of food. Like Liebig, with whom he kept up correspondence for many years (in German), he tried to promote the use of phosphates in agriculture and in food products. From these studies originated the idea of using calcium dihydrogenphosphate as the acid ingredient in baking powder. In 1856 Horsford obtained a patent (U.S. Patent 14,722) for the new baking powder acidulant. A laboratory to develop the product was established by Horsford and George F. Wilson in Seekonk (now East Providence), Rhode Island, in 1854. This laboratory, expanded into a chemical factory, the Rumford Chemical Works, put the product on the market in 1859, "a pulverulent phosphoric acid, commonly known as Horsford's Cream of Tartar Substitute, for bread-raising", as it was called in the early accounts of the company (Rumford Chemical Works, 1939). The unknown writer, maybe Horsford himself, continued: "In the invention of this chemical, the main desire was to avoid all fermentation, and to give to bread made of bolted flour a *deficient* ingredient which would add to the nutritional qualities of the bread." Horsford left Harvard in 1863 to become a full-time entrepreneur in the Rumford Chemical Works. This was probably related to the increasing demand of the Union Army for baking powder during the American Civil War (1861–1865).

The Massachusetts Institute of Technology

In 1861 another institute of higher learning was founded in the city of Boston itself; the Massachusetts Institute of Technology (originally usually referred to as Boston Tech, after it moved to Cambridge in 1916 the abbreviation MIT came into use).

Because of the Civil War the Institute did not open until 1865, when the first courses began with 15 students. The first Professor of Analytical Chemistry at MIT was Charles William Eliot (1834–1926). Eliot had graduated from Harvard in 1853, was Assistant Professor of Chemistry at his alma mater for some years, and had just returned from a two-year tour of the more prominent European universities to acquaint himself with European methods of higher education.

Eliot's stay at MIT was only of short duration: he was appointed President of Harvard in 1869, where he almost immediately began a complete overhaul of the curriculum, putting much more emphasis on science, on practical instruction methods, and on the learning of foreign languages. Professor of General and Industrial Chemistry in the first years of MIT was Francis Humphreys Storer (1832–1914). Storer[2] had studied chemistry with Horsford at the Lawrence Scientific School (Harvard), obtained his B.S. in 1855 and then studied abroad for two years in Germany and France. Upon his return in 1857 he established himself as a practising chemist in Boston until he was appointed to MIT in 1865. Because of lack of textbooks he wrote, together with Charles Eliot, two manuals for the students, one on inorganic analysis and one on qualitative analysis. Later a "Cyclopedia of Quantitative Chemical Analysis" followed. In 1870 Storer became Professor of Agricultural Chemistry at Harvard, an appointment probably influenced by his friend, colleague and future brother-in-law, Charles W. Eliot, the President of Harvard University.

Some information on the chemistry courses given at MIT by Professors Storer and Eliot during the years 1867 and 1868 is presented in Figure 4,1. The influence of Liebig's School is evident, but a prominent place is also given to the industrial uses of the science of chemistry, and the study of the German language is a requirement.

I

The Instruction in Chemistry. In the chemical laboratories provision is made for teaching General Chemistry, Qualitative Analysis, Quantitative Analysis, Assaying, Determinative Mineralogy, the Use of the Blowpipe, Metallurgy, and Industrial Chemistry. The Department occupies five ample laboratories, a chemical lecture-room, and a recitation-room, besides rooms for apparatus, balances, and storage.

In the first year, instruction is given in Inorganic General Chemistry by a weekly exercise which combines a recitation and an illustrated lecture, and by a weekly lesson in the laboratory, where every student is provided with a desk and the necessary apparatus, and is required to perform, under the supervision of the professors, a large number of experiments selected to illustrate the laws of chemical action and the properties of all the important chemical elements. In the second year, a systematic course of instruction in Qualitative Analysis is given, by laboratory practice and oral and written examinations. Every student works in the laboratory twice a week during the greater part of the year. Towards the close of the year instruction is given in the Elements of Organic Chemistry. Manuals, specially prepared for the purpose, are used in aid of the laboratory instruction given to the classes of the first and second years.

In the third and fourth years the principal subjects of study are Quantitative Analysis, Assaying, Mineralogy, the Use of the Blowpipe, Metallurgy, and Industrial Chemistry. Competent students are encouraged to undertake special researches, and are assisted in bringing them to useful results.

II

III. — Course in Practical Chemistry.

1. Chemical Analysis, Quantitative,—embracing the Analysis and Commercial Testing of Ores, Metals, Alloys, and Mineral Materials; Soda-ash, Bleaching-salt, Saltpetre, Paints, Drugs, and Manures; Drinking and Mineral Waters.
2. Lectures on Industrial Chemistry; on the Manufacture of Glass, Pottery, Soda-ash, Acids, Soap, Gas, etc.; on the Arts of Dyeing, Calico-Printing, Tanning, Brewing, Distilling, etc.
3. Descriptive and Determinative Mineralogy.
4. Lectures on Structural and Systematic Geology.
5. Drawing,—of Apparatus, and of Machinery and Plans for Chemical, Dyeing, and Print works.
6. Physics continued.
7. Logic, Rhetoric, and the History of English Literature.
8. Extended study of French and German, for the first of which Spanish may be substituted.

III

7. *Excursions for the Inspection and Study of Machines, Processes of Manufacture, Buildings, Works of Engineering, Geological Sections, Quarries, and Mines.* In aid of the practical studies of the School, and as a means of initiating students into the actual details of the professions for which they are preparing, they will be required from time to time, in the progress of the course, assisted by one or more of their teachers, to make visits of inspection to machine-shops, engines, mills, furnaces, and chemical works, and to important buildings and engineering constructions which are within convenient reach.

With a like view, and under the same direction, they will be expected to spend such parts of vacations as may be assigned, in excursions for observation and practice.

Fig. 4,1. Chemistry courses at Massachusetts Institute of Technology, 1867–1868 & 1868–1869. Reproduced with permission from MIT.

I Catalogue 1868–1869, p.22.
II Catalogue 1867–1868, p.19.
III Catalogue 1867–1868, p.25.

Developments after the Civil War

Although the universities had trained only a few students in the methods of experimental chemistry until the middle of the century, many practical chemists were at work in the United States at that time: some in apothecary shops, others as owners of small chemical enterprises, and still others as independent consultants, offering their services to an ever larger variety of manufacturing enterprises. These men exchanged their ideas, wrote popular articles about practical applications of chemistry, gave courses for laymen and, in general, did much to promote the use of chemical products and methods in everyday life. (An interesting account of the activities of these pioneer chemists is given by Haynes, in his history of "American Chemical Industry", Vol. I, Ch. 14.) In many cities these practical chemists, including university professors who also tended to be more practical than theoretical, convened more or less frequently and kept each other abreast of their activities and of new developments in the United States and abroad.

Through the activities of these early chemical entrepreneurs the basis was laid for an independent chemical industry. The Civil War provided a large impetus for accelerated growth. The requirements of the Union Armies stimulated the production of many materials needed for warfare, for transportation, and for the provision of one and a half million soldiers.

After the conclusion of the war the American Nation entered a period of impressive changes. It became an important industrial nation, comparable to Great Britain, and in a few decades would surpass the Mother country. The West now became really opened; a vast railroad system was built, farming became a mechanized operation, making the United States by the end of the century the major grain producer in the world.

The new industries, of which the steel and the mining industries were the most important, were greatly encouraged by a shield of high tariffs against competition from England. Of even greater advantage was the huge internal market created by the excellent transportation system; the shortage of labor which stimulated the introduction of labor saving methods; the availability of enormous amounts of coal to supply the energy needed; and not least of all, the democratic spirit of the people, who disliked the class society of Europe with its remnants of feudalism and who were very receptive to use new techniques and new products for their material advancement.

The chemical industry did not fully participate in these developments. Certainly, the industry and the population needed chemical products, which gave the chemical industry a sound basis, but investors shied away from the chemical industry; they preferred to invest in railroad companies and other enterprises engaged in more “classical” fields of activity. Chemistry was still considered as something alien, if not bizarre; its methods were incomprehensible, dangerous and often accompanied by bad fumes and obnoxious residues. The ordinary investor stayed away from it; the financing of a typical chemical enterprise usually was in the hands of the owner or of a small number of partners (Haynes, 1954, 239).

Baking Powder

The development of the baking powder industry is a typical example of the way in which Americans accepted new products and new methods.

It is not known who first had the idea to use a chemical leavening agent instead of yeast for the raising of bread. The immigrants from Europe, of course, had all brought their national recipes for bread and pastry making, using brewers' yeast or a sour dough process of one kind or another. In Europe the use of pressed yeast, an offspring from the distillery industry, had made its entry on the bakery market. In the new still scarcely populated country people had to make their own bread, and as yeast was difficult to obtain housewives began to try combinations of sour milk and soda (sodium bicarbonate) as leavening agents. (Soda had become a “popular” chemical with the housewives in the beginning of the nineteenth century to produce hard soaps.)

The combination of soda and sour milk worked reasonably well, but the price of the soda, called *saleratus* (sal aeratus = aerated salt) was much too high, in 1840 still around 2.75 $/kg. In 1846 John Dwight and Austin Church installed a factory for the manufacture of saleratus in America, their production was so successful that by 1850 the price of baking soda had come down to 0.10 $/kg. Housewives and bakers all over the United States readily accepted the new chemical, a big help particularly to settlers in remote areas in the West. They used it together with vinegar or sour milk to get well-raised bread, as good and soft as that which the citizens of the cities ate. In the beginning of the 1850's the use of cream of tartar (potassium bitartrate) as a more convenient acid ingredient was promoted. The first, or one of the first, companies to put a mixture of cream of tartar and sodium bicarbonate on the market was the firm of Preston and Merrell in Boston (Rumford, 1939).

In the same manner as baking soda had become a chemical “commodity”, now cream of tartar manufacture increased rapidly to a comparable status:

1874: four US producers make 300 tons; imports are 1,370 tons
1881: total US production is 3,000 tons
1900: total US production is 5,000 tons

Compared to these figures the production and imports of tartaric acid were of minor importance. In 1871 131 tons were imported, in later years, partly thanks to high tariffs, imports dropped to almost zero. National production of tartaric acid based on imported raw materials (argols and lees) rose gradually and reached 1,215 tons by 1900.

The most important producer of cream of tartar and tartaric acid was the firm of Charles Pfizer. This enterprise had been founded in 1849 by two recent German immigrants, Charles Pfizer and his cousin Charles Erhart, to manufacture pharmaceuticals and fine chemicals. During the Civil War the company had set up a refining operation in its Brooklyn plant to manufacture cream of tartar and tartaric acid for use in bakeries, by beverage manufacturers and by housewives (Mines, 1978). In 1870 a large plant was built to meet the increased demand of the baking powder formulators for cream of tartar.

The successful introduction of cream of tartar as an acidulant in baking powder induced a number of chemists to look for substitutes. One of those was Horsford, who developed a dry, "pulverulent" acid phosphate as a substitute. Another inventor attracted by the idea of making a cream of tartar substitute was a young chemist educated at MIT, Charles E. Avery. Avery tried acid calcium lactate as an acidulant; the course of events following the discovery of what he thought to be a valuable product led him to build the world's first lactic acid factory to provide for the raw material.

Most successful in providing the acid ingredient of baking powder were, however, the producers of "burnt alum", a product based on sodium aluminum sulfate[3]. These products came into production in the early 1880's and soon seized a large share of the baking powder market. Consequently the price of cream of tartar dropped to about 0.45 $/kg and remained that low until World War I.

Charles Ellery Avery, the First Manufacturer of Lactic Acid

Who was Charles Avery, the man who tried to make a cream of tartar substitute on the basis of acid calcium lactate? Not much is known about him in readily accessible publications, but from a number of obscure sources, some of which were hidden under half a century of dust, it has been possible to obtain an impression, still rather blurred, of this chemist and inventor.

According to an obituary written by Charles' old playmate Goodwin Palmer, which appeared in a Littleton, MA, newspaper in January 1917, his father Alden Avery was a successful builder and real estate dealer in the city of Boston. Charles Avery, born in 1849, graduated with honors from high school and, in 1867, entered the Massachusetts Institute of Technology "then just started and occupying a few rooms in the building on Summer Street". Charles Avery was enrolled as a non-degree candidate and, therefore, did not receive a degree when he left MIT two years later[4]. (Some parts of the MIT chemistry curriculum of those years are shown in Figure 4,1).

Avery was a diligent student and "showed marked proficiency in research work". (In his second year Charles Avery carried out some research work for his teacher, Professor Storer.) After leaving MIT Avery was elected an Associate Member of the Society of Arts. This Society was one of three proposed entities comprising the Massachusetts Institute of Technology, as was decided in its incorporation in 1861. The purpose of the Society of Arts was to provide a forum for lectures and discussions of developments in the pure and applied sciences. The election of Charles Avery shows that at the age of only 20 years, he was already considered to be a well-educated student of "science".

The first job Avery had was with the Boston Dyewood & Chemical Company. Here he was engaged in the handling (processing) of logwood and black oak bark. Before the advent of synthetic coal-tar dyes, textile dyers and printers used natural coloring materials, of which indigo and madder were the best known. After the discovery of the New World extracts from logwood and brazilwood were found to be quite suitable as well. The logwood extract became very popular because it was the first dye with which satisfactory black colors could be obtained on silk and wool. Chemically it is haematoxylin, a flavone derivative (Karrer, 1947). On extraction black oak bark yields a chemically related compound, also used as a mordant dye in the textile industry of the production of yellow, orange, and red dyed or printed goods (See also Intermezzo 2). Through these researches Avery became interested in dyestuffs and "spent some time investigating the aniline dyes, but did not follow it up closely" (Palmer, 1917).

After this excursion into the field of textile colors Avery was engaged for short periods in analytical work (for the milk inspector of Boston), in the sewage problems of a paper mill, and he was associated with the state assayer[5]. He also lectured at the Massachusetts Dental College. Then, probably in 1879, one of his business friends suggested that he develop a cream of tartar substitute. Avery first tried to make an acid phosphate based on apatite (Horsford used bone meal), and he claimed that he had found a fairly good process (Avery, 1881). When he realized, however, that lactic acid could be produced more economically than phosphoric acid, and that, moreover phosphoric acid was considered by some to be "objectionable", being a constituent of various "urinary deposits", he decided to make lactic acid the subject of further study.

In 1880 Avery developed, in a laboratory in the house of his landlord, first, a process for making acid calcium lactate (U.S. Patent 235,615, issued December 21, 1880) and second, a process for the manufacture of lactates by fermentation (U.S. Patent 243,827, dated July 5, 1881). The Boston Directory of 1881 records Avery as living at the same address as his parents, so presumably his landlord was his father. The baking powder patent was assigned to his mother, Lucinda M. Avery, the fermentation patent to a new company, the Avery Lactate Company, incorporated in Portsmouth, New Hampshire.

In the paper presented to the Society of Arts, Avery had already sketched some of the economics of the venture that he had in mind (February 1881):

> With regard to cheapness, little needs to be said. With a yield of lactic acid equaling fifty per cent of the weight of corn employed, with meal at one cent per pound, chalk or whiting at seven-eights of a cent, and sulphuric acid at one and three quarter cents, coal being assumed at US$ 6 per ton, the costs of materials for one pound of lactic acid is about four and one quarter cents. By shrewd location of the factory, and good buying in large quantities, these expenses may be largely reduced. The operations of manufacture are not unlike those of the sugar refinery, and those are known not to be excessive.
>
> \- *Charles Avery, 1881*

Land was bought in 1882 in the small town of Littleton, 50 km north-west of Boston. The factory was built in the same year and production was started probably at some time in 1883. Why did Avery, or his financers, choose a site in Littleton? What was so shrewd about this choice?

Littleton, site of Avery's lactate factory

Littleton, originally part of Concord Village (where the first battle of the Revolutionary War took place in 1775) was in 1882 a small farming town with a population of about 1060. The town had good railroad connections with the Boston and Lowell areas and other centers of industry and commerce in Massachusetts and New Hampshire (Map 4,1). Of the three railroads which served the town the Boston-Fitchburg Line was the most important; it had a station at West Littleton – the Depot – which had special facilities for handling freight and mail.

Littleton was not without industrial experience. The white settlers (the first had arrived around 1650) had taken with them the usual crafts needed in such an isolated spot in the hilly woodlands. Gradually they had chopped down and burned the forests to obtain arable land, producing timber for their own use and for sale, and potash for export to England. (Soda was still a rare and expensive chemical.) Lime from local limestone deposits soon followed, and later vinegar, probably produced by fermentation of sweet cider. At the time that the Avery plant was built, saw and grist mills run by water power were still in operation (potential suppliers of corn meal?). In 1864 a factory for canning and pickling fruit (and cider?) had been established at the Depot; as a matter of fact this was the first industry especially built in that area which had been set aside for industrial purposes when the railroad was built (Town of Littleton, 1964).

The lactate factory was erected at a cost of US$ 100,000[6]. The building was of a sturdy construction with very thick walls made of brick. It had many floors and a large elevator "for the conveyance of barrels of Acid from the upper floors". In 1911 the plant had a fire room (boiler house) equipped with four boilers with a total capacity of 390 hp; 22,000 tons of coal were yearly used to fire these kettles. (This is an unbelievably large amount of coal, even if one assumes that in 1911 the plant generated its own electricity. The figure is probably wrong by a factor of about ten.) "Other rooms are the [filter] press room, the laboratory, cooper and barrel room and office". The young writer forgot the rooms in which the fermentation vats were placed, the purification department (Avery used animal charcoal to remove color and odor), the room in which the crystallization of the acid calcium lactate originally took place, and the evaporating chamber. It most certainly was a large plant for the time. It also must have been quite impressive too; the outside was all of brick, the plant had many floors and the inside of the building was full of special equipment, such as large wooden vats and heavy flywheels which kept a number of centrally driven shafts in smooth operation (Figure I4,1, 205 of Merck's lactic acid plant provides an example of drive shafts and flywheels in operation.) For the first ten years the plant certainly had no electricity (the generation and use of alternating current

Fig. 4,2. Aerial photograph of the site of the Avery factory. Courtesy: Lone Star Co., Littleton.
Since 1942 a concrete products plant occupies the site of the Avery Lactate Company at Littleton. The building behind the large conveyer in the foreground is located at the site of the original lactic acid plant. The road opposite the plant disappearing into the woods was called Lactate Road (now New Estate Road). The Boston-Fitchburg railroad can just be seen at the lower right hand edge of the picture.

did not come to the industry until the 1890's). Illumination was most probably provided by kerosene lamps.

The First Years of the Avery Company

The process developed by Charles Avery was, in the light of the history of (technical) microbiology, a well advanced process and one certainly suitable for the industrial production of lactic acid. Avery's lecture at the Society of

Arts (Avery, 1881) shows that he not only possessed the latest information about the chemistry and laboratory preparation of lactic acid, but also that his knowledge of fermentation processes was ahead of what was written in generally used textbooks[7].

Avery completely discards the ideas (of Liebig and his followers) that some nitrogeneous organic substance in a state of putrefaction has to be present to make fermentation possible. Also Pasteur's view that "albuminoid matter was non-essential, that it could be substituted by inorganic salts", did not find his approval:

> It is substantially here that I depart from the beaten track. I find albuminoid matter, though not, scientifically speaking, essential, is essential in a technical point of view ... I prefer them as fresh and free from putrefaction as possible ....
> To obtain this fresh albuminoid matter at no further expense, and in large quantity, I leave in the glucose solution made from the corn meal [by hydrolysis with sulfuric acid] the albuminoid residue. And from this substitution [of Bensch's rotten cheese] I obtain the advantage of healthfulness, a crude product free from the products of putrefaction, and hence more easily purified; and, lastly, the fermentation takes place in three or four days, instead of requiring ten days to a month, as by other processes.
>
> - *Avery, 1881*

Avery did not report whence he obtained his bacterium (lactic ferment). He invented a process, however, "of purifying and preserving a lactic ferment by adding said ferment to a medium specially favorable to its growth and less favorable to the growth of other ferments" (U.S. Patent 330,815, application date November 2, 1883, granted November 17, 1885).

Important conditions, favoring the growth of a "pure" culture of lactic ferment, are a high temperature (35–45°C, the higher the better), the use of fresh cultures for inoculation to suppress the development of butyric fermentation, and addition of sufficient carbonate of lime to neutralize the acid formed. The patent also specifies that

> impregnation [inoculation] of each solution from the preceding one [should be] effected at the point of full height of fermentation, as evidenced by the evolution of the carbonic-acid gas ...
>
> - *Avery, 1883*

Avery's ideas about preparing pure cultures by selecting conditions favoring the growth of the desired culture while at the same time suppressing the growth of other microorganisms can be seen as an extension of the work of Pasteur who already in his early investigations of lactic and alcohol fermentations had stressed the point that the conditions of the medium play a great part in the predominant development of any given ferment (Nicolle, 1961, 43).

Charles Avery Turns Seriously Ill

The patent granted in 1885 was one of Charles Avery's last published contributions to the new enterprise. Goodwin Palmer wrote in the 1917 obituary:

> On the completion of this work Mr. Avery suffered a complete breakdown from over study and strain and for over two years was in Florida trying to "come back". While here he suffered from a stroke, from the effects of which he never fully recovered. On his first visit to the factory which bore his name, he was carried in a chair ....
>
> *- Palmer, 1917*

At another place in the obituary Palmer writes that at the early age of thirty-two Avery's work was done. Which means that he turned ill in 1881 and probably did not come back to Boston until the fall of 1883.

Meanwhile the factory had been opened and started; the technical management proceeded according to the plans and policies initiated by Charles Avery. In 1883 patents were applied for by Charles Oliver Thompson[8] of Terre Haute, IN, by Charles Nelson Waite[9] of Medford, MA, and by George Austin Marsh[10] of Littleton.

Acid Calcium Lactate for Baking Powder

Thompson's patent (U.S. Patent 290,284) is in part an improvement over Avery's original acid lactate patent; he adds about 10% of commercial pure, white glucose to the hydrolyzed corn meal, which helps in obtaining a higher yield. For our story, however, the most important part of the patent is the description of the preparation of the acid calcium lactate itself, a description which is lacking in the Avery patent[11]. To a dilute (warm) solution of calcium lactate, obtained by crystallization and washing of the product of the fermentation, half the equivalent amount of sulfuric acid is added. The sulfate of lime (gypsum) formed is filtered off, the filtrate concentrated and thereupon placed in a cold chamber to crystallize. The temperature in this chamber should not exceed 45°F [7°C].

... acid calcium lactate crystallizes substantially solid, the slight amount of mother-liquor still remaining with the crystals being readily pressed out therefrom and the acid crystals are now ready for use.

- Thompson, 1883

It is not known whether cold chambers were installed in the factory. Refrigeration machines had already been invented and some were in use to cool perishable goods during ocean shipments, but workable machines for industrial use were not yet in general use. On the other hand, ice from ice farming was readily available, ice was still cut from the rivers and lakes of New England on a large scale in those years. (Burke, 1978, who gives an interesting picture of this peculiar "industry", tells that in some years more than 100,000 tons of ice were shipped from New England harbors.)

It seems fairly certain that acid calcium lactate was not a success in the baking powder market. Thompson's improvements of Avery's manufacturing process probably were of not much help either. A patent obtained by George Marsh in 1885 (U.S. Patent 328,419) bears testimony to this:

This invention has for its object to provide a method of preventing the change and reaction which takes place when acid calcium lactate is mixed with bicarbonate of soda to form baking-powder. The reaction causes the active properties of the baking-powder to be given off in the form of a gas, which, escaping, makes the mixture useless. It has therefore been necessary heretofore to put up the said ingredients in separate packages and mix them when they are to be used.

- Marsh, 1885

Marsh added 5% of lard or tallow to the dried acid calcium lactate, passed the crude mixture through a fine sieve, and then added 20% of powdered starch or flour. The acid calcium lactate, now having a protective waterproof coating, was thereupon mixed with the sodium bicarbonate, and packed in tin boxes. As a further precaution Marsh advises to give the bicarbonate a similar pretreatment.

The fate of the baking powder venture has been well pointed out by the high school student in 1911:

At the beginning of the factory ... a baking powder was tried but was not uniform or well dried.

It is not really possible to judge how much competition from cream of tartar and exsiccated alum also contributed to the failure.

The Fermentation and Purification of Lactic Acid

Before further turning our attention to the commercialization of the new product, a summary of the technical aspects of the production process is warranted. This summary is based on Avery's lecture before the Society of Arts and on the patent applications filed by the Avery company until 1885 (See Exhibits 4,I and 4,II).

EXHIBIT 4,I
AVERY'S FERMENTATION PROCEDURE
lit. see text

raw material	corn meal hydrolyzed with 2–4% of sulfuric acid (16 hours at room temperature, then at boiling temperature until the iodine reaction with starch becomes negative). Or: corn meal or starch converted with diastase. Glucose may have been added.
concentration	of sugars 12-14% (estimate)
inorganic nutrients	not specifically added
organic nutrients	organic albuminoid matter from the flour; or the matter supplied by the action of diastase on corn flour
neutralization by	fine powdery carbonate of lime added in the beginning
fermentation temperature	35–45°C
fermentation vessels	wooden vats of about 4 m^3 or larger
sterility of operation	??
mixing	from time to time
acidity	slightly acidic (remains constant because of the presence of excess chalk)
fermentation time	3–4 days
control of end	the broth solidifies (crystallization of calcium lactate)
yield	about 90% based on starch content of the meal = about 80% of the theoretically possible yield
type of bacterium	lactic yeast [?]
cultivation method	inoculum is taken from previous fermentations in their most active stage
control of purity	??
amount of inoculant	2–10%

EXHIBIT 4,II
AVERY'S PURIFICATION PROCEDURE
lit. see text

treatment of broth	the mass of calcium lactate, resembling thick mortar, is dissolved in hot water
filtration	following the dissolution of the lactate
concentrating	yes
crystallization	no details
purification	the solution is treated with animal charcoal to remove odor and color; after crystallization the calcium lactate is pressed free from adhering liquid. This procedure is repeated twice (in the manner of Bensch, Exhibit 2,II)
conversion to acid	no details
filtration	butyric acid is distilled off [probably by blowing steam through the solution]
concentrating	in the first years the dilute solution was probably not concentrated but sold as such in a concentration of about 10% (as vinegar substitute)

*****

The most essential process improvements of Avery over Bensch were:

- the use of a culture of bacteria instead of rotten cheese
- fermentation at a higher temperature to give the lactic acid bacteria a better change of overgrowing other, undesirable microorganisms (infections)
- inoculation from previous fermentations at their moment of maximum growth, that is in their most vigorous stage

In hindsight, it is evident that the use of cheap corn meal must have been very detrimental to the quality of the acid calcium lactate due to the high percentage of non-fermentable material it contained. The same would apply to the production of lactic acid and most certainly so when the acid was intended to be a vinegar substitute. The patent of Thompson (U.S. Patent 290,284) showed that the Avery staff were aware of this problem: by replacing part of the corn meal with glucose a higher yield, also meaning less contamination, could be obtained.

When the fermentation was conducted in the right manner and no mishap took place in the form of a butyric infection, the product obtained from the fermentation contained dextrins (10% or more ?), soluble and insoluble proteins and cell material from the corn meal (5 to 10% ?), a small amount of unfermented sugars and soluble and insoluble inorganics from the whiting and the corn meal. Due to the low solubility of calcium lactate most of it would precipitate spontaneously towards the end of the fermentation, in particular when the temperature would drop to about 35°. The broth had then first to be heated up to dissolve the calcium lactate again, followed by filtration of the hot solution to separate from the real insolubles.

Thereupon the crude solution of calcium lactate was purified in the manner described in Exhibit 4,II, a rather cumbersome and inefficient procedure. Calcium lactate precipitated from the clear filtrate would still be very impure. This was due to the mortarlike consistency of the precipitated crystal mass, as a consequence of which it easily retained 100% of the mother liquor, including of course the impurities contained in this adhering and adsorbed liquid. (Roughly calculated not more than two thirds of the impurities were removed in each crystallization step, hence at least two crystallizations were needed to produce, with considerable losses, a product of acceptable purity.)

Following the (partial) purification of the calcium lactate it was converted with sulfuric acid into lactic acid and gypsum. Most of the gypsum would precipitate, but about 0.2% would remain in solution. By means of filtration the solids, gypsum and some other material precipitated by the action of the sulfuric acid, were removed and a dilute solution of lactic acid (of about 20% strength) was obtained.

The remark "butyric acid is distilled off" tells us that infections of the fermentation process occurred frequently. Such infections were not only detrimental to the yield, but, more seriously, they gave rise to the very unpleasant smell of butyric acid, which is almost impossible to remove by distillation. Avery's remark in this respect should be taken with more than a grain of salt.

It can, however, be assumed with a reasonable degree of certainty that part of the lactic acid production of the company was acceptable in odor, color and taste to be sold as a vinegar substitute. (The idea to replace vinegar Avery had already expressed in his 1881 lecture to the Society of Arts.)

The need to further purify the acid was soon felt. In 1884 a patent application was filed by Charles N. Waite for a process in which lactic acid was purified by distillation with steam (U.S. Patent 321,925). This was a highly impractical method while the lactic acid was a hundredfold diluted by the large amounts of steam (See Holten, 1971, 37). When the distillation is carried out at reduced pressure better results can be obtained. Waite mentions the possibility of using a partial vacuum but gives no details.

Another patent application for the purification of lactic acid was filed by Waite in 1890 – after the dissolution of the Avery Lactate Company a few years earlier (U.S. Patent 455,078). This was a modification of Bensch's method in which the purification was carried out by recrystallization of zinc lactate. Instead of liberating the lactic acid from the zinc salt with the aid of hydrogen sulfide, Waite converts the zinc salt first with lime into soluble calcium lactate and insoluble zinc oxide, and thereupon recovers the lactic acid in the usual way with sulfuric acid.

This method, although rather laborious, was much better and was put into use in a somewhat modified form a few years later by the firm of E. Merck of Darmstadt, Germany, to prepare a pharmaceutical grade of lactic acid (Ch. 5, 133).

No indication has been found that the method was ever put into commercial practice in the United States.

Lactic Acid in Soft Drinks

After the failure of acid calcium lactate as baking powder, other uses for the new acid had to be developed if the company wanted to survive. In his 1881 lecture to the Society of Arts Charles Avery had expressed the view that in foods and drinks a "set of uses" might be found.

> In this field the industrial improvements have not been of so striking a nature [as in the manufacture of oxalic acid], and, as a result, the food acids, citric and tartaric, remain high in price.
>
> \- *Avery, 1881*

The first use of the acid itself proposed by the company was that of a mordanting aid in the dyeing of wool and silk, replacing tartaric and citric acid (Compare Intermezzo 2). Charles Waite applied for a patent on the combined use of oxalic and lactic acid "in conjunction with some suitable metallic salt [the mordant]" for the "fixing of colors to animal fibers" (Waite, U.S. Patent 290,294). The patent application was made in September 1883, about a month after Waite became employed by the Avery Lactate Company. At that time Thomas Shepard Nowell was President of Avery Lactate; he held office in Boston (Boston Directory, 1883).

In 1884 Avery Lactate began to promote (dilute) lactic acid as a vinegar substitute, viz. the reproduction of an advertisement in the Boston Directory of 1884 (Figure 4,3). The idea to use a food acid in lemonade drinks was old, as a matter of fact, Scheele had already mentioned it.

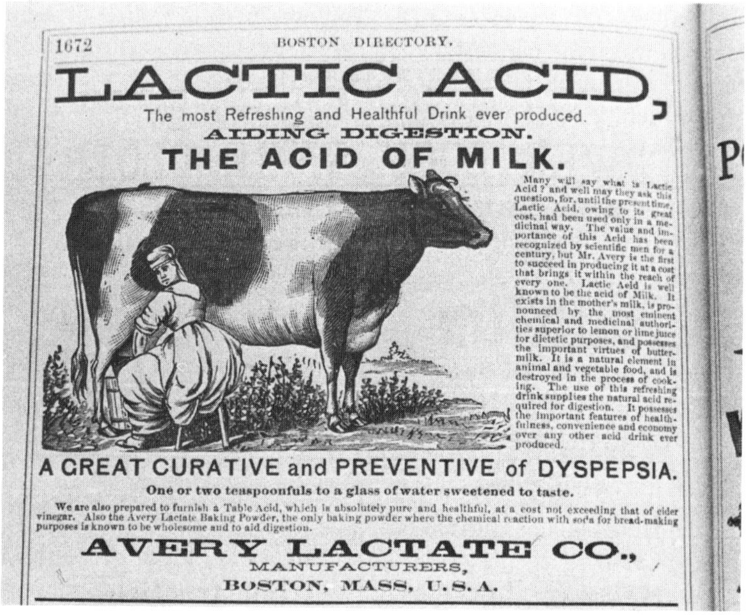

Fig. 4,3. Advertisement of Avery Lactate Company in the Boston Directory of 1884. Courtesy: New England Historical Genealogical Society.

When citric acid and tartaric acid had become articles of commerce, these acids were also applied in the preparation of lemonade drinks and the like in Germany from about 1840 onward, and not much later also in the U.S.A. (Riley, 1958).

Housewives made these lemonades themselves. They tried many mixtures of syrups, vinegar, sometimes together with fermentation juices, and plant and root extracts. Root beer and ginger ale are beverages surviving from those experiments.

The idea of Avery Lactate to introduce its dilute lactic acid into this market received more emphasis in succeeding years. In 1885 the trade name "Lactart" was registered, a "pure vegetable acid for household use", which together with honey provided a healthy drink. In 1886 the company advertized it as "Professor Avery's Food Acid", and in 1887 even as "Lactart Champagne".

Fig. 4,4. Bottle of Avery's "Lactart" food acid. Photo: Forrest K. Marcy, Littleton. Courtesy: Littleton Public Library.

Under the management of Thomas Nowell the Avery Lactate Company was a progressive company, one is tempted to believe. It was for instance quick to adopt at least one of the miraculous inventions of the century, the telephone; in 1885 its address in the Boston Directory included the telephone number (No. 217). This was only one year after the first intercity telephone line in the world, between New York and Boston, had been established.

Nowell also took care of the registration of U.S. Patents of the company in Great Britain. In May 1885 he applied personally for a new patent in the U.S.A. (U.S. Patent 346,824, granted in 1886). This patent related to

the production of an acid base for aerated [carbonated] and other beverages and tonics, which may be used by druggists in connection with soda water fountains, as well as by physicians in preparing a cooling acid drink, and by the public generally.

It consists of lactic acid containing from one to two per centum of lactate of lime, with traces of less than one fourth of one per centum each of vegetable gluten, vegetable coloring matter, gum and sulphate of lime, combined with water and sugar of plain syrups, to which may be added any desired fruit syrup, in any particular case. The lactate of lime serves to take off the sharp edge of a chemically pure lactic acid, and the gluten, gum, etc. named serve to give it a distinctive color and odor and, as they are present in such minute quantities, are incapable of doing any injury to the human stomach, and at the same time render the acid more palatable than the pure acid would be.

- Nowell, 1885

Since the 1850's soda-fountains had become very popular in the larger cities of the States not only because of the wide selection of flavored soft drinks they offered but also because of their social function: offering a meeting place for people different from those frequenting the places where hard drinks (alcoholic liquors) were sold. The history of soft drinks goes back to, among others, Benjamin Silliman Sr., who as the young Professor of Chemistry at Yale College in 1807 had produced a drink for public sale, combining "fixed air" and fruit juice (Riley, 1958, and American Can Company, 1972).

Around 1860 druggists in the United States began to install so-called soda fountains in their stores to sell effervescent waters and attract more public, more buyers. Fountains serving 1,500 glasses or more a day were not uncommon in New York City by 1870 (Riley, 1958).

Again the Avery company was not successful; they did not sell enough of the "pure vegetable acid" to keep their creditors at bay. In 1889 the company disappeared from the Boston Directory, "for tax reasons" as another source mentions[12]. But they must have sold some lactic acid to keep the company solvent from 1883 to 1887. The picture of a bottle still in the possession of a Littleton family, shows how lactic acid was sold to housewives as a vinegar replacement (Figure 4,4). Perhaps it was also delivered to producers of soft beverages such as the Waco, Texas, producer R.S. Lazenby, who brought "Dr. Pepper's" on the market in 1885.

Lactic Acid in the Mordanting of Textiles

As previously mentioned, Charles N. Waite had, in 1883, secured a patent on the use of lactic acid in the mordanting of animal fabrics (wool and silk). A few years later, in 1886, he obtained a second patent on a mordanting process, this time on the fixing of aniline and other dyes on cotton by means of antimony lactate. (About the history of these processes, and the development of artificial dyes, see Intermezzo 2.)

The oldest record of the actual use of lactic acid in mordanting appeared in the *Technology Quarterly* (of MIT) in 1890 (Norton and Tuttle, 1890). The authors had successfully replaced cream of tartar by the acid lactate of potash in a mordanting recipe for wool. As to the source of lactic acid they remark: until very lately lactic acid was too expensive to be

able to compete in price with tartaric and citric acids. Furthermore the cheaper grades of acid have not been colorless nor free from impurities capable of working injury in coloring. It is only within a few months that an acid sufficiently pure for use in coloring has been produced at a cost which will enable it to be obtained at a reasonable price. Its use has become practicable, and the experiments described ... were designed to throw light upon its effect. The samples used in this investigation contained about 43.5 per cent of lactic acid and were nearly colorless. The impurities present were sulphate of lime and a faint trace of iron.

- *Norton & Tuttle, 1890*

The sufficiently pure acid that the writers had used had been prepared according to patents of Charles N. Waite. The question arises as to who the actual producer was, and if the production had been performed on a commercial scale or only on laboratory scale. By whom and where?

Maybe by Waite himself who then owned a chemical plant at Newton, MA (Haynes, 1954, 277).

A comprehensive review of the use of lactic acid in wool dyeing was published in the English *Journal of the Society of Dyers and Colourists* in June 1895 (Whitaker and Denison, 1895). These authors wrote about lactic acid:

The product has recently been introduced in the United States by the Avery Lactart Company as a dark brown liquid, with a smell of sour milk.

The concentration of the acid was 24.1%, the specific gravity 1.145. From these data and other information given in the paper it can be concluded that Avery lactic acid contained about 50% of dry solids, half of which was lactic acid. The article says:

the lactic acid contained a large amount of organic impurities, probably mannitol and sugars or albumen, ... [The results, even with this very crude lactic acid, were quite promising:] If lactic acid can be introduced at a sufficiently low price there is no doubt it will prove a very valuable assistant to the chrome mordanting bath.

It will require a good many tests by practical dyers before the advantages or disadvantages of the use of lactic acid in place of tartar [cream of tartar] can be thoroughly understood, and we should be pleased to hear the experience of our district dyers on the use of this product.

Whitaker and Denison, 1895

Probably in 1893 or 1894 the lactic acid factory at Littleton was put into operation again, after having stood idle for about five years. The factory must have been renovated to some extent and furnished with some new equipment. The company also acquired a new name: Avery Lactart Company. Names of the new (?) owners have not been found. A few years later Alan Avery Claflin[13] is listed as chemist of the company. His father, Frederick Alan Claflin, brother-in-law of Charles Avery, was then treasurer, a position he also held with the old Avery Lactate Company and which he would keep until his death in 1907. So it seems that the Avery Claflin family had taken over the plant. [They may have bought it in the state of bankruptcy, perhaps from the proceeds of the estate of Charles Avery's father, a well-to-do businessman who had died at the age of 80 in 1892.]

In 1897 Alan A. Claflin presented a paper to the New York Section (of the American Chemical Society ?) on the manufacture and applications of lactic acid (Claflin, 1897). He refers to the work of his predecessor, Charles E. Avery, who had seven years' experience in the manufacture of lactic acid on a large scale (1881–1888 ?). "The applications of lactic acid, while limited, are increasing", Claflin wrote:

The majority of all that is produced is used by the woollen dyer as an assistant in mordanting with bichromate of sodium. [The advantages of lactic acid over tartaric and oxalic acid are greater reducing powder, greater solubility of the acid itself and of its salts.]

Claflin then continued with the following statistics:

In 1894 the dyehouses of the United States and Canada used about 400,000 lb lactic acid, while none was used in England and Europe; in 1895 there was a domestic consumption of 1,000,000 lb, and a foreign consumption of 500,000 lb; in 1896 the domestic consumption was 1,200,000 lb and the foreign 1,000,000 lb.

- Claflin, 1897

These figures probably relate to actual amounts of 25–30% lactic acid. According to a reference in Chemical Age (Chemical Age, 1932) the Avery Chemical Company produced in 1897 about 180 t of technical lactic acid, probably calculated as 100%. Still other data were given by the Littleton

High School student in 1911:

> The [lactic acid] industry has grown from 60,000 pounds in 1896 to 300 and 350,000 pounds, the largest trade being carried on in 1904 and 1905.

This means either that Claflin did blow up his figures, or that by 1896 there already was a second producer in the U.S.A.

Lactic Acid in the Tannery

In his 1897 New York lecture Claflin also drew attention to the use of lactic acid in tanning. This application, he said, was attributable to the researches of J.T. Wood, published after 1885, and of E. Andreasch, which appeared in "Der Gerber" in the mid-1890's. These researchers had discovered appreciable concentrations of lactic acid in deliming baths.

As a matter of fact, Braconnot had already isolated lactic acid from sour baths in 1832 (Ch. 1, 14). The first proposal to use lactic acid for the deliming operation was made by Georges Jacquemin, also of Nancy, France, in 1891, in an addition to his French Patent 204,874).

Before continuing Claflin's story, a few words on some aspects of the history of leather tanning will help in understanding the big success of the application of lactic acid in the tanneries. Most of this historical information has been borrowed from a book by Joseph Turney Wood on "The Puering, Bating and Drenching of Skins" (Wood, 1912).

Depending on the type of leather, the following steps were involved in the preparation of the hide for the tanning process (du Pont, 1922):

Cleaning. Soaking of the raw hides in water to remove foreign material and dirt and to soften the hides.
Unhairing. After sufficient soaking in solutions of various chemicals (sulfides and lime) the hairs can be scraped off.
Liming. Treatment with a solution of lime causes swelling of the fibers and loosening of the fleshy material on the underside of the skin.
Washing and deliming. To remove the flesh from the hide and to dissolve the lime which penetrated into the hide.
Bating or puering. A treatment with manure to make the hides more supple.
Drenching. The bating process leaves the hides in a slightly alkaline condition (amines from the dung treatment). The hides are now soaked in a fermented bran mash to remove the alkali. This action also causes some swelling.
Scudding and washing. To remove compounds and particles loosened in bating and drenching.
Plumping. To attain further swelling of the hide.

The most primitive as well as filthy parts of the preparatory process were the treatment with dog turd or with chicken manure, respectively called puering and bating. Drenching, the treatment with a sour bran drench must have been more acceptable, although the same drench was used for a long time and usually fouled considerably.

J.T. Wood, who entered the leather trade as an apprentice in 1885, had set himself the task of studying the various aspects of the bating and drenching processes. These studies extended over a period of about 20 years, and concerned the chemistry, the microbiology and the physical phenomena taking place in the hides. Wood established that in the manure treatment, the internal structure of the skin was loosened through the action of bacteria. A result of this was that the hide lost its original stiffness, and "came down" when spread out on a flat surface. Wood tried to obtain the same result by using enzymes. When this failed, he proposed the use of bacterial cultures. These were produced by a company in Frankfurt am Main for some time, but were not really accepted by the trade. The real breakthrough came in 1908 when Otto Röhm of Darmstadt, Germany, introduced Oropon, a combination of pancreatic enzymes with inorganic salts. In subsequent years, Röhm improved the formula to meet the requirements of the tanneries, while his friend Otto Haas was active in introducing the product first in Germany and then in the United States. Oropon became a tremendous success. It also marks the beginnings of two important chemical enterprises: Röhm und Haas A.G. in Germany, and Rohm and Haas Company in the United States of America (Hochheiser, 1986).

The drench or sour bate seems to have been developed in the eighteenth century. The original idea was that lime remaining in the skin after the first washing could be removed with any acid obtainable at a reasonable price, which was then sulfuric acid from the new lead chamber process. This proved to be a failure; with dilute sulfuric acid the hides very often were damaged beyond repair. Then a sour bran was introduced – a fermented mixture of about 1% of bran in water, usually applied at a temperature of 30°C or slightly higher. This bate was used repeatedly and kept at strength by the occasional addition of bran. Wood found that such bates contained from 1 to 2 g of acids – mixtures of lactic and acetic – per 1,000 ml, which after repeated use would increase due to the neutralizing action of the lime dissolved from the skin. He also found that artificial cultures, inoculated with cultures of bacteria obtained from drenches, could produce acid concentrations up to 11 g per 1,000 ml when calcium carbonate was present. With artificial bates of such a composition, the deliming and bating operation would give results in $1^1/_2$ hours, comparable with those normally achieved in about 14 hours. The results of Wood's researches were published in the early 1890's, the tanneries began to use artificially soured bates in 1896.

The first successful experiment with lactic acid was reportedly performed in 1896 by a sheepskin tanner, who was already using commercial lactic acid in coloring. As reported by Claflin this tanner first experimented with mixtures of lactic acid and acetic acid, the two acids present in the sour bran drench. Soon he changed to dilute solutions of about 1% lactic acid alone, which were found to be as effective and somewhat safer to use. Claflin gave an account of what happened after these first trials (Claflin, 1901).

> The success met with the sheepskin tanners induced Theodore Haight, of the firm of Hall, Haight, and Co., Ballston, N.Y., very extensive tanners of heavy upper leather, to make experiments on a large scale with lactic acid. The success of the experiments of Mr. Haight was such that he abandoned the use of other bating agents and adopted lactic acid permanently. With this result the practicability of the lactic acid bate may be said to have been demonstrated and it has been since then only a question of for what kind of stock the lactic acid bate is especially adaptable. With the experience of about 200 tanners at my disposal, I will give my opinion, subject to the criticism of many who have had far wider experience in this field of chemistry, where the laboratory helps us so little. For bating steer and cow hides for heavy upper leather, belting, carriage stock, etc., lactic acid is well suited; the slightly increased first cost is far more than compensated by the increased yield of finished leather, and the greatly diminished risk of over or under bating, and lessened demand for skilled supervision. For sheepskins the same is true, except that for these, where a depleting action is desired, the bate must be used at a higher temperature. For calfskins the advantage of lactic acid is disputed by the manure bate, all dependant on that mysterious operation called "bringing down". Many, especially some large chrome calf makers, find that by using a warm lactic acid bate the skins are brought down sufficiently, and the bate adaptable for their use. For goat and kid skins the manure bate must be used, but is wisely followed by a lactic acid drench. To summarise: the advantages claimed for lactic acid by its adherents are cleanliness as opposed to filth, economy as opposed to waste, and accuracy as opposed to uncertainty.
>
> *- Claflin, 1901*

Lactic acid also was used in the tanning bath iself. Andreasch had shown that lactic acid was present in old sour tanning liquors and that those liquors gave the best results. The function of lactic acid here is, Claflin says, to cause swelling of the hide, opening up the structure, making it more accessible to the tanning fluid, so that the tanning takes place more evenly throughout the leather. Claflin continues (Claflin, 1901):

> ... many hides intended for sole and other very heavy leathers are not bated at all, but

put directly into the handler liquors. These handler liquors are old sour liquors with a minimum of tanning material and a large amount of free non-tanning acids present. The presence of this acid necessarily removes a large amount of the lime, but it also has another function – that of plumping the hide. This plumping is extremely important in the manufacture of sole leather, and of more or less consideration in the making of all leather. Plump stock means heavy leather and a profit, while depleted stock may mean the reverse. In the old days of very slow tanning, liquors were weaker, and the plumping in a large measure took care of itself. Lately, tanners seeking for quicker methods, have used strong liquors, and these have made the necessity of plumping more evident.

- Claflin, 1901

Claflin then recommends the use of commercial lactic acid in the tanning liquor at an acid concentration of 0.8% maximum. The commercial success of the application of lactic acid in the leather industry was very impressive indeed: The first large scale trials were made in 1896; the sales of technical lactic acid (of 20–25% concentration?) amounted to about 300 t in 1897. By 1930 the total volume of lactic acid used in tanning in the United States had increased to between 3,200 and 3,600 t. of the 50% grade (Garrett, 1930).

Notes

[1] Dictionary of American Biography; Dumas Malone, Ed.; Chas. Scribner's Sons: New York, 1932, Vol. 9, 236.
[2] Dictionary of American Biography; Dumas Malone, Ed.; Chas. Scribner's Sons: New York, 1932, Vol. 18, 94.
[3] Burnt alum, also called exsiccated alum was prepared by gentle heating of sodium alum. Chemically it is composed of basic aluminum sulfate and sodium sulfate (Morrison, 1904).
[4] Private communication from the MIT Institute Archives and Special Collections, 1986.
[5] Publications from these period concern:

a. The analysis of arsenic (Strecker, 1869, 871).
b. The determination of silicate (Strecker, 1869, 879).
c. An improvement in the use of filter paper (Strecker, 1869, 989).
d. Report on the Miller's River Nuisance (Avery, 187?).

[6] This figure and more details come from a sketch written by a student of the Littleton High School in 1911.
[7] Avery mentions a textbook by Miller. William Allen Miller (1817–1870) studied medicine at Birmingham University, attended a short course at Liebig's laboratory in Giessen in 1840, became Assistant lecturer of Chemistry, and in 1848 was appointed Professor of Chemistry at King's College, London (Gillespie, 1974).

The textbook Avery is referring to possibly was the fourth, 1869, edition, third part, "Organic Chemistry", of Miller's "Elements of Chemistry". Although Miller appears to present the latest ideas about for instance the formulation of organic compounds, he seems a bit confused when he discusses lactic acid fermentation:

According to the observations of Pasteur, there is a special lactic ferment, consisting of microscopic vegetable cells, which are much smaller than those of beer yeast. These are developed during the decomposition [sic] of other animal matters, such as moistened bladder, and muscular and albuminous tissues, at a particular stage of their decomposition after exposure to air [sic] for a certain time.

- Miller, 1869

[8] Charles Oliver Thompson (1836–1885). In 1868 appointed Principal of the Worcester, MA, Institute of Industrial Science, where he held the Chair of Chemistry and "was charged with the development of a scientific and practical course of instruction which had no recognized type in this country" (Appleton's Cyclopedia, 1889, VI, 88). In 1883 Thompson was called to head the Rose Polytechnic Institute at Terre Haute, IN. He was the author of numerous papers on technical instruction.

[9] Charles Nelson Waite graduated in Civil Engineering from MIT in 1876. In 1877 he accepted a position at the Manchester Mills, an important textile enterprise at Manchester, NH. According to the MIT class alumni handbook of 1883 he resigned as chemist from the company in August 1883, "having become interested in the manufacture of lactates, for use in dyeing, and as a substitute for the tartrates used in domestic industry. He expects to go to Paris in a few months, as Resident Agent for the Avery Lactate Co." (Information received from the Librarian, MIT, 1988).

Waite obtained many patents, until 1885 in the name of Avery Lactate, in later years usually in his own name. From about 1900 onward the subject of his inventions changes from lactic acid and its applications to the field of viscose manufacture.

[10] George A. Marsh (1858–1942) received his bachelor degree in science from the Worcester Polytechnic Institute in 1882. His thesis was on "Sources and Uses of a Glucose". (His teacher supposedly was Charles O. Thompson.) Went to work for the Avery Lactate Company in April 1883, where he became assistant superintendent. He stayed with the company until August 1885 when he accepted a position as assayer and chemist at a Pennsylvania Lead Company plant near Pittsburgh, PA. (From an obituary in the possession of the Worcester Polytechnic Institute.)

[11] Avery had most likely followed the method given by Engelhardt and Maddrell in Liebig's Annalen (Engelhardt and Maddrell, 1847):

This composition [of acid calcium lactate] was prepared by evaporating to dryness a solution of calcium lactate and lactic acid in the proportion called for by the formula.

[12] Thomas Nowell's name disappeared from the Directory in 1886, in the following years C. van Brunt was president. Frederick Alan Claflin, a brother-in-law of Charles Avery, was treasurer from 1886, or earlier.

[13] Alan A. Claflin (1873–1954) studied chemistry at MIT, obtained his B.S. degree in 1894 and started his professional career with the Avery Lactart Company in the same year. In 1918 he founded a consulting firm in Boston, specializing in products used in the textile and leather industries (American Dyestuff Reporter, January 18, 1954).

Books Consulted

The New Encyclopaedia Britannica; 15th ed.; Encyclopaedia Britannica: Chicago, 1987.

Webster's New Collegiate Dictionary, 1974; G. & C. Merriam: Springfield, MA.

American Can Company, 1972. *Carbonated Beverages in the United States.* Historical Review; American Can: Greenwich, CT.

Burke, J., 1978. *Connections*; Book Club Associates; MacMillan: London.

Gillespie, C.C., Ed., 1974. *Dictionary of Scientific Biography*; Charles Scribner's Sons: New York.

Haynes, W., 1954. *American Chemical Industry, Vol. 1.; Background and Beginnings*; D. Van Nostrand: Toronto; reprinted by Garland: New York, 1983.

Karrer, P., 1947. *Organic Chemistry*; Elsevier: New York.

Mines, S., 1978. *Pfizer ... An Informal History*; Pfizer: New York.

Riley, J.J., 1958. *A History of the American Soft Drink Industry*; American Bottlers of Carbonated Beverages: Washington, DC.

Veysey, L.R., 1965. *The Emergence of the American University*; U. of Chicago: Chicago.

References

Appleton's Cyclopedia of American Biography, **1889**. Wilson, J.G., and Fiske, J., Eds.; D. Appleton: New York, Reprinted by Gale Research Co., 1968.

Du Pont, **1922**. *Du Pont Lactic Acid; an Essential Material in Leather Manufacturing Processes*; Du Pont Brochure.

Rumford Chemical Works, **1939**. *Eighty Years of Baking Powder History*, Rumford Chemical Works: Rumford, RI.

Town of Littleton, **1964**. *250th Anniversary of the Town of Littleton, Massachusetts*; Official Program [of Celebration].

Avery, C.E., **187?**. *Report on the Miller's River Nuisance* (listed in the National Union Catalog).

Avery, C.E., **1881**. *Proceedings of the Society of Arts, Meeting 264*, February 10, 1881.

Claflin, A.A., **1897**. *J. Soc. Chem. Ind. London*, **16**, 516.

Claflin, A.A., **1901**. *J. Soc. Chem. Ind. London*, **20**, 210.

Engelhardt, H., and Maddrell, R., **1847**. *Ann. Chem.*, **63**, 118.

Garrett, J.F., **1930**. *Ind. Eng. Chem.*, **22**, 1153.

Hochheiser, S., **1986**. *Rohm and Haas, History of a Chemical Company*; U. of Pennsylvania; Philadelphia.

Holten, C.H., **1971**. *Lactic Acid*; Verlag Chemie: Weinheim, Germany.

Miller, W.A., **1869**. *Elements of Chemistry: Theoretical and Practical*; 4th Ed., Part III, Organic Chemistry; Longmans, Green, Reader and Dyer: London.

Morrison A.C., **1904**. *The Baking Powder Controversy*; The American Baking Powder Association: New York.

Nicolle, J., **1961**. *Louis Pasteur. A Master of Scientific Enquiry*; Hutchinson: London.

Norton, L.M., and Tuttle, H.C., **1890**. *Technology Quarterly III, (MIT)*, 287.

Palmer, G., **1917**. *Obituary of Charles Avery*; in a not identified Littleton, MA, newspaper; January 1917.
Strecker, A., **1869**. *Jahresber. Fortschr. Chem.*; J. Ricker: Giessen.
Whitaker, T., and Denison, J.R., **1895.** *J. Soc. Dyers Colour.*, June 1895, 98.
Wood, J.T., **1912**. *The Puering, Bating and Drenching of Skins*, E. & F.N. Spon: London.

TEXTILE COLORING AND LACTIC ACID

England had emerged from the Napoleonic wars as the leading industrial nation and the most important industry of this nation was the textile industry. Textile exports were more than five times the home consumption and were by far the largest item of commerce. Thanks to the new processes for making cheap cottons the calico industry had dwarfed the formerly much larger wool industry, viz. Table I2,1.

Table I2.1. British export of textile goods.

	1769	1829
woolen products	£ 3,900,000	£ 5,400,000
cotton products	£ 200,000	£ 37,300,000

Source: The New Encyclopaedia Britannica, 15th ed., 1987, Vol. 18, 779.

Textile products were manufactured and exported in almost every stage of processing. Finishing operations also became highly mechanized, a development which had great bearing on the infant chemical industry. Soda, bleaching chemicals and other chemicals were used in ever increasing amounts.

Of particular interest for the trade was the dyeing of the cloth: dyed and printing fabrics, of course, provided the highest added values. These also were the operations where the highest chemical knowledge, more appropriately the highest chemical skills, were required to produce attractive and lasting results. [Textile dyeing and printing of woolen yarns and fabrics were already well developed industries in Great Britain and on the Continent before the Industrial Revolution made cotton textiles cheap commodity products.]

Attractive colors, however, were very expensive and only a few were available: red from madder (Turkey red), blue from indigo, and violet from Mediterranean mollusks (Tyrian purple). In the seventeenth and eighteenth century this range of colors was extended with dyes from a number of tropical and Northern American woods: black and blue from logwood

(Haematoxylon campechianum), yellows from fustic (*Morus tinctora*) and quercitron (*Quercus nigra*), and reds from various types of redwood.

Like madder most of the wood dyes did not adhere well to wool or cotton fiber. The problem of insufficient adhesion – leading to poor mechanical and washing fastness – had in practice already been solved in ancient times by forming insoluble precipitates of the coloring matter with inorganic substances. Alum, potassium aluminum sulfate, was precipitated onto or preferably inside the textile fiber with the aid of some acid generating substance like tannin; the dye would then in turn react with this precipitate to form a colored lake locked up in the fiber as well. This process was called mordanting (See for instance Berzelius, 1839).

Around 1800, chemists and the chemical industry turned their attention to the production of better mordants and mordanting recipes to improve the (wash)fastness of the precipitated color, which depended much on the manner in which the precipitation was effected. The brilliancy and hue also appeared to be influenced by the mordanting process.

In the same period small chemical enterprises came into existence in the neighborhood of textile finishers to supply them with chemicals for their bleaching, dyeing and printing operations. An example of such an enterprise was the firm of Charles Kestner mentioned in Chapter 2, 34. This company was one of the early producers of a mordanting acid – tartaric acid (1809). It also was the company which discovered racemic acid, by-product of the tartaric acid manufacture (1826) and subject of historic investigations by Pasteur.

More "exotic" salts than alum, some of them coming from the shops of the alchemists, also found use as mordants in textile dyeing. Berzelius mentions sugar of lead [lead acetate], iron acetate, tin salts and gallnut extracts. A combination of alum with tartar yields better results than alum alone, he adds to his summary of mordanting agents.

The first antimony salt used for mordanting was tartar emetic, discovered in the seventeenth century; Glauber produced it (in 1648) by boiling antimony oxide with tartar (Ullmann, 1923).

The new wood colors introduced in the eighteenth and early nineteenth century required special mordanting recipes to achieve acceptable fastness. A great many materials were tried for this purpose, Berzelius names for example copper and bismuth salts.

Another new and widely applicable chemical which would become an interesting mordant was potassium bichromate. The mordanting action of this product was discovered in the 1840's. Not much later it was established that bichromate only acted as a mordant in the reduced state[1]. The woolen yarn or fabric had to be boiled in a solution of the chromate together with a suitable reducing organic acid and, in some cases, in the presence of the dyestuff as well. The extraordinary fastness of chrome dyes was first recognized in 1853 by Camille Koechlin of Mulhouse (Alsace), one of the foremost textile chemists – colorists – of the time.

Chrome Mordants Need an Organic Acid

The preferred organic acid was tartar – potassium bitartrate -which sometimes was used in the crude form (red argol) in which it was procured from the vinery.

In the 1850's and 1860's, when the demand for cream of tartar [which is a specially selected or recrystallized argol] rose strongly because of the application in articifially raised bakery products, tartar prices moved upward and textile colorists began to look for substitutes. In 1854 Bolley (Wagner, 1855) investigated the use of oxalic and lactic acid; he believed that lactic "could offer good services provided ... a cheap and prolific manufacturing method could be found."

On the next page of his book Wagner summarizes the results presented in a British patent (of 1854) by Gatty and Kopp. In this patent lactic acid is proposed as a substitute of tartaric acid, citric acid and lemon juice. The replacement of cream of tartar by the sodium salt of (double) lactic acid is another "invention" claimed in this patent. [Evidently based on the theory that lactic acid was dibasic, see Intermezzo 1, 64.]

Dyers and printers also learned to fix one of the best dyestuffs – madder – with chrome, in particular on wool. In this case potassium bichromate was added to the suspension of madder (or to garancine, a dye made by treating ground madder with sulfuric acid). A brownish red was obtained with excellent fastness. When bichromate was replaced by alum and tartar a yellowish red of comparable quality was produced.

The action of bichromate on logwood dyeing was tried by many colorists. It seems that good results were obtained, but the blacks were inferior in beauty and fastness to those of the old logwood and copperas black recipe (copperas = ferrous sulfate heptahydrate). For economic reasons logwood-chrome blacks nevertheless attained some importance in the 1860's (Crookes, 1874; O'Neill, 1878). Charles Avery studied these processes while working with the Boston Dyewood & Chemical Company, one may assume (Ch. 4, 87).

The Dyeing and Printing of Cotton

Compared to wool, cotton is chemically inert, which entails even more problems in dyeing when reasonable wash fastness is desired.

Only a limited number of dyes was available for the coloring of cotton, of which indigo and madder were superior. Indigo derives its fastness from the special chemical way in which the color is precipitated in the fiber.

Madder was fixed onto the fiber by means of a mordanting process, called Turkey red dyeing. This was a very elaborate process taking more than twenty operations and a period of time of up to thirty days (Verbong, 1986). Mordanting in this case was done with alum and tannin or tannic acid, after a thorough pretreatment of the fabric with 'turned' (rancid) olive oil, soda and sheep dung.

The First Synthetic (Coal Tar) Dyes

Then came, in 1856, the accidental discovery of a synthetic dye by William Henry Perkin [Sr.], (1838–1907). Perkin, a student of August Wilhelm von Hofmann[2] tried to make quinine, but obtained a black tarry product, which on dilution assumed a violettish color, mauve. Perkin tried the color on silk and obtained such interesting results that he decided to build a manufacturing plant for the production of the new dyestuff. This decision marks the beginning of the synthetic dyestuffs industry.

Production of the new dye, called mauveine, began in 1858, and, from the first day, met with much success. This invited French competitors to imitate Perkin's product. Within a year they also succeeded, with the "help" of poor patent protection, in putting mauveine on the market.

The discovery of the first aniline dye also initiated the search for other synthetic dyes based on aniline. A second dyestuff, fuchsine, was discovered by the French chemist Verguin (1858). The search for new aniline dyes now became feverish: By 1862 about 50 companies in England, France, Germany and Switzerland were engaged in the production of aniline dyes for the textile industry (Homburg, 1983). Although aniline dyes provided a much desired extension of the range of colors, and their brilliancy was a tremendous improvement over most of the old natural dyes, color and wash fastness was rather inferior.

The new aniline dyes were basic dyes which required fixation techniques different from those the industry was accustomed to. This applied in particular to the coloring of cotton, volume and pricewise the most interesting textile material.

Colorists soon learned that with one of the old mordants, tartar emetic, good results could be obtained in the dyeing of cotton with the new aniline dyes (Crookes, 1874; O'Neill, 1878). O'Neill provides as an example a patent of Lloyd and Dale of 1861, in which the use of tartar emetic together with tannic acid is claimed for the fixation of aniline colors.

Dammer's Handbook of Chemical Technology (Vol. 5, 1898), states that tannin-antimony mordants are the best for the fixation of basic dyes on cotton. The colors are less brilliant than those obtained with the traditional Turkey red mordanting process, but show better wash fastness. Loewenthal (Loewenthal, 1898) also calls tannin-antimony mordants the most often used ones for the dyeing of cotton with basic dyes. Both authors mention tartar emetic as the preferred antimony salt for these processes.

Textile Coloring Demands Much Chemical Skill

Due to these developments (from about 1800 to 1870) the printing and dyeing of textiles had become quite sophisticated operations and the number of chemically trained people increased at pace with the skills required to put good quality textiles on the market. During this period a new type of highly specialized chemist evolved: the textile colorist.

In dyeing evenness, uniformity of shade is of prime importance; in printing the contours of the print and the (lack of) penetration of the dye into the fabric play a more important role. In the larger works colorists had a full time job in the formulation of new methods and recipes for the ever increasing demands of the trade, and for the introduction of the new dyestuffs into the textile finishing plants.

Only a few years after the discovery of the first aniline dyes, three large factories were established in Germany to manufacture dyes and chemicals in greater quantities and on a more scientific base:

> Chemische Fabrik Meister, Lucius & Co. (now known as Farbwerke Hoechst) in 1863,
> Farbenfabriken Bayer also in 1863, and
> Badische Aniline und Soda Fabriken in 1865.

The rapid expansion of these enterprises (and of other chemical works as well) was only possible because of the large number of qualified chemists emerging from German universities and technical institutes, many of which were molded after the famous school of Liebig at Giessen. Many foreign chemists also received additional training in Germany; quite a few Germans moved to France and Great Britain to accept positions as colorists in the textile finishing industry.

The new German dyeworks (Farbwerke) embarked within a decade on very ambitious research programs, aimed at the synthesis of two of the best natural coloring substances: alizarin (madder) and indigo. The synthesis of alizarin was accomplished by Carl Gräbe (1841–1927) and Carl Liebermann (1842–1914) in 1868. Production of synthetic alizarin began in 1869, simultaneously by BASF and Hoechst. An indication of the revolution the synthesis of alizarin must have caused, is given by the development of its price: in 1870 a kg of natural madder cost 80 Mark, in 1888 alizarin, the synthetic but fully comparable product was sold for 8 M/kg.

Production had hardly begun when the search for alizarin derivatives already started. Many of these alizarin dyestuffs found their way to the textile dyers and printers because of their superiority in light fastness over aniline colors. On wool synthetic alizarin dyes became very popular.

The German dyestuff manufacturers were also very active in developing dyeing and printing recipes. These recipes were presented to customers all over the world in beautiful brochures containing samples of yarns or tissues dyed and printed with the new dyestuffs.

As most of the colors were not really substantive, mordanting recipes also had to be elaborated. For the coloring of wool with alizarin type dyes madder mordanting recipes with chromate and an organic acid could be copied with minor modifications only.

For this and other mordanting applications textile colorists between 1850 and 1885 quite frequently recommended the use of lactic acid. Which advice often was followed by a remark that the high price of lactic acid would prevent such use!

Until the development of more substantive dyes a great number of aniline dyes was applied in the coloring of cotton. In this case tartar emetic was the preferred mordant. The first proposal to replace the antimony salt of tartaric acid by the corresponding lactate came from Charles Waite (Ch. 4, 97, and below).

For other dyes special recipes were developed, induline dyes for instance were precipitated onto the fiber with a combination of tannin and acetic acid.

Induline Dyes

Induline dyes are large complex molecules with basic functions. They give blue colors, resembling that of indigo, and show high resistance to washing and have good light fastness (Garçon, 1902). As said the lakes have to be formed with tannin and acetic acid, a volatile acid with a pungent smell.

A. Schlieper of Schlieper und Baum, well-known calico printers at Elberfeld (a city now incorporated in the city of Wuppertal, FRG), proposed in 1885 to add lactic acid to induline printing paste, instead of the acetic acid recommended by the manufacturer (Farbwerke Hoechst, vorm. Meister, Lucius und Brüning):

> ... that substance [lactic acid] always produces a homogeneous print of excellent beauty and deep intensity which can not be achieved with acetic acid. The high price of lactic acid will, however, certainly prevent its use in the industry.
>
> *- A. Schlieper, according to Garçon, 1902*

The Introduction of Lactic Acid in Mordanting

The patents of Charles Waite – US Patent 290,294 of 1883 on the use of a combination of four parts of lactic acid with one part of oxalic acid in the mordanting of animal fabrics, and U.S. Patent 341,294 of 1886 on the use of antimony lactate as mordant for cotton yarn or cloth – brought lactic acid again to the fore. The first successful experiments still had to await the introduction of an acid of sufficient purity and a reasonable price (Norton and Tuttle, 1890). In April 1890 Georges Jacquemin (Ch. 5, 134), Nancy, France, demanded a patent on the manufacture of lactic acid by fermentation (French Patent 204,874). The fifth claim of this patent reads:

> The application of lactic acid and of lactates is claimed in general, and in particular those of the acid sodium and potassium lactates for the same uses as tartaric acid, tartrates and acid tartrates ... in the dyeing and printing of fabrics ...
>
> *- Jacquemin, 1890*

Whitaker and Denison present some information on the gradual acceptance of lactic acid for wool dyeing operations between 1892 and 1895 in the United States and Great Britain (Whitaker and Denison, 1895).

An early report in a German textile journal on the possible use of lactic acid in textile coloring appeared in the Färber-Zeitung in 1890/1891 (p. 376). The journal gives a summary of the paper of Norton and Tuttle, and reproduces a remark from the British Journal "The Dyer and Calico Printer":

When the information about the value of the new lactic acid [of Waite] is correct ... then certainly an important and useful invention has been made.

Some four years later the use of lactic acid in chrome mordanting again was brought to the attention of German wool dyers. In the October 15, 1895, issue of the Färber-Zeitung questions were raised about the supply situation and the use of lactic acid to replace tartar. The answer was:

Lactic acid in the form of a clear red-yellowish solution of 20–21° Bé [probably about 50% by weight] is supplied by C.H. Boehringer Sohn, Chemische Fabrik at Nieder-Ingelheim am Rhein.

An editorial in the next issue of November 1 stated that lactic acid appears to be the best reducing acid for chrome mordanting, but the price is still too high. A more complete answer is given in December, 1895, by E. Hoffmann (Hoffmann, 1895, see also Hoffmann, 1896). The author reports on the fixation of alizarin colors as well as of "wood colors" on wool. For both dyes lactic acid is to be preferred. "The price is not as high as usually is stated. German factories [plural!] supply 50% lactic acid at a price of 60–65 M/100 kg."

The story of lactic acid use in textile coloring will be continued in Chapter 5, 150.

Notes

[1] In the reduction step hexavalent chromium is reduced to chromium trioxide, which forms the lake to which the dye adheres. When the bichromate solution has been allowed to fully penetrate the fiber the precipitated chromium oxide will be much more finely divided inside the fibers than can normally be achieved with other mordants. Such was even more the case when stronger acids such as tartaric acid and oxalic acid were applied for special prints like reserves and etches. The price of tartaric acid, about one U.S. dollar per kg in the 1850's, was strongly influenced by the availability of tartar, which changed much with the wine harvest. Consequently the intensity of the search for substitutes of tartar and tartaric acid also changed with those fluctuations in price.

[2] Hofmann (1818–1892), one of the best students of Liebig, had been appointed Director of the Royal College of Chemistry when this institute was founded in 1845. He returned to Germany in 1864.

Literature consulted

The New Encyclopaedia Britannica; 15th ed.; Encyclopaedia Britannica: Chicago, 1987

Bäumler, E. **1968.** *A Century of Chemistry*; Econ.: Düsseldorf.

Dammer, O., **1895–1898.** *Handbuch der Chemischen Technologie*; Ferdinand Enke: Stuttgart.

Homburg, E., **1983.** *J. Soc. Dyers Colour.*, **99,** 325.

Molenaar, L., and Kooiman, P., **1986.** *Chemie en Samenleving. Van Kleurstof tot Kunstmest*; Natuur en Techniek: Maastricht.

Osteroth, D., **1985.** *Soda, Teer und Schwefelsäure*; Rowohlt: Reinbek by Hamburg.

References

Berzelius, J.J., **1839.** *Leerboek der Scheikunde*; Dutch translation of the third edition; Mulder, G.J., Ed.; van den Heuvel: Leyden, Vol. 4, 373.

Crookes, W., **1874.** *A Practical Handbook of Dyeing and Calico-Printing*; Longmans, Green and Co.: London, 542 and 543.

Garçon, J., **1932.** *Encyclopédie Universelle des Industries Tinctoriales*; Vol. 31, Société Industrielle de Mulhouse; Macon, Protat Frères: Paris, 681.

Hoffman, E., **1895.** *Färber-Z.*, 1895/1896, 65.

Hoffman, E., **1896,** *Färber-Z.*, 1895/1896, 217.

Loewenthal, R., **1898.** *Die Färberei der Spinnfasern*; Otto Spamer: Leipzig, 32.

O'Neill, C., **1878.** *The Practice and Principles of Calico Printing*; Palmer and Howe: Manchester, 274 and 277.

Norton, L.M., and Tuttle, H.C., **1890.** *Technology Quarterly III, (MIT)*, 287.

Ullmann, F., **1913–1923.** *Enzyklopädie der technischen Chemie*; Urban und Schwarzenberg: Berlin, Vol. 12, 70.

Verbong, G., **1986.** *Jaarboek voor de Geschiedenis van Bedrijf en Techniek*; Utrecht, Vol. 3, 183.

Wagner, J.R., **1855.** *Jahresber.Fortschr. Chem. Techn.*; Otto Wigand: Leipzig, 343.

Whitaker, T., and Denison, J.R., **1895.** *J. Soc. Dyers Colour.*, June 1895, 98.

CHAPTER 5

INDUSTRIAL PRODUCTION OF LACTIC ACID COMES TO GERMANY

Since the middle of the nineteenth century pharmaceutical lactic acid and a few of its salts were commercially available through pharmacists in Germany and France (see Ch. 2, Table 2,I).

E. Merck, Darmstadt, seems to have been one of the early producers. A short history of the company (Löw, 1952) mentions lactic acid as one of the company's products as early as 1842, which is amazingly soon after the first publication of a convenient method to produce larger quantities of the acid by fermentation (Boutron and Frémy, 1841a and b). In a footnote to the German translation of this paper Liebig remarks:

> Recently we have heard of a chemical manufacturer who, following our directions, produces lactic acid by saturation of waste water from the preparation of starch with lime.
>
> *- Liebig, 1841*

Considering later events it can be surmised that Merck bought crude lactic acid as calcium lactate which it then converted in its own works to pharmaceutical lactic acid. Outside the pharmaceutical field interest in organic acids also rose. The textile industry had been using tartar and tartaric acid as mordanting aids since the beginning of the century, and with the advent of artificial dyestuffs in the 1850's and 1860's, became very interested in other acid ingredients as well (Intermezzo 2). Lactic and citric acid were both subjects of studies to replace tartaric acid. Due to high production costs they were discarded time and again for this type of technical application.

Tartaric Acid Production

As long as raw materials for tartaric acid, argols and lees from the vineries were in good supply, temporary shortages of these chemicals could be and usually were met by increasing production facilities, or by installing new plants.

Demand kept rising, however, and by the middle of the 1890's total world production had increased to the large volume of 5,000 t/a (Dammer, 1998). Largest producing countries were the still leading industrial nation Great

Britain, and the United States of America, where a substantial use of tartaric acid had developed for use in baking (cream of tartar) and in soft drinks.

German tartaric acid production then was about 1,500 t/a. The most important producers were Joh. A. Benckiser of Ludwigshafen, founded in 1823; Goldenberg, Geromont & Cie of Winkel near Wiesbaden, founded in 1875; and Fabrique de Produits Chimiques de Thann et de Mulhouse, a continuation of the chemical works of Charles Kestner (Ch. 2, 34). A fourth producer, probably smaller than either of these three was C.H. Boehringer Sohn of Nieder-Ingelheim near Mainz, founded in 1885. It is this company which would succeed in (partly) replacing tartaric acid in textile coloring by lactic acid.

Albert Boehringer Acquires a Chemical Plant at Nieder-Ingelheim[1]

The tartrates plant at Nieder-Ingelheim (see Map 5,I) had been bought in 1885 by Ernst Boehringer, owner of the chemical enterprise C.F. Boehringer Söhne at Mannheim[2], for his younger brother Albert (1861–1939).

Fig. 5,1. Albert Boehringer. Courtesy: C.H. Boehringer Sohn, Ingelheim.

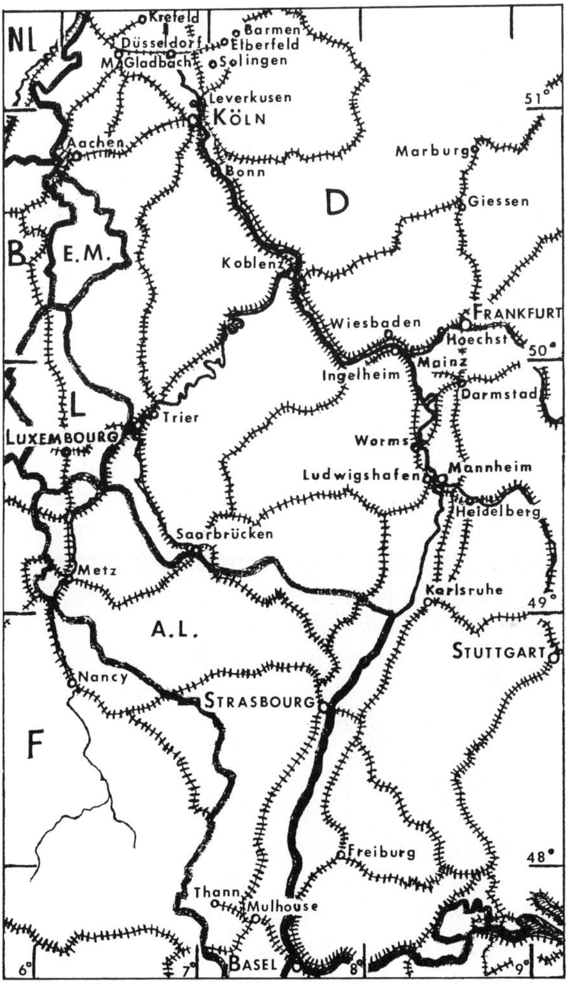

Map 5,1. The Rhine basin.

– region of important textile industries

– birthplace of the German chemical industry

I-I-I-I-I-I railroads;

A.L. = Alsace-Lorraine, incorporated in the German Empire in 1871; returned to France in 1919;

E.M. = Eupen and Malmédy, territory ceded to Belgium after World War I.

The Boehringers came from a Swabian family of industrialists, their grandfather owned an apothecary in Stuttgart and had founded together with a colleague (apothecary) by the name of Engelmann in 1817 an industrial enterprise engaged in the manufacture of ether, chloroform, santonin, and potassium iodide. As traders they were even more important, especially in morphine, quinine and tartaric acid. Albert Boehringer's father Christoph Heinrich moved with the company to Mannheim in 1870, where the aniline factory of Engelhorn and Clemm was acquired. This factory had been vacated because the newly founded company "Badische Anilin und Soda Fabriken" in which Engelhorn and Clemm participated needed more space and facilities than the city of Mannheim was prepared to give. BASF moved to Ludwigshafen on the other bank of the Rhine.

When Christoph Heinrich Boehringer died in 1882, his eldest son Ernst took over the reins of the family enterprise, then called C.F. Boehringer Söhne[2]. The second son, Albert, had apparently inherited too much of the Swabian love for work and independence to be content with a second position in the family enterprise. When Ernst was informed that at Nieder-Ingelheim a factory producing cream of tartar was for sale, he stepped in and bought the company for his brother. This happened with Albert's consent, the payment of 40,000 Mark was effected with Albert's money. The date of the purchase was July 31, 1885. The company was named "Albert Boehringer, Chemische Fabrik": a few years later it was renamed "C.H. Boehringer Sohn, Chemische Fabrik" in honor of the deceased father of the owner.

Nieder-Ingelheim, A Rural Village Attracts Industrial Enterprises

The communities of Nieder-Ingelheim, Ober-Ingelheim and two other villages that today comprise the city of Ingelheim were rather small at the time. The total number of inhabitants was only 6,590. The basis for growth was provided in 1859 when the railroad from Mainz to Köln was finished and Ingelheim as a station on this line became connected with the growing German railroad system. In 1863 the first industrial enterprise was opened, a portland cement works.

A small factory of agricultural chemicals (bone meal) was erected by D'Avis and Klein in 1867. A few years later it only manufactured salts of tartaric acid from argols and lees. In 1885 this plant was owned by Philipp

Adam Schneider, who was dissatisfied with the financial results and had offered it for sale.

Fig. 5,2. The works of C.H. Boehringer Sohn in 1908. Courtesy: C.H. Boehringer Sohn, Ingelheim.

The conditions under which Albert Boehringer then started his enterprise are almost unbelievable for the modern reader: *Electricity* had not yet been introduced in Ingelheim. As a matter of fact the first electric power station in Germany was built in Berlin in the same year 1885, three years after Edison had installed the world's first power station in New York. The larger cities of Europe and North America were still illuminated with gas, the use of which had been invented in Great Britain in the beginning of the century. The village of Nieder-Ingelheim, however, did not have a gasworks. One must assume that the factory of Boehringer as well as most private houses were lit with kerosene lamps. This changed for the better after the installation of a gas works in 1891. Street illumination with electricity was installed at Nieder-Ingelheim in 1899, most private houses were then also connected to the electric grid.

The company had its own *steam* supply, however. Schneider had even obtained a permit for the installation of a second boiler, which probably came into operation in 1882. Steam was used for two purposes: for the supply of heat to the vessels in which the products were being treated and also to supply power to equipment in the plant and in the workshop. For this reason the manufacturing plant and the boiler house were adjacent buildings. Around the turn of the century the Boehringer plant became equipped with its first locomotive engine and generator with a capacity of 160 kW (Figure 5,2).

The *plant* was a modest one compared to the huge factories BASF, Hoechst and Bayer had erected. Even the works of future competitor E. Merck in Darmstadt were much larger (Figure 5,3).

Fig. 5,3. The factories of E. Merck in 1892.
Courtesy: E. Merck, Darmstadt.

But it already employed 28 people when it was acquired, a number which grew steadily after the lactic acid plant came into full operation in 1898. By 1910, about 150 people were employed[3]. Even then almost all work was done by hand.

Cream of Tartar Production by D'Avis & Klein

The chemist of the company, Dr. Ferdinand Klein, had developed a new process for the manufacture of chemically pure cream of tartar from lees. Considering the technological knowledge and possibilities at that time he probably used an acid process (see Dammer, 1898):

Lees were boiled in lead lined wooden vessels with dilute hydrochloric or sulfuric acid. After boiling for several hours the bulk of insoluble potassium bitartrate would have been converted into soluble tartaric acid. The liquid was left standing for a while to allow for the settling of insolubles. The solution of tartaric acid was separated from the sediment by decantation or filtration. The extraction was repeated to recover most of the tartaric acid.

It can be supposed that Klein obtained cream of tartar by adding potassium carbonate to the acid extractions, cream of tartar – potassium bitartrate – would be the precipitate and could be filtered off, dried and sold (see also Friedburg, 1882).

Tartaric Acid Production by Albert Boehringer

Albert Boehringer who had finished a practical course in pharmacy and had studied chemistry at Munich University for some time, decided very soon after the "take-over" to manufacture tartaric acid itself instead of the salt. From his brother, he probably knew that the acid was of more importance commercially than cream of tartar.

The process implemented for the production of tartaric acid also made use of the acid extraction process. The extracts were treated with milk of lime to precipitate tartaric acid as the calcium salt. The calcium tartrate was separated by filtration, and reconverted into the acid with sulfuric acid. In a second filtration the gypsum was removed; the filtrate, a rather pure solution of tartaric acid, was concentrated in a lead lined vessel to about 30° Bé, filtrated again to remove residual gypsum and thereupon concentrated to about 43° Bé in order to obtain large, regular crystals. Crystallization had to be carried out very slowly. This was done in leaden, well insulated vessels and took 5 to 8 days. Boehringer used strings on which the crystals grew, a process known from the production of sugar candy. In this way large crystals were obtained, which bought the best price in the market.

The description of the process is given to provide the reader with an impression of what Albert Boehringer was up to when he bought the tartar works, and what he had to do to make an acceptable quality of tartaric acid. But he managed the problems, about a year later the first deliveries of tartaric acid were being made!

Other conditions required to achieve an economical production were:

vacuum pans to concentrate under mild conditions to prevent browning of the solution and of the final product due to traces of sulfuric acid;

a system to recover tartaric acid from the mother liquors of the first crystallization;

centrifuges to obtain a good separation of fine crystals which are formed in some of the crystallizations;
good analytical equipment and procedures.

It is not known how much tartaric acid Boehringer produced in these first years. As previously mentioned total German production, including that of the Alsace, in 1896 was about 1,500 t/a. There are good reasons to believe that Boehringer's share of the market in those years was not more than about 10%. The market share of the company almost certainly was too small to allow for a profitable production. A number of times Boehringer seems to have been on the verge of quitting the market. Between 1888 and 1898 five (small) German producers of tartaric acid actually stopped production (Gehe & Co., 1898). In this period raw material prices, although fluctuating, on average did not decrease substantially, the price of the final product, however, went down consistently, probably due to heavy competition from the major (international and German) producers who had the advantage of economy of scale.

Only U.S. prices of crystalline tartaric acid have been found for this period (Haynes, Vol. 1, 1954):

1871	1.10 $/kg
1881	0.93 $/kg
1891	0.73 $/kg
1901	0.62 $/kg

Gehe reports a German price of 2.60 M/kg in 1890, equivalent to 0.70 $/kg (Gehe & Co., 1890).

Boehringer Is Interested in the Manufacture of Citric Acid

Albert Boehringer must soon have felt the necessity to manufacture other products. In 1887 for instance, he concluded an agreement with his brother on financial cooperation in the production of citric acid. One must assume that the brothers had the "classical" production method in mind, a method based on extraction of the acid from lemon juice (Scheele's method). The project was not realized.

It is not known whether lactic acid also was considered, but it is not unlikely. Ferdinand Klein, the company chemist of d'Avis and Klein, who probably stayed on as chemist for one or two years after Albert Boehringer's take over, almost certainly was interested[4]. A few years later, in 1888, Klein, then living at Tyrnau (Trnava, near Bratislava, now in Czechoslovakia) approached E. Merck, Darmstadt, with a process for the manufacture of lactic acid by alkaline decomposition of beet sugar. Klein had modified a process for the preparation of half kilogram quantities of lactic acid, developed by Professor Heinrich Kiliani (1855–1945) of the University of Munich in the very same years Albert Boehringer attended chemistry lessons in that city (Kiliani, 1882). In a letter to Merck, Klein proposed to carry out this chemical manufacturing process in an autoclave under the influence of lime (instead of alkali) (Private information from Merck's Archives, 1987).

E. Merck, Darmstadt, Sells and, Maybe, Manufactures Pharmaceutical Lactic Acid

It is unlikely that Merck made any practical use of the proposal of Ferdinand Klein. As said before, it can be assumed that the company at that time already manufactured pharmaceutical lactic acid from carbohydrate raw material by means of the Bensch process. The oldest archives of the firm of Merck, Darmstadt, which actually yield information about lactic acid preparation, are from 1893. In that year one of the production chemists of the company, Dr. Carl Krauch[5], spends part of the summer in Isny, a small town in the South of

Fig. 5,4. Carl Krauch in front of the dairy Eisenharz, 1893. Courtesy: E. Merck, Darmstadt.

Fig. 5,5. The dairy Eisenharz at Isny, Württemberg, 1893. Courtesy: E. Merck, Darmstadt.

Württemberg, near the Austrian border, to assist in the production of calcium lactate (Figures 5,4 and 5,5).

In three letters (of 1893) to one of the directors of the company he tells about the problems he has in establishing calcium lactate production in the local cheese factory Eisenharz. From these letters a clear picture emerges of the know-how of the Merck company at the time. The second letter, of July 12, 1893, is the most interesting one. Krauch reports in this letter about the unexpected arrival on the previous Sunday of three officials who had received complaints about the smell of the plant and now were staging an official investigation. The delegation, consisting of the local burgomaster, the head of the county and the county medical officer were shown the premises and the works and, on request, were given a description of the process (Figure 5,6):

I told him [the county doctor] that we produce a solution of calcium lactate by fermentation of whey in the presence of chalk, this solution is clarified by boiling, the precipitate containing excess calcium carbonate and protein is deposited on Wunderlich's dung-hill [Wunderlich was the owner of the cheese factory annex piggery], the liquid is evaporated and the solid residue dispatched to Darmstadt for further processing to lactic acid."

- Krauch, 1893

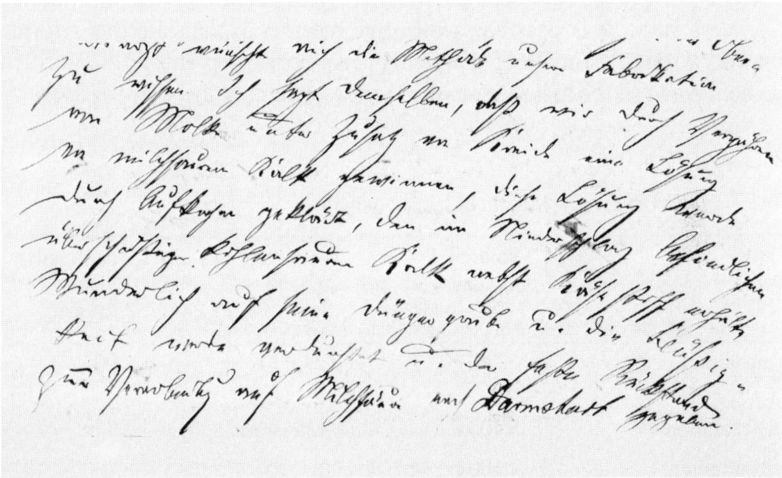

Fig. 5,6. Krauch's description of the fermentation process in the dairy plant Eisenharz, 1893. Courtesy: E. Merck, Darmstadt.

At the end of the inspection, Krauch reports, the officials were convinced that the smell of the piggery was worse than that of the fermentation process.

The three letters of 1893, together with an additional letter from the summer of 1896, do give sufficient information on the process to allow for a more detailed description:

6000–7000 l of whey are received daily for the fermentation, which is carried out in 1200 l vats at a temperature of about 30° and normally takes from 3 to 5 days. The end of the fermentation is controlled by a reaction on sugar [Fehling's reaction?], thereupon the solution is boiled to flocculate milk proteins and excess chalk [as well as the biomass]. No information is given how the separation of the flocculate from the calcium lactate solution is effected. The clear solution of calcium lactate was then concentrated in open iron pans, probably by the direct heat of flue gases, to a concentration of 9 to 10 Bé (about 14 to 15%).

The concentrated solution was left standing to crystallize and then passed over a small hand-operated hydraulic filter press (capacity about 30 l "crude concentrate" per pressing, yielding about 8 kg of filtercake). In 1893 the filtercake was dried in the open air. One of the reasons Dr. Krauch visited the plant in that year was to install a drying room.

In the 1896 letter Dr. Krauch reports on yield determinations of a number of process steps. It is possible, with some reserve, to translate this information into a rough estimate of the yields per process step and of the purification achieved (on the assumption of a lactose content of the whey of 5%).

Table 5.I. Yield in the preparation of lactic acid at Isny.

process step	volume per day	Ca-lactate content (kg)	cumulative yield (m/m)*
fermentation	6.000 l	(270**)	90
flocculation and separation	5.400 l	435	72
concentrating	2.100 l		
filtrating			
wet filtercake	500 kg	260	54
drying of the cake	235 kg	235	49

* m/m = molar yield
**lactose

At Darmstadt the dry filter cake with an estimated purity of 98% was converted at boiling temperature with a solution of zinc sulfate into zinc lactate and calcium sulfate, which precipitated and was filtered off. The hot solution of zinc lactate was cooled to room temperature to crystallize. The crystals were separated, washed and dried. The molecular yield of zinc lactate on calcium lactate was only about 60% (Krauch), the calculated content of impurities: less than 0.05%. Pharmaceutical lactic acid was prepared from zinc lactate by a modification of Bensch's method (Exhibit 2,I) proposed by Kolbe (Kolbe, 1860). Merck extracted the product of the reaction of zinc lactate with hydrogen sulfide

$$(CH_3 . CHOH . COO)_2Zn + H_2S \text{ (gas)} \rightarrow$$
$$2CH_3 . CHOH . COOH + ZnS \text{ (precipitate)},$$

a pulpy mass of lactic acid and insoluble zinc sulfide, repeatedly with diethyl ether. Later manufacturing reports of Merck state that the ether extraction was carried out in closed, enameled "Schütteltrommeln" (shaker barrels). The ether extracts containing nearly all the lactic acid were collected; the ether distilled off after water had been added. The final concentration step (to a density of 1.21, that is about 90% by weight) was performed on a steam bath in open enameled or porcelain pans.

The oldest manufacturing report which has been saved is from 1898/99. It gives as the yield of pharmaceutical lactic acid on zinc lactate 61% by weight, which amounts to a molecular yield of 90%. The overall molar yield calculated on the lactose content of the whey was therefore only 0.49 x 0.60 x 0.90 = 26%.

When Did Lactic Acid Fermentation Begin in Darmstadt?

In his letter of July 20, 1893, Krauch writes:

> Probably as a consequence of the peculiarly high sugar concentration the whey (here in Isny) ferments much more energetically than the Darmstadt whey. Temperatures which are conditional in Darmstadt, are almost forbidden here. ... When we have ... a temperature between 20 and 25°C in the fermentation room, and the temperature of the fermenting whey reads between 27 and 29, or a maximum of 30°C, we have the right conditions to ferment the whey completely in 3 days. For the Darmstadt whey a 5 to 10°C higher temperature is required.
>
> *- Krauch, 1893*

This statement, together with the fact that at Isny Krauch is evidently the only man who knows the ins and outs of the process, supplies convincing proof that the company of E. Merck as early as 1893 possessed considerable experience in the manufacture of lactic acid by fermentation. They also had the know-how to produce calcium lactate and lactic acid at the Darmstadt works.

Production of crude calcium lactate was cheaper in Isny than in Darmstadt, however, and the first location was therefore preferred. Whey could be had at only one third Pfennig per liter against a half Pfennig per liter in Darmstadt and, moreover, contained a higher percentage of sugar of milk.

It also fermented more readily.

The complicated way in which Merck manufactured pure lactic acid made the product too expensive for technical use. The sales prices of somewhat larger amounts of Merck's acid in the 1890's were probably over 4.50 M/kg (Güstrow, 1905), while tartaric acid could be bought at 2.60 or less.

The initiative to produce technical lactic acid in a more economical manner came from the French chemist Georges Jacquemin, son of a Professor of Pharmacy, who was well-versed in the new dyestuff chemistry and also was well acquainted with Louis Pasteur.

Jacquemin Obtains a Patent for Lactic Acid Production

In 1890 George Jacquemin (1862–1925) applied for a patent for a practical process to manufacture technical lactic acid (French Patent 204,874). The opening sentence of the patent reads in translation:

> The production of lactic acid ... by fermentation is described in the handbooks of chemistry and has been the subject of many special studies. However, it is a fact – as remarked by Mr. Dublaux in his summary of the state of the art in 1887 (in Frémy's Encyclopédie Chimique) – that all processes are long and uncertain.
>
> \- *Jacquemin, 1890*

It is therefore not surprising, Jacquemin continues, that lactic acid and its salts have not left the domain of pharmaceutical preparations and will not be found in the industry at large. The product could potentially attain an industrial importance comparable to that of other organic acids, however. To clear the way for industrial production, Jacquemin said, he had carried out a series of experiments, from which he had derived a suitable manufacturing method.

When Jacquemin's method (Exhibit 5,I) is compared with Avery's, only small differences are apparent. Jacquemin seems to have taken more sterility precautions and his bacterial culture is airborne, while Avery's original culture has probably been taken from sour milk. From the description of the process it is evident that Jacquemin knew the literature, including the patent literature, quite well. It may therefore be assumed that he had read some of the patents of the Avery company on the manufacture of lactic acid by fermentation.

EXHIBIT 5,I
JACQUEMIN'S FERMENTATION PROCEDURE
lit. Jacquemin, 1890, Dammer, 1898

raw material	fully saccharified wort, enriched with sugar, maltose or glucose
source of raw mater	malt from a brewery
concentration	?
sterilization	the enriched wort is boiled and cooled under sterile conditions
inorganic nutrients	supplied by the malt
organic nutrients	supplied by the malt
neutralization by	pure sterilized chalk added in the beginning
fermentation temperature	45°C
fermentations vessels	vats or wine casks
sterility of operation:	the vessels are closed, with special provisions for the escapement of gases
or	the vessels are covered with two layers of filtercloth
mixing	2 times a day with filtered air blown through a sparge ring
acidity	slightly acidic
fermentation time	5–6 days
control of end	termination of the generation of gas
yield	??
type of bacterium	airborne in Pasteur's manner
cultivation method	in Pasteur flasks
contr. of purity	microscopically
amount of inoculant	??

Who Was George Jacquemin?

Georges Jacquemin (Figure 5,7) was born in 1862 at Strasbourg as the son of Eugène Theodore Jacquemin, Professor of Chemistry at the École Superieur de Pharmacie in the same city. The father had studied chemistry at the University of Strasbourg with Louis Pasteur and had maintained excellent contact with his former Professor after his departure to Lille, and later to Paris (Jacquemin, 1900).

When Georges was seven years old the Franco-German War of 1870 broke out. In a short campaign the French army was overrun by the German invasion force. Under the terms of the Treaty of Frankfurt (1871) France had to cede Alsace and part of Lorraine to the new German Empire. Many inhabitants of the ceded territories emigrated to France; one of the cities that was particularly popular among these immigrants was Nancy.

Fig. 5,7. Georges Jacquemin. Reproduced from "Les fermentations rationnelles", Jacquemin, 1900.

In 1873 the French parliament took the initiative to transfer the Medical Faculty and the School of Pharmacy from Strasbourg to Nancy. The government executed the decision of the parliament within a year, and the first courses in medicine and pharmacy at Nancy were given in 1873/1874. Eugène Jacquemin was one of the members of the staff of the School of Pharmacy who came to Nancy in 1872 or 1873. A provisional laboratory with room for 30 students for practical instruction opened in January 1874 and by the summer of that year the transfer of the Strasbourg facilities was more or less complete (Revue médicale de l'Est, January 1, 1874, 5). The young Jacquemin studied chemistry at the Nancy School of Pharmacy in 1885; his position is given as laboratory attendant.

In 1885 the first article of Georges Jacquemin was published in the Revue médicale, a short communication on the analysis of mineral water. Not much later he published a paper about reactions of cyanogen and urethane, a subject related to his father's thesis at Strasbourg ("Considérations générales sur le cyanogène et ses composés, suivie d'observations particulières sur les prussiates").

In 1886, on the advice of his father, Georges Jacquemin begins with the study of the ferments of wine (wine-yeasts). In a certain sense he considers this work a continuation of the studies of Pasteur and the Institut Pasteur on the fermentation processes of the distillery and brewery industries. Pasteur, Jacquemin writes in 1900, encouraged his investigations. In his book "Les fermentations rationnelles" of 1900 he reproduces a letter (Figure 5,8) from Pasteur to Eugène Jacquemin of March 2, 1888, which reads in translation:

Dear Sir,
My health prevents me indeed from assisting in the meetings of the Académie.
I hope you will have enclosed with the copy in the possession of Monsieur Berthelot a summary of the work of Monsieur your son, reduced in accordance with the rules, for insertion in the Comptes-rendus.
I have read it with great interest. The route on which Monsieur your son is embarking is very fertile.
Please receive my highest respect.

- L. Pasteur

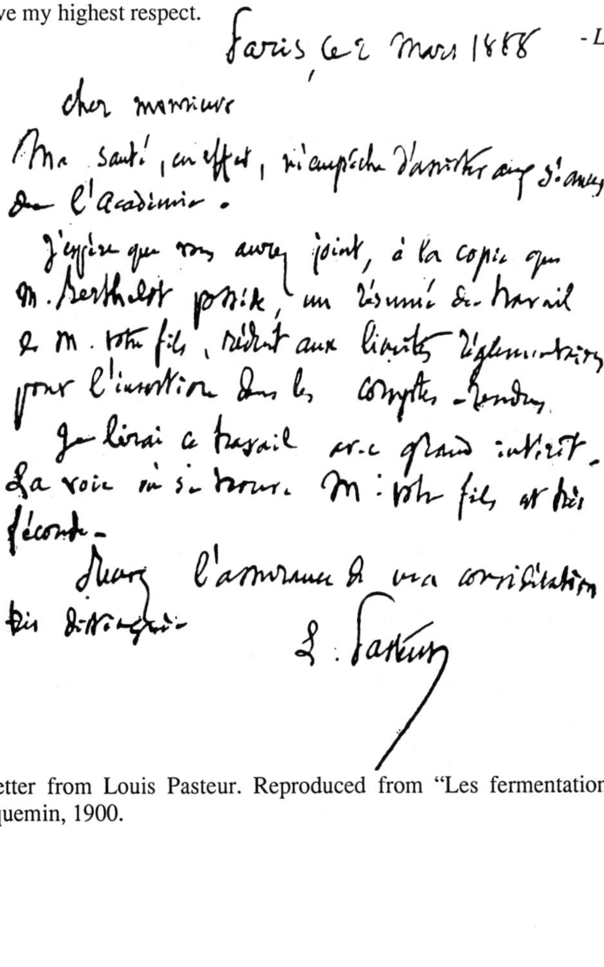

Paris, le 2 Mars 1888

Cher Monsieur

Ma santé, en effet, m'empêche d'assister aux séances de l'Académie.

J'espère que vous avez joint, à la copie que M. Berthelot possède, un résumé du travail de M. votre fils, réduit aux limites réglementaires pour l'insertion dans les Comptes-Rendus.

J'ai lu ce travail avec grand intérêt. La voie où se trouve M. votre fils est très féconde.

Agréez l'assurance de ma considération très distinguée

L. Pasteur

Fig. 5,8. Letter from Louis Pasteur. Reproduced from "Les fermentations rationnelles", Jacquemin, 1900.

The paper on the use of a pure culture of wine-yeast (*Saccharomyces ellipsoïdeus*) in the fermentation of malt was presented to the Académie Française by Berthelot on March 5, 1888, and subsequently published in the Comptes rendus of the Académie.

In the course of further study Jacquemin arrives at the important conclusion that some kind of controlled initial acidity often is rather helpful in maintaining the right type of fermentation. [As a matter of fact the basis for this conclusion had already been laid by Pasteur's studies in Lille on the fermentation of beet juices.]

Gradually Jacquemin then turns his attention to the fermentation of wine. He observes that wine growers in the southern regions of France often add tartaric acid to their wines to redress low acidity. Would it not be advisable, he contemplates, to bring about a mixed fermentation by adding a pure culture of a lactic ferment to a selected yeast [strain]? In view of the great practical difficulties of delivering to the viticulturist two pure cultures, that is one of the lactic acid bacterium and one of the special yeast, Jacquemin gives up this idea. First of all the conditions which favor the growth of the two microorganisms are too different to make a combination possible. Secondly even the most careful adjustment of fermentation times and temperatures by the winemaker cannot sufficiently be trusted to ensure the right fermentation conditions for the wine fermentation process. He then continues (Jacquemin, 1900, 45).

> I also had the idea to use lactic acid in wine making in another way, either by substituting the addition of tartaric acid by that of lactic acid made with my process in a special factory, or by the addition of calcium lactate to replace the whiting now applied in the clarification of the wine. I have patented my process for the manufacture of lactic acid and lactates with pure ferments (cultures) on April 14, 1890; below I give the description of my system as presented to the Société Chimique de Paris in the beginning of 1891 [Follows the exact text of French Patent 204,874 (1890)].
>
> - *Jacquemin, 1900*

In an addition to the patent presented on August 3, 1891, Jacquemin claims a number of applications, all resulting from successful experiments he had carried out himself (Jacquemin, 1900):

lactic acid and lactates, particularly the acid lactates of sodium and potassium, can be used instead of tartaric acid, tartrates and bitartrates;
to replace benzoic acid and other organic acids which transform some coal-tar dyes [into soluble adducts];
for the deliming of skins in the tannery;

to replace vinegar wholly or in part for culinary uses, and for the preparation of vegetable preserves, e.g. gherkins;
in the future lactic acid could be introduced to dissolve fish glue, which is used to clarify wine, beer and other fermented drinks.

In 1900 Jacquemin wrote about the fate of his lactic acid patent:

When produced on a large scale it should be possible to offer lactic acid of a medium concentration [60%] for a sales price below 0.60 francs per kg, which would destine dilute lactic acid to replace vinegar of an average price of 0.10 francs per kg. A trial to apply my process was performed in a chemical works in the Alsace, but it was discontinued because I had no possibility to go there and direct the manufacture myself. I was at the time very much engaged in my research on selected yeast [strains] and the foundation of the "Institut La Claire"[6]; for this reason I have not been able to devote to lactic acid the time needed to transfer my laboratory process to the industry.

It is a pity that this process for cheap, pure lactic acid has not made headway in France; without doubt such a new industry would have been very prosperous provided the required energy would have been mobilized to develop the many applications of the acid. Presently lactic acid is hardly used due to the high price caused by obsolete manufacturing processes. As if to demonstrate the regrettable fact that not one French producer finally has accepted my system, I now have received confirmation of the recent beginning of the production of lactic acid in Germany. This product is sold at about 50 francs per 100 kg of medium concentration; this tends to prove that my studies have not gone unnoticed outside France and that my research has brought profit to others than myself.

- Jacquemin, 1900

The company Jacquemin is referring to in 1900 was undoubtedly C.H. Boehringer Sohn of Nieder-Ingelheim. One wonders which was the chemical firm in the Alsace (then Germany!) which tried to manufacture lactic acid according to Jacquemin's patent. Could it have been the "Fabriques de Produits Chimiques de Thann et de Mulhouse"? The company was quite prominent in the production of auxiliary chemicals for the textile industry, and in the very same years was engaged in trials to make citric acid by fermentation.

Jacquemin lost his interest in lactic acid in 1893 when he stopped paying the yearly taxes to maintain his patent rights. From then on he devoted most of his time to the development and commercialization of pure yeast cultures for wine manufacturing.

Wehmer Makes Citric Acid by Fermentation

In 1893 a short communication from the hands of Carl Wehmer appeared about citric acid fermentation in the Bulletin de la Société Chimique de France (Wehmer, 1893a). In the same year he applied for patents in Great Britain (British Patent 5620, granted in 1893) and in Germany (German Patent 72,957, granted in 1894); the "Fabriques de Produits Chimiques de Thann et de Mulhouse" was registered as the owner of both patents.

In the communication to the "Bulletin" Wehmer reported that he had already begun the experiments in 1890, and now was able to produce citric acid on a 10 kg scale.

Who Was Carl Wehmer?

Carl Wehmer, born in 1858, lost his father at an early age. (Biographic information from Jahn, 1935, and Koch, 1936). His mother hardly had the means to let her son study at the university, but Carl got the opportunity nevertheless. He was an excellent student, in 1890 he majored in botany and chemistry at the University of Göttingen. In 1891 he was admitted as private teacher at the Technical College of Hanover, a position which earned him hardly any income. During his studies he had become interested in plant metabolism. In Hanover he started those studies using fungi as model substances for the higher plants which he intended to study later. But he never came to do so. For his entire life, Wehmer remained a student of mycology, the study of fungi (then also including other microorganisms). The first metabolic product obtained by Wehmer was oxalic acid. This acid is present in many plants and in molds as well, which sometimes produce it in appreciable quantities.

The next metabolic acid of importance he discovered was citric acid. Two species of fungi which he denominated *Citromyces* – but today are included in the genus *Penicillium* – were able to produce interesting quantities of citric acid when grown on 10% sugar solutions. Wehmer immediately recognized the practical importance of his findings and applied for patents. He also got into contact with the important producer of tartaric acid and other textile chemicals "Fabriques de Produits Chimiques de Thann et de Mulhouse". The owner of the company, Scheurer-Kestner, a descendant of Charles Kestner, the discoverer of Biot's racemic acid some fifty years earlier, showed much interest and went into some kind of a consultancy agreement with Wehmer.

Over a period of five or six years a large number of trials were made by the plant chemist G. Wyss under the guidance of Carl Wehmer. In 1893 the first amounts of fermentative citric acid were offered to the market. The project, however, never surpassed the trial, pilot plant scale. (For the main features of the process, see Exhibit 5,II.)

EXHIBIT 5,II
WEHMER'S FERMENTATIVE PREPARATION OF CITRIC ACID
lit. Wehmer, 1893a, 1893b, 1894, 1933

raw material	sucrose or glucose
source of raw material	??
concentration	10 to 20%
sterilization	yes, reasonable [?]
organic nutrients	none [not required]
inorganic nutrients	ammonium, potassium, magnesium phosphate, sulfate and nitrate
neutralization by	calcium carbonate
fermentation temperature	ambient, minimum 10°, maximum 50°C
fermentation room	a constant temperature chamber kept reasonably free from infections; sterile air has to be introduced to provide sufficient oxygen
fermentation time	8 to 14 days. Intermittently some mixing has to be done [by hand ?], because the fermentation only takes place at the surface and chalk has to be mixed in to convert citric acid to calcium citrate
acidity	probably slightly acidic
control of end	??
yield	low
type of bacterium	*Citromyces [= Penicillium]*
cultivation method	a 5 to 10% sugar solution containing 1 to 2% of citric acid is left standing at the air. Spores of colonies formed after some days on the surface are collected to prepare pure cultures.
control of purity	The control of the purity is done microscopically (in German patent 72,957 Wehmer presents drawings of the 2 types of fungi he employed).
amount of inoculant	??

Two factors seem to have been responsible for the termination of the trial productions (Foster, 1949):

1. The technology for large scale pure culture fermentation under sterile conditions was not available. Even the fermentation pans presented problems: they were made of very expensive lead and because of their weight were difficult to handle. Wehmer (Wehmer, 1933) estimated the investment costs for a 100 t/y plant at 650,000 Mark, a huge amount of money for a product with a market price of only 2.20 M/kg.
2. Wehmer had not succeeded in finding conditions that would give his *Penicillium* fungi a selective advantage over other contaminating microorganisms. In particular the acidity of the medium was well within the range that is favorable for the growth of many yeasts and bacteria.

In one respect the discovery of citric acid producing molds was a success for Wehmer: it provided him for many years with a sound financial basis for the continuation of his scientific work. His biographer, Jahn (Jahn, 1935), adds to this: following the studies concerning the production of citric acid by molds, Wehmer worked on lactic acid fermentation. The technical success of these investigations enabled him to establish a household; he married in 1902.

The work on lactic acid fermentation to which Jahn is referring are studies Wehmer did as consultant of Albert Boehringer. The latter had asked his assistance in or shortly before 1895 for help in the development of a reliable process for the manufacture of lactic acid by fermentation.

Boehringer Tries To Make Citric Acid

The most complete account of how Boehringer ventured into lactic acid has been given by A. Heusner, who wrote in the 1970's a history of the Boehringer enterprise. About the events of the 1890's leading to the commercial production of lactic acid, while searching for a method to make citric, Heusner wrote (translated):

> The first experiments to produce citric acid by fermentation were conducted in 1893. The direct inspiration not only came from Wehmer's paper [in the Bulletin de la Société Chimie de France], but also from direct contacts with Wehmer as well as with the firm in Mulhouse. [The writer is inclined to believe that Wehmer, in search for a profitable application of his invention had not only approached the "Usines de Thann et de Mulhouse" to which Heusner refers here, but other tartaric acid manufacturers as well, including Boehringer.]
>
> The experiments … can at best be called dilettantish.

We suppose that the first experiments were performed in wooden vats. Growing microorganisms, however, develop quite a bit of heat, and wood being a kind of thermal insulator does not conduct well enough to dispose of the heat so generated. As a consequence the temperature rose to values above 40°, killing all the fungi. Their position in the broth was taken over by lactic acid bacteria which were abundantly present and some of which grow optimally at temperatures between 40 and 50°C.

- Heusner, 197?

The chemists of the company could prove that the product they had obtained was lactic acid instead of citric acid. In fact, it is more proper to say that it predominantly contained lactic acid, because as Heusner writes it was a malodorous dark oily product, probably being a mixture of lactic acid, polymeric lactic acid, other organic acids and residues from the starting materials.

The company chemists had almost certainly made another important mistake in the fermentation process of which they, however, could not be aware: they supplied a nutrient rich substrate solution in which lactic acid bacteria could grow! The molds capable of producing citric acid do not require organic nutrients, but lactic acid fermentation is not possible without organic nutrients in substantial amounts. According to Dreher (Dreher, 1898) molasses was used in the first trials, soon a change to mashed potato starch was made (the mashing process is explained in Intermezzo 3, 160).

Both shortcomings of the process – submerged, that is bulk fermentation instead of surface fermentation as practised by Wehmer in the Thann factory, and the unintentional supply of organic nutrients – might be related to the need Boehringer felt to use existing equipment (the vats of tartaric acid manufacture) for the fermentation and the cheapest raw material, molasses or potato starch as substrate. Both mistakes were understandable; the company was short of means, the owner was very thrifty and the chemists knew nothing (or next to nothing) about the microbiological requirements of the fermentation process.

Of course, the real problem was that the science of microbiology had not yet advanced enough to prepare citric acid by fermentation. Even Wehmer did not really know how to choose conditions that would give his molds a selective advantage over other – contaminating – microorganisms. Only many years later, in 1917, would Currie discover that under strongly acidic conditions, that is in the absence of chalk, *Aspergillus niger* – a mold also tried by Wehmer – could be a very efficient producer of citric acid (Ch. 7, 230).

Boehringer Decides To Manufacture Lactic Acid

The decision to scale up the primitive laboratory preparation of lactic acid to industrial production was apparently reached in 1895. The most crucial step taken in that year was the engagement of Wehmer to advise on the fermentation procedure. One of the first things he did was to teach the responsible chemist how to prepare industrially useable cultures of lactic acid and how to maintain and work with those cultures (Dreher, 1898). The first allegedly pure culture put into use was probably one supplied from Lindner's collection in the Institute for the Fermentation Industries in Berlin. From 1896 onwards this institute delivered pure cultures of *[(Lacto)-bacillus delbrueckii]* on a regular basis (Henneberg in Ullmann, 1920; see also Wehmer, 1898; and Dellweg, 1974, who writes that commercial cultures of *Lactobacillus delbrueckii* and of *Lactobacillus bulgaricus* were shipped regularly).

Thanks to this application of the most recent advances in the preparation of pure cultures for the industry, Boehringer succeeded in establishing reliable large scale fermentation methods for the new chemical, lactic acid. This had certainly not been possible on the basis of Bensch's method in which foul cheese was used as inoculant.

The few data available on the quality of Avery's product (Whitaker and Denison, 1895; Hoffmann, 1896; see also this Chapter, 148) suggest that Boehringer had also mastered the fermentation of lactic acid better than the Avery Company. To a large extent this can be attributed to the eminence in the science of microbiology achieved in Germany as well as to the involvement of the excellent and very industry minded institute of Delbrück. This opinion does not imply in any way a belittlement of the important role of Wehmer, nor that of the anonymous workers at the Nieder-Ingelheim factory of Albert Boehringer.

Trial production came on stream towards the end of 1895: about 4 tons per month (Ullmann, 1953) of a rather impure product. Drums containing this material were collected on the yard, awaiting customers (Heusner, 197?). In the ensuing years the technical staff of the company worked out, and partially installed, acceptable procedures for the large scale production of technical grade lactic acid (50% and 80%).

On February 15, 1898, the company applied for a building and operation permit. The handwritten [sic !] request reads in translation:

Explanatory Note Relating to the Lactic Acid Factory

As essential raw materials carbohydrates (sugar etc.) will be used. These materials will, with the aid of purely cultured lactic acid bacteria, be subjected to fermentation; a faint smell as is common to all fermentation industries (e.g. distilleries) cannot be prevented from occurring in the process. The calcium lactate obtained in the fermentation will be converted with sulfuric acid under the formation of lactic acid liquor and gypsum. The lactic acid liquor will be concentrated in vacuum stills, thereupon decolorized and purified. The gypsum will be completely washed out, it is the sole waste product[7]. Waste waters are not formed in the manufacture.

Though Boehringer had been prudent in the description of possible environmental problems, the authorities in Bingen were not to be circumvented. On May 23, 1889, Boehringer was granted a permit for the manufacture of lactic acid, under the proviso that "in case the plant would increase the offences to the population through (the emission of) malodorous smells, the owner would be obliged to install at his own costs such equipment as (needed) to eliminate these offences in the best possible way." The reason for this special condition was obvious: the existing industries were already causing too much pollution, a straw cellulose plant had turned the little river flowing through Nieder-Ingelheim into a badly smelling sewage canal while the cement works offended the population even more with its regular emissions of dust (Heusner, 197?).

In 1902 a permit was obtained for an enlargement of the factory; this happened again in 1908, 1914 and 1916 (Figure 5,2 shows the plant at about 1908). Production rose from more than 200 t/y in 1901 to a figure between 300 and 500 t in 1907. During the Great War a level of 1000 t/y was reached (Schäfer, 1907; Ullman, 1953).

Boehringer's Production Method

A rather complete description of lactic acid manufacture has been given by Wilhelm Henneberg in the first edition of Ullmann's "Enzyklopädie der Technischen Wissenschften" (Henneberg, 1920):

As fermentative lactic acid is produced by certain species of bacteria, a study of these species and their conditions of life had to be carried out first before the manufacture of lactic acid on an industrial scale could be successful. It appeared very fortunate that the distilleries and yeast manufacturers also applied lactic acid bacteria in the acidification of the yeast substrate used in alcohol fermentation. These bacteria belong to the same species, later named *Bacillus delbrücki* [and today *Lactobacillus delbrueckii*], which is used in lactic acid fermentation. Hence the knowledge and results acquired in the yeast and alcohol manufacture could be applied successfully in lactic acid manufacture. Pistorius had established that for the fermentative acidification of the wort a mimimum temperature of 45°C was required. Delbrück showed the (lactic) fermentation to be least susceptible to infection at 50°C. At this temperature the lactic acid bacteria used had a definite advantage over other bacteria and would prevent their growth [See Ch. 3, 56]. First Lafar [in 1893] and independent of him Leichmann [in 1896] cultivated "natural pure cultures" of lactic acid bacteria and applied these successfully in the distillery. Wehmer introduced the same cultures in the industrial production of lactic acid at Nieder-Ingelheim. Kownatzki[8] introduced equipment (1,000 and 2,000 l vessels) for the scaling up of the pure culture, in 1902 in Nieder-Ingelheim and in 1903 in Niederstriegis. [In Niederstriegis, a small town in Saxony, near Dresden, a competing lactic acid factory had been built by the Chemische Fabrik Knab & Lindenhayn[9].]

The details of the fermentation and purification procedures given by Henneberg probably relate to the Boehringer manufacture. The main features of the processes have, with a few minor alterations based on other sources (Schäfer, 1907; Dreher, 1898; private information from Boehringer-Ingelheim, 1986), been summarized in Exhibits 5,III and 5,IV.

EXHIBIT 5,III
BOEHRINGER'S FERMENTATION PROCEDURE about 1908 (?)
lit. see Exhibit 5,IV

raw material	potato starch converted at 65°C with 5 to 10% malt or malt sprouts into "sugar" (conversion about 80%)
concentration	10-11% of sugar [maltose]
sterilization	pasteurization at 80°C
inorganic nutrients	the malt provides enough, sometimes the fermentation is accelerated through the addition of ammonium phosphate and sulfate
organic nutrients	supplied by the malt
neutralization	[sterilized?] chalk is added at the beginning
fermentation temperature	50°C
fermentation vessels	wooden vats of about 10 m^3; originally of 600 l
sterility of process	the "sugar" solution is sterilized by boiling; the covered fermentation vessels are cleaned with soda or dilute sulfuric acid, and steamed before every use
mixing	every two hours the partially deposited chalk is stirred up
acidity	slightly acidic
fermentation time	8–10 days, sometimes less

control of end	Fehling's reaction on reducing sugars should be negative or almost negative
yield (estimates)	on dextrose 100% [?] on potato starch 74–80% on corn 65%
type of bacterium	*L. delbrueckii*
cultivation method	in 2–5 l flasks under sterile conditions using the same substrate and nutrients as in the manufacturing process
source of bacterium	the Institute for the Fermentation Industries; scaling up by factors of about 10 to 20
control of purity	microscopically (Henneberg give sketches of *Lactobacilli* in various stages of growth as well as of possibly infecting microorganisms [10])
amount of inoculant	the more the better [probably 5–10%]

EXHIBIT 5,IV
BOEHRINGER'S PURIFICATION PROCEDURE about 1908(?)

lit. Henneberg, 1920; Dreher, 1898; private information Boehringer-Ingelheim, 1986

treatment of broth	extra lime is added to make it slightly alkaline, the broth is then boiled to precipitate proteins and bacteria
decantation	the mass is left to settle for several hours, then filtrated or decanted, the residue – the sludge – is first passed over a sieve to remove the "sprouts" and then filtered through a filter press
concentrating	in open pans to a concentration of 30–35%
acidification	with the equivalent amount of sulfuric acid (methyl violet as indicator)
filtration	through a filter press, the cake of gypsum is washed in the press
purification [11]	dissolved iron and copper are removed with yellow prussiate salt, arsenic and lead with barium sulfide, excess barium is precipitated quantitatively with sulfuric acid
decolorization	color and [some of the] organic contaminating material is removed through treatment with a large amount of activated carbon (charcoal). In between these chemical treatments the solution is passed over filters, if required
concentrating	to 50 or 80% in a copper vacuum still (at a pressure of less than 100 mm Hg); originally a leaded iron still was used

Applications of the New Product

BAKING POWDER

The first patent ever granted to the Boehringer company was a patent on the use of acid calcium lactate together with sodium bicarbonate in a baking powder formula (German Patent 87,235 of 31 October 1895). Almost exactly the same "invention" as the one which formed the backbone of Avery's decision to build the world's first lactic acid plant in Littleton, 11 years earlier. The application was as unsuccessful as Avery's: the acid calcium lactate turned out to be a sticky product, difficult to handle and not having the desired keeping qualities when mixed with bicarbonate.

TEXTILE INDUSTRY

Already in 1895 Albert Boehringer had decided to consult external experts about possible uses for his new acid. The first experts mentioned (by Heusner, 197?) were Dr. Carl Dreher[12], Dr. F. Düring and Dr. P. Fuchs. The first application investigated was, quite naturally for a tartaric acid manufacturer, the replacement of tartaric acid and tartrates in the textile industry. (On the history of the use of organic acids in the textile industry see Intermezzo 2.) Carl Dreher, then superintendent of a wool dyeing works and one of the advisers contacted, wrote in May 1896 that he had carried out experiments with lactic acid for almost two years, first with 30% acid from the Avery Lactart Company of Boston, and more recently with 50% acid from C.H. Boehringer of Nieder-Ingelheim (Dreher 1896a).

In the mordanting process technical lactic acid, Dreher wrote, can replace cream of tartar provided the quantity of sulfuric acid is slightly increased and the main oxidation is performed at 50° R (=62.5°C) instead of at boiling temperature. The substitution at equal weight makes the recipe a bit cheaper: Avery's 30% lactic acid costs about 85 Mark per 100 kg, cream of tartar 90 Mark. Boehringer's acid is still cheaper, it has a concentration of 50% and is supplied at the same price as the Avery material. These technical lactic acids

represent yellowish-brown to dark syrupy liquids. They are produced by fermentation, and retain from the fermentation process polluting substances like sugars, dextrins and components from molasses. These substances are responsible for the color. Also present are small amounts of other organic acids like for instance butyric acid, and sometimes traces of sulfuric acid. In wool dyeing the pollutants are not bothersome, it is only the concentration of the acid which counts. In printing and finishing only light grades of lactic acid should be used.

- Dreher, 1896

As is clear from Dreher's account the idea to use lactic acid in the mordanting of wool goes back to the activities of the Avery company. In the fall of 1896 Dreher presented a second article on the merits of lactic acid in the chrome mordanting of wool to the Färben-Zeitung (Dreher, 1896b). This paper is very positive about Boehringer's lactic acid:

While initially from all sides brown to dark and very impure lactic acids containing only 25 to 30% lactic acid were introduced on the market[13], the firm of C.H. Boehringer Sohn at Nieder-Ingelheim is now supplying an almost chemically pure, faint yellowish product in a concentration of 50% w.w. (or 600 g lactic acid per l) at a price of 80 M/100 kg. The product is guaranteed free from iron, sulfuric acid and dextrins.

-Dreher, 1896

The quality changes Dreher notes between the first material he received from Boehringer (in the first months of 1896?) and the second sample (almost a year later) probably represent the result of the introduction of good culturing methods by Wehmer. It can also relate to a change in raw material from molasses to potato starch. The fact that Dreher had in the meantime entered the employment of the company might have influenced his judgement as well.

Heusner says that the names of the chemists employed by Boehringer in those years are not found in the archives of the company. It has been recorded, however, that besides a generalist, one analytical chemist was engaged. It is not unlikely that this analytical chemist was Dr. Wilhelm Klapproth, born in 1878 and in 1905 registered as a Nieder-Ingelheim member of the Verein Deutscher Chemiker.

Another chemist employed by Boehringer in the late 1890's was Dr. Johannes Schäfer who had studied with Professor Kiliani at the Technical University (Hochschule) of Munich. His thesis (Erlangen, 1896) deals with the properties of the natural dye quercitron. Maybe he entered Boehringer's service in 1896 or 1897.

Carl Dreher's stay at Boehringer was only short, he left the company within two years, and not in a friendly way.

Dreher Sets Up a Consultancy Business

Carl Dreher left Boehringer in 1897 or 1898 to establish his own (chemical) consultancy in his home city Freiburg. In the summer of 1898 he offered his services (for the manufacture of lactic acid) to the firm of E. Merck, Darmstadt. In October 1898 this matter was discussed with Dr. Carl Krauch, superintendent of Merck's lactic acid and butyric acid plant. (Krauch's report on this meeting has been saved in Merck's Archives.)

Dreher claims to possess practical knowledge of methods to produce (chemically) pure lactic acid for a price of 1.25 M/kg and technical lactic acid for 0.35 M/kg. After some questioning Krauch finds out that the first assertion is not based on actual knowledge. ("He doesn't understand anything about the manufacture of this product", Krauch reports).

Krauch first interrogated Dreher about his situation with Boehringer, however. The employment, Dreher said, had been terminated because of financial matters. These have now been resolved and Dreher is free to exploit "the lactic acid method." The information Dreher further gives about the manufacture helps Krauch in his advice to engage Dreher as consultant for the establishment of technical lactic acid production by E. Merck. An important point in support of his opinion is Dreher's practical knowledge about the use of technical lactic acid:

> Dreher is well-informed about the sales and applications of technical lactic acid. The sales are in the hands of representatives who are in particular active in industrial centers like Leipzig, Mühlhausen [Mulhouse in the Alsace] and Barmen [now part of the city of Wuppertal]. The dyestuffs industry also buys much technical lactic acid, the general usage of the acid Dreher considers to be very large. He suggests that an annual sale of about 50,000 kg would be within easy reach.
>
> *-Krauch, 1898*

Small scale trials on the basis of Dreher's know-how were run by Merck in the next years. The first year of a sizeable output was 1902/1903 with 9,557 kg (50%). In the same year the company produced 17,000 kg pharmaceutical lactic acid (most of it of 90% concentration), 2,000 kg calcium lactate and 5,000 kg ferrous lactate.

Boehringer's First Successful Derivatives

On September 4, 1896 Boehringer demanded a patent on the manufacture of double salts of antimony lactate with calcium lactate. The preamble of the patent (German Patent 95, 828, granted in 1898) goes to great lengths to prove that the new method of manufacturing these antimony salts is superior to the one claimed by Charles Waite (see I.2, 117). The new product which was registered under the trade name "Antimonin" competed successfully in Germany and also internationally[14] against tartar emetic in the mordanting of cotton (Düring, 1900).

Another derivative introduced in the same years was a 50% solution of an equimolar mixture of lactic acid and potassium lactate. Fuchs calls the product acid potassium lactate (trade name "Lactolin"), although as he admits, lactic acid as monobasic acid cannot be expected to form acid lactates. The existence of acid calcium lactate nevertheless supports the view that "Lactolin" also can be considered a double salt of potassium lactate and lactic acid (Fuchs, 1897).

Dreher reacts vehemently to this article. He apparently feels offended that his recipes for the mordanting of wool with lactic acid are replaced by recipes with a partially neutralized solution of lactic acid, which have no advantage for the practical dyer, and are based on the theoretical impossibility of the existence of double salts of lactic acid (Dreher, 1898b).

Dreher's reaction was countered by E. Kielmeyer, who painstakingly pointed out the advantages of "Lactolin":

- it slows down the reaction of chromate with lactic acid, enabling the dyer to obtain a more evenly dyed product;
- the product is of consistent quality due to Boehringer's superior manufacturing method of lactic acid, "[other] German, American and English lactic acid liquids [sic !] being of the most variable consistencies, colors and densities."
- the fact that imitations of "Lactolin" are traded widely is considered further proof of its superiority.

Boehringer also objects to Dreher's article and advises every practitioner that the "applicability of a new mordant cannot be decided in the laboratory". The large scale acceptation by the industry is proof of the advantages of "Lactolin" over other mordanting aids (Boehringer, 1899). This is also substantiated in a final article on the subject by Dr. Sigmund Kapff (Kapff, 1900).

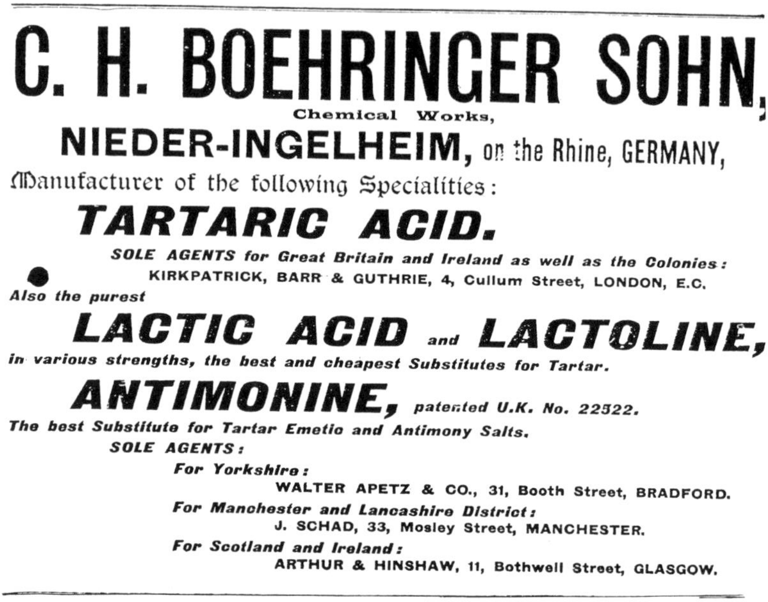

Fig. 5,9. Advertisement of Boehringer, 1899. Reprinted from The Dyer and Calico Printer, *19* (December 1899).

Notes

[1] Most of the information relating to Albert Boehringer and his company has been collected through the intermediary of Dr. Philipp Bennecke of Boehringer Ingelheim Zentrale GmbH. Particularly helpful were reports written by Dr. A. Heusner in the 1970's on the history of production processes applied by the company.
[2] Today this important pharmaceutical/chemical company is known by the name Boehringer-Mannheim. "Our" company, C.H. Boehringer Sohn of Nieder-Ingelheim today is usually referred to as Boehringer-Ingelheim.
[3] In 1881 BASF already employed over 2,000 people; the Hoechst company was of a comparable size.
[4] In 1886 and 1887 Ferdinand Klein, in the function of director of a chemical works at Nieder-Ingelheim is mentioned as a member of the Verein zur Wahrung der Interessen der Deutschen Chemischen Industrie. In 1886 a son, Ferdinand (Ferry) Klein, was born at Nieder-Ingelheim. (From his dissertation of 1915, registered in the National Union Catalog.)

The next chemist about whom a few details are known is Dr. H. von Knapp, educated at the University of Strasbourg. The list of members of the Deutsche

Chemische Gesellschaft of 1890 mentions him as an employee of Albert Boehringer. In 1892 he appears to have another employer.

5 Carl Krauch (1853–1934) started his career in an apothecary, studied pharmacy and chemistry from 1874 till 1877 and entered the employment of E. Merck in 1882. He served first as analytical chemist and later became production manager. In the 1890's Krauch was responsible for the production of tanning preparations, lactic acid and butyric acid. He retired in 1919 (Information from Deutsche Apotheker-Biographie (Hein & Schwarz, 1975), corrected by Merck).

6 L'Institut La Claire was founded in the fall of 1891 to produce and sell pure cultures of wine yeasts for the wine growers and the distillery industry.

7 The amount of gypsum formed in the reaction of calcium lactate with sulfuric acid is about equal to that of the amount of lactic acid (100%). As wet filter cake the volume is about double that of the lactic acid.

8 Dr. Erwin Kownatzki, born in 1872 in Tilsit, Ph.D. Erlangen in 1900, was then assistant of Henneberg, the head of the botanical laboratory of the Institute for the Fermentation Industries in Berlin (Henneberg, 1901). He supposedly was later employed by Boehringer (Heusner) and died before 1931 (Ullmann, 2nd ed., Vol VII, 1931, 587).

9 The plant at Niederstriegis, a small town about halfway between Leipzig and Dresden, Saxony (now GDR) belonged to the firm of Knab & Lindenhayn, established by Julius Lindenhayn and Heinrich Knab in 1881 to manufacture dye wood extracts. When synthetic dyestuffs had replaced most natural dyes by the end of the century Lindenhayn decided to take up lactic acid production. After the introduction of pure cultures he succeeded in improving the quality. (From a short historical note supplied by Byk Gulden Lomberg Chemische Fabrik GmbH, Konstanz, 1986.)

10 Until the appointment of Dr. Rauch in the 1920's the company chemists only had fragmentary knowledge of microbiology and of microbiological procedures. It remains doubtful if they were able to "control" the "purity" of the *Lactobacillus* properly. The process relied primarily on the application of "natural culturing" conditions and on the regular renewing of the culture with fresh material from the Fermentation Institute in Berlin.

11 Originally Boehringer only made a technical quality of lactic acid (50%). By improving the chemical purification they succeeded between 1900 and 1910 in upgrading the product to a much better quality suitable for the food industry. This quality usually was supplied in 80% concentration.

12 Carl Dreher was born in Freiburg (Baden). In 1888 he obtained the doctor's degree at the University of Freiburg. Thereupon he spent some time at the Textile Institute of Krefeld (Königliche Webe-, Färberei- und Appreturschule), then entered the employment of BASF at Ludwigshafen (the technical dyeing department, probably including customer service). In 1894 he became superintendant of a large wool dyeing works. In 1896 he assumed a position with the Boehringer company and moved to Nieder-Ingelheim. (Information reported by Carl Krauch to the Merck management from an interview with Carl Dreher, October 7, 1898.)

13 Besides Boehringer only one other German producer (or trader?) is known in 1895 with certainty:

P.H. Paulcke, Chemische Fabrik, Leipzig (communication to Färber-Zeitung, November 1895). It is rather likely. however, that Knab und Lindenhayn in

Niederstrieger also began production in the 1890's, see note 9.
In England the first recorded production is by Ernest Hickson in Bradford, Yorkshire. In, or shortly after 1894, Hickson "began the manufacture of lactic acid which was then chiefly employed for the mordanting of wool with bichromates" (Fox, 1987).

[14] During the first years of World War I when England had put an embargo on German chemical products, the American industry was hit by a severe shortage of chemicals, dyestuffs and intermediates. An American steamer; the Ogeechee, trying to break the blockade with a number of essential chemicals, carried a substantial amount of the apparently vital product Antimonin (Haynes, Vol II, 27).

References

C.H. Boehringer Sohn, **1899.** *Färber-Z.*, **1899**, 149.

Boutron-Charlard, A.F., and Frémy, E., **1841a.** *Ann. Chim. Phys.*, [3], **2**, 257.

Boutron-Charlard, A.F., and Frémy, E. **1841b.** *Ann. Chem.* **31**, 181.

Dammer, O., **1898.** *Handbuch der Chemischen Technologie*; Ferdinand Enke: Stuttgart, Vol. IV; Weinsäure.

Dellweg, H., **1974**. *100 Jahre Institut für Gärungsgewerbe und Biotechnologie zu Berlin*, 1874–1974, Festschrift: Berlin.

Dreher, C., **1896a.** *Färber-Z.*, **1895/1896**, 233.

Dreher, C., **1896b.** *Färber-Z.*, **1895/1896**, 477.

Dreher, C., **1898a.** *Färber-Z.*, **1898**, 21 and 277.

Dreher, C., **1898b.** *Färber-Z.*, **1898**, 248.

Düring, F., **1900.** *Färber-Z.*, **1900**, 319.

Foster, J.W., **1949.** *Chemical Activities of Fungi*; Academic Press: New York, 333

Fox, M.R., **1987.** *Dye-Makers of Great Britain*, **1856–1976**; Imperial Chemical Industries: Orpington, Kent.

Friedburg, L.H., **1882.** *J. Am. Chem. Soc.*, **4**, 292.

Fuchs, P., **1897.** *Färber-Z.*, **1897**, 133.

Gehe & Cie, *Marketing Reports in Färber-Z.*, **1889/90, 1893/94, 1895/96, 1898.**

Güstrow, **1906.** *German Patent 171,835.*

Haynes, W., **1954.** *American Chemical Industry*; Vol. 2; D. Van Nostrand: New York; reprinted by Garland: New York, 1983.

Hein, W.H., and Schwarz, H.D., **1975**. *Deutsche Apotheker-Biographie*; Wissenschaftliche Verlagsgesellschaft: Stuttgart.

Henneberg, W., **1920**. *In Ullmann, F., Enzyklopädie der technischen Chemie*; Urban und Schwarzenberg: Berlin, Vol. 8, 129.

Hoffmann, E., **1896**. *Färber-Z.*, **1895/96**, 65, 217.

Jacquemin, G., **1891.** *Bull. Soc. Chim. Fr.*, **5**, 294.

Jacquemin, G., **1900.** *Les Fermentations Rationnelles*; Edg. Thomas: Malzéville-Nancy.

Jahn, E., **1935.** *Ber. Dtsch. Bot. Ges.*, **52**, 233.

Kapff, S., **1900.** *Färber-Z.*, **1900**, 149.

Kielmeyer, A., **1899.** *Färber-Z.*, 17 and 34.

Kiliani, H., **1882.** *Ber. Dtsch. Chem. Ges.*, **15**, 136 and 699.

Koch, R., **1936.** *Zentralbl.Bakteriol. Parasitenk. Infektions-krankheiten*,II, **93**, 417.

Kolbe, H., **1860.** *Ann. Chem.*, **113**, 242.
Löw, C., **1952.** *Die Chemische Fabrik E. Merck Darmstadt*; E. Merck: Darmstadt.
Revue médicale de L'Est, **1874**. **1**, 1,3.
Schäfer, G., **1907.** *Chemische Zeitschrift*, **6**, 177 and 189.
Ullmann, F., **1953.** *Enzyklopädie der technischen Chemie*; Vol. 12, Milchsäure, 525.
Wehmer, C., **1893a.** *Bull. Soc. Chim. Fr.*, **9**, 728.
Wehmer, C., **1893b.** *British Patent 5620.*
Wehmer, C., **1894**. *German Patent 72,957.*
Wehmer, C., **1933**. *Chem.Ztg.*, **57**, 313.
Whitaker, T., and Denison, J.R., **1895**. *J. Soc. Dyers Colour.*, June 1895, 98.

INTERMEZZO 3

LACTIC ACID PRODUCING BACTERIA

THE DISCOVERY AND FATE OF *LACTOBACILLUS DELBRUECKII*

During the second half of the nineteenth century the identification and nomenclature of microorganisms became generally felt as a prerequisite for further advances in the science of bacteriology. Until then only very general names were used; Leeuwenhoek spoke of animalcules; Cagniard-Latour thought of yeast globules as being plants; Schwann named them fungi; and Pasteur distinguished in his first papers about fermentation between lactic yeast and the yeast of beer.

Classification of microorganisms began with the extension of the system devised by the Swedish botanist Carl Linnaeus (1707–1778) for plants and animals to the submicroscopic sphere of living nature. In the linnaean system two names were applied, the first name denotes the name of the genus to which a group of related species belongs, the second name defines the particular species itself. An early example of this system applied to bacteria is given by Lister, who used in 1877 for his isolation of a lactic acid forming bacterium the name *Bacterium lactis*. Bacterium from the Greek word "bakterion" for rod, lactis indicating the source of the bacterium.

It was soon discovered that a distinction between rodlike bacteria, spherical bacteria (cocci) and bacteria that occur in bunches or clusters (staphylococci) did not suffice. More morphological criteria were developed, most of which were based on particular staining techniques with which microorganisms were made microscopically visible.

By 1890 it had become evident that bacteria exibited great diversity with respect to infectious behavior, nutritional requirements and the production of metabolites like for instance toxines, various organic acids and alcohols. It was a logical step to apply these functional (physiological) properties together with morphological criteria for an improved classification system.

One of the functions thought useful in the classification of lactic acid forming bacteria was the property of forming one of the enantiomeric forms in the pure state, or the exact racemate.

Some Lactic Acid Bacteria Form Pure Stereoisomers

Around 1880 it had been discovered by a number of investigators that some lactic acid producing bacteria exuded an enantiomeric form of lactic acid only, while other strains made the racemic mixture:

Richard Maly of Innsbruck, Austria, reported in 1874 on a series of studies in which various carbohydrates were fermented with the mucous membrane of swine stomach at temperatures from 30 to 40°C (Maly, 1874). He found that in one of his experiments only sarcolactic acid [L(+)-lactic acid] was produced, whereas in the other experiments mixtures of fermentation acid (inactive acid) and stereoisomers developed. Maly made microscopic studies of the fermentation liquids and observed that in the absence of living (rodshaped) bacteria no lactic acid would be formed (See also Nencki and Sieber, 1889). [The conclusions about the isomerism of the lactic acid molecule were based on the water contents of the hydrates of the zinc lactates.]

The first report on fermentative production of levorotatory lactic acid came from the Austrian army physician Franz Schardinger (Schardinger, 1890). In a study aimed at discovering the causes of anthrax in horses ten watersprings were searched for bacterial contamination, using the methods of Koch, and "applying in all the experiments exactly the rules of the bacteriological technique". Although Schardinger did not find the cause of the anthrax infections, he discovered a new bacterium consisting of small rods which could convert dextrorotatory sucrose into levorotatory lactic acid [D(-)-lactic acid]. In the introduction to his paper Schardinger refers to the theory of Le Bel and van 't Hoff about the existence of (geometrical) isomers of carbon compounds with asymmetrical C atoms. It being established that inactive lactic acid consists of equal portions of the two isomers, Schardinger believed that separation probably could be effected with the aid of a microorganism. So, when he found the bacterium, which he called *Bacillus acidi laevolactici*, he exposed a solution of inactive calcium lactate to the new bacterium to see whether it would ferment and consequently destroy the dextrorotatory molecules and leave the levorotatory intact. This, however, did not happen.

Schardinger's experiment was based on the theory that during a "normal" fermentation process racemic lactic acid was formed. Certain bacteria, it was thought, could convert one of the stereoisomers further but lacked the ability to also digest the other form. This other form would become the pure endproduct. [This was also one of the methods Pasteur had applied to separate the isomers of tartaric acid.]

A few years later, investigations by Lafar (Lafar, 1896), and Leichmann (Leichmann, 1896) proved that it was an intrinsic property of pure cultures of lactic acid forming bacteria to produce either one of the enantiomers or racemic acid only. [Another complicating factor in the experiment of Schardinger was the fact that his bacterium was heterofermentative, which means that it did not solely form lactic acid, but in addition also produced other chemical compounds.]

Lactic Acid Bacteria in the Distillery Industry

The reasons for the studies of Lafar and Leichmann which led to the isolation and discovery of *Lactobacillus delbrueckii* were found in the distillery industry. It is more than coincidental that this was the very same industry which called upon Pasteur to solve certain fermentation problems in the 1850's; a request which led to his discovery of "lactic yeast".

The same topic – lactic fermentation in the distillery – later in the century had become one of central interest at the Berlin Institute for the Fermentation Industries, Delbrück's Institute. Some elements of the distillery process should be considered in some detail to appreciate the researches of Lafar and Leichmann as well as those of a number of later workers in this field (process details mainly taken from Dammer, Vol III, 1896).

Malting in the Distillery Process

The first step in alcoholic fermentation is the germination of barley grains. This has the purpose of producing enzymes capable of converting starch into lower sugars, which form a good substrate for the fermentation.

Barley grains are soaked in water for a period of about 40 to 100 hours, depending on the temperature, which should be kept between 11 and 15°C. Thereafter the soaked and swollen kernels are spread on a threshing floor, the malting floor, in a layer about 40 cm thick. This layer, which should be kept at a temperature between 7 and 10°C, is turned over twice a day to achieve a uniform germination of the grains. The germination is accompanied by the evolution of heat, which is dispersed off by further spreading out the layer. The germination process takes about 7 days. At the end of the germination process the grain kernels have formed sprouts, in fact the course of the process can be followed by the length and uniformity of the sprouts. The amount of diastase [a mixture of amylases, enzymes capable of breaking down starch molecules to lower sugars, of which maltose is the most important constituent] has then reached its optimal value. The product carries the self evident name *malt sprouts* or green sprouts.

Other cereals can also be used to produce these enzymes, the choice usually depending on local availability and tradition. Germination times and conditions have to be adjusted to attain the optimal content of [amylases].

Recently, Dammer says in 1896, process equipment has been introduced in which conditions of temperature and humidity can be much better controlled, allowing for the production of green malt during the whole year, which, the writer says, also lowers (manual) operating costs. The cleanliness of the operation remains, of course, as critical as when germination takes place on the malting floor. But with good design the requirements can be met.

Mashing in the Distillery

In the production of alcohol and yeast, green malt is used directly to convert (other) starch containing raw materials into lower sugars suitable for the ensuing alcoholic fermentation.

The sprouts are first broken in a crushing mill to achieve a high rate of extraction in the following mashing step:
2.5 to 3% of the crushed green malt is then added to a solution of gelatinized starch. The suspension is kept at a temperature of about 55°C, the optimal temperature for the action of the [amylases] on the starch. In practice, Dammer writes, a higher temperature towards the end of the reaction is required to make the destructive fungi originating from the malt and the air harmless. [The literature of that time often uses the term fungi when in present vocabulary the word microorganisms is used.]

The hydrolysis of the starch takes from less than an hour to about two hours, the solution then contains about 68% of maltose and 32% of dextrins. The mash is thereupon cooled from about 70°C to a temperature below 20°C to prevent or slow down infection processes. Clean equipment and clean air are other prerequisites for a good fermentation.

The Inoculation

The mash is now ready for alcoholic fermentation. This is induced by inoculation with (a suspension of) yeast. At the time a specially prepared so-called "artificial" yeast was applied. The term artificial relates to the special way in which this yeast had been prepared:
Originally, of course, all yeast was brewer's yeast, easy to separate from the

beer because it rises to the surface at the end of the brewing process (or sinks to the bottom in the later "Pilsener" process). In the distillery the mash, usually being more viscous and often also containing more suspended material, does not allow for an easy separation of the yeast from the alcoholic solution. It had therefore become a regular practice to inoculate the next batch with a certain volume of a previous batch. This meant some loss of alcohol and of capacity, it also increased the danger of infection. In the beginning of the nineteenth century a fairly simple method was found to "breed" the yeast separately, "artificially". From then on the distillery had its own supply of good, [hygienically safe] yeast.

Lactic Acid Bacteria in the Production of "Artificial" Yeast (German: Kunsthefe)

About 10% of the sweet mash is used in the preparation of artificial yeast (the description of the process comes from Lafar, 1896). This part of the mash is only cooled to about 50°C and kept at that temperature for a period of at least half a day. Usually spontaneous acidification takes place caused by lactic acid bacteria which have survived the heat treatment, or have accidentally been introduced by the stirring spoon [sic!] or have fallen into the mash from the air. Such an acidification had been found to be a desirable pretreatment because of the better results obtained with yeast [cultures] from the ensuing yeast fermentation. Not every acidification guaranteed good results, however, as Lafar writes:

> One can expect that the acidification often will be insufficient or of the wrong kind [butyric acid!]. Such evils will manifest themselves particularly in the beginning of the working season (after the summer) when the lactic acid bacteria ... which do not endure drying conditions very well, are almost extinct from the yeast manufacturing room.
>
> *- Lafar, 1896*

Some distillers applied special measures to cope with this situation. Leichmann (Leichmann, 1896), another student of bacteriological processes in the distillery, wrote:

> The distillers took and still take special measures to prepare the first yeast of the new season. A few days before the start up they let sour (spontaneously) a freshly prepared sweet mash at a temperature of 50°C. From this acidified mash a new sweet mash was inoculated [supposedly after one day of souring and with an amount of about 10% of

the volume of the fresh mash]. This inoculation procedure was repeated until a strongly acidic mash [with a lactic acid concentration of about 1%] was obtained, which was thereupon applied as inoculant in the normal production.

- Leichmann, 1896

The new method, although quite an improvement over the old practice still gave erratic results. Lafar, who does not mention the use of "culturing" by some distillers, says that until presently the distillers have relied only on good luck. This, he continues

can now be shoved aside in one stroke when one leads [the souring] in the right direction by adding to the sweet mash at a temperature of 50°C a sufficient quantity of **a pure culture of a selected strain of lactic acid bacteria.**

- Lafar, 1896

Lafar (Lafar, 1896 and 1897) succeeded in isolating and culturing such a bacterium from fresh yeast from a distillery in Lietzen (Mark Brandenburg, today G.D.R.). In the winter of 1894/1895 this culture was tested in practice in a distillery in East Prussia (now part of the Soviet Union). For the start-up of the fermentation in the following campaign only the new culture was used.

The accepted procedure was as follows:

The sweet mash prepared at a temperature between 60 and 70°C is cooled to 50°C, inoculated with about 10% of a mash soured with the "pure culture" and subjected to acid fermentation at that temperature for about half a day. Then 10% of this acidified mash, called "Muttersäure" [= mother acid] with a lactic acid content of 1.0–1.1% is set aside to inoculate sweet mash for the production of Muttersäure for the next day. The remaining bulk of the volume is heated to 70°C to kill most of the bacteria, cooled to 20°C and inoculated with yeast to form in about half a day a suitable "Mutterhefe" [= mother yeast]. In this way the inoculant for the normal batchwise conversion of sweet mash into alcohol is obtained as a relatively pure culture.

The Origin of *Lactobacillus delbrueckii*

Lafar not only prepared a pure culture of the acidifying bacterium, he also gave it a name: *Bacillus acidificans longissimus*, and described the morphology:

"The cells of the new species have characteristic forms of growth: elongated rods and threadlike cells."

Leichmann (Leichmann, 1896) observed similar phenomena. He referred to the studies of Delbrück and collaborators who had found that to achieve a pure lactic fermentation of the sweet mash the temperature had to be kept at or above 50°C. Microscopic examination of a mash acidified under these conditions

> shows a very characteristic picture, known to every practician: luxuriously developed, slim, non-motile rods of varying lengths, often connected in pairs or strings, in which the separate rods have sharply defined contours; usually the strings do not form a straight line, but are bended to some extent.
>
> \- *Leichmann, 1896*

Leichmann compared the new bacterium with the bacterium he had obtained from milk which had soured spontaneously at 50° C. This bacterium showed very similar behavior and appearance. Another remarkable similarity between the two bacterial species was that both of them produced **optically active, levorotatory lactic acid** [now indicated as D(-)-lactic acid]. The only difference Leichmann discovered was the inability of the bacterium isolated from the soured distillery mash to ferment lactose. For the two closely related bacteria he proposed the names:

- *Bacillus lactis acidi* for the bacterium isolated from milk soured at 50°C, and
- *Bacillus Delbruecki* for the bacterium isolated from distillery mash at the same temperature.

The distillery bacterium was named after Delbrück to honor the scientist who had been so involved in the elevation of the distillery practice, and, in particular, had contributed so much to clarify the conditions under which pure cultures of lactic acid producing rods perform in the manufacturing of yeast.

Leichmann concludes his article with a quotation of historical interest for our subject from Märcker's Handbuch der Spiritusfabrikation (1894):

> ... it should be pointed out that recently lactic yeast [sic !] has been cultured artificially for the purpose of acidifying cream in the dairy industry, and that equipment for pure culturing has been installed (for instance at the laboratory for dairy sciences of the "Landwirtschaftliche Versuchsstation für Schleswig-Holstein" in Kiel). The same could be done in the (distillery) practice which should make it possible to obtain a pure lactic fermentation right in the first mash (of the season).
>
> It is known, however, that not one, but a number of lactic acid producing fungi [sic !] exist, different species and strains as is the case with alcohol producing yeasts. It might possibly be of interest to advise the yeast culturing department of the "Verein der Spiritusfabrikanten" [Delbrück's Institute for the Fermentation Industries] to breed and deliver the particular strain of lactic acid fungus which could effect a

vigorous and pure lactic fermentation at the start up of the distillery season.

- Leichmann, 1896

Two important conclusions can be drawn from the experiments of Lafar and Leichmann:

1. Some species of lactic acid forming bacteria are resistant to temperatures above 50°C, may be even to temperatures of about 60°C. As the great majority of other bacteria proliferating in the green malt or present in distillery air and equipment do not survive at these temperatures, it is possible to breed – culture – lactic acid forming bacteria selectively at about 50°C.
2. Lafar and, independently, Leichmann, discovered the same, or closely related bacteria in soured distillery mash. Both scientists find predominantly (or exclusively?) levorotatory lactic acid as the metabolic product of their bacterium.

Henneberg Defines *Lactobacillus delbrueckii* More Exactly

The suggestion given in Märcker's book in 1894 to the Institute for Fermentation Industries to supply the distilleries with pure cultures of lactic acid forming bacteria was only followed up about ten years later.

Much earlier, in 1895 or 1896 Boehringer's new lactic acid factory had, at the request of Carl Wehmer already received a pure culture of *Bacillus acidi lactici* from the Institute.

About a decade later Dr. Wilhelm Henneberg, who had joined the Institute in 1897, reported the results of extensive microbiological studies of lactic acid bacteria found in distillery mashes, in milk and in beer (Henneberg, 1901). The report confirms and completes the earlier investigations of Lafar and Leichmann, and supplies a rather detailed description of the bacterium:

BACILLUS DELBRUECKI (LEICHMANN)

Lafar named it *B. acidificans longissimus*.
Bacillus of the lactic distillery mash and of the lactic yeast mash.
The species grows in the form of separate small, whitish colonies, only to a very limited extent on the agar surface, much better in a stab culture (in agar).
...

The cells are thin and elongated 2.8–7 micron: 0.4–0.7 micron, separate or twinned, or united into straight or angularly bent lines.

...

No clustering of cells nor sporeforming does occur.

Artificial nutrient substrates have a very unfavorable influence. The species is very demanding with respect to quantity and quality of the nitrogen containing nutrient. Of the natural nutrients yeast water, yeast extract, mash and wort (without hops) appeared to be very favorable.

The reaction of the substrate should be slightly acidic only, a small amount of (excess) acid will already slow down or even stop growth.

The most favorable growth temperature is about 45°C, growth is good in the range from 30 to 48°, moderate between 22 and 29°, no growth takes place below 18°.

The cells are killed within one minute at 68 to 70°C, in 5 min. at 65–68° when in liquid form, in the dry condition at 93 to 96°.

Lactic acid is formed from levulose, dextrose, galactose, cane sugar, malt sugar, dextrin, – not from arabinose, xylose, rhamnose, lactose, ...

...

The lactic acid produced is not (further) digested.

Kowatzki[1] (see Ch. 5, 153) has determined that dextrose, cane sugar, malt sugar and dextrin are converted into levorotatory lactic acid.

...

The species is found quite generally on germinated barley, green malt and kiln-dried malt; in the mashing process carried out at 50°C, it becomes the predominant species. For this reason its use is recommended in the distillery practice being the easiest "natural culture."

- Henneberg, 1901

Henneberg's last sentence already contains the germ of major future confusion regarding the identity of *Lactobacillus delbrueckii.*

Since the days of Henneberg the name *L. delbrueckii* has been adopted for the bacterial culture isolated by means of "natural culturing" from malt sprouts. Whether such a strain actually produced pure D(-)-lactic acid or some combination of enantiomers, or pure L(+) was not considered to be of any practical importance. As, moreover, the determination was difficult, hardly anybody paid attention to the optical properties of the product.

French Pharmaceutical Lactic Acid Must Be Racemic

One exception was found to the general lack of interest in the optical properties of commercial lactic acid. The Pharmacopoeia Gallica specified that lactic acid had to be optically inactive, that means the composion should be racemic. For this reason Merck studied in 1912 the optical properties of its

production in quite some detail (Krauch's 1912 Manufacturing Report gives the details; source: Merck's Archives):
The specification in the Pharmacopoeia Gallica that lactic acid should be optically inactive proved almost impossible to comply with. As the reporter wrote: only synthetic lactic acid will be racemic, the fermentation product will always contain one of the isomers in excess, hence the final product will always show some optical activity. This will be made worse when the acid is concentrated to the high density of 1.24 [kg/l] specified in the Ph. Gallica[2].

The conclusion of Merck's study of the optical activity of its pharmaceutical lactic acid was that the ratio of the two enantiomers was not constant. This means that Merck's fermentation was run with a mixed culture of lactic acid bacteria.

Confusion about the Enantiomer Produced by *L. delbrueckii*

In a 1926 study of the factors influencing the degree of enantiomorphism, commercial samples from Merck and Du Pont were evaluated for comparison. Both contained a slight excess of the D(-) modification, as determined by the rotation of the zinc salt (Pederson et al., 1926). The authors report that some pure cultures produce dextro- or levorotatory lactic acid almost exclusively. They also found indications that mixed cultures with an organism which does not produce lactic acid may produce the enantiomer opposite to the one produced by the pure culture alone.

The theme of the above study was continued for a number of years. A 1935 paper from the same research group of the University of Wisconsin defines the conditions under which it is believed to be possible to produce pure L(+)-lactic acid commercially (Tatum and Peterson, 1935). In most cases pure strains of *L. delbrueckii* type bacteria at a fermentation temperature of 37°C produced L(+)-lactic acid, the same enantiomer was prepared by fermenting at 30°C with *Streptococcus lactis* and *Lactobacillus casei*. [In the ninth 1986 edition of Bergey's Manual it is stated that all *L. delbrueckii* subspecies produce D(-)-lactic acid only, while three of the four subspecies of *L. casei* produce the pure L(+) enantiomer ! (Kandler and Weiss, 1986)]

How can one be sure that Tatum and Peterson indeed had identified their *L. delbrueckii* strains correctly? The fermentability of a range of sugars was usually one of the most important criteria. In an extensive review of the literature Briggs and Briggs point out that sugar fermentation reactions often give variable results and are therefore not always reliable (Briggs and Briggs,

1954). The summary of the paper refers to much confusion in the literature "caused by too much regard to names and sources of cultures, and too little to their stable and characteristic properties."

A review on the taxonomy (= systematic classification) of the *Lactobacilli* of 1962 (Sharpe, 1962) gives an example of the confusion when identification of *Lactobacilli* as *L. delbrueckii* is based only on their failure to ferment lactose: "... more than 50% of the strains of *L. delbrueckii* investigated in the authors laboratory have proved on further examination to be *L. casei* var. *alactosus* (in Bergey's ninth edition this subspecies is designated as *L. casei* subsp. *casei*, a producer of L(+)-lactic acid).

Bergey's eighth edition (1974) about the confusion relating to *L. delbrueckii*:

"At the time the neostrain was isolated, all known culture collections named *L. delbrueckii* were incorrectly designated; they were typical strains of *L. casei* subsp. *rhamnosus* [Buchanan and Gibbons, 1974]."

Now two questions remain regarding the identity of the bacteria used in commercial lactic acid manufacture.

The first one is the optimum temperature, usually above 50°C, sometimes even 55°. This is some 5° above the optimum or maximum usually given for *L. delbrueckii*, and about 10° higher than the maximum reported for *L. casei* subsp.*rhamnosus*. The reason for the high temperature is found in the selective methods applied, the principle of "natural pure culturing" formulated by Delbrück around 1890: try to find conditions favoring the growth of the wanted species which are at the same time highly unfavorable for the growth of possibly contaminating bacteria.

The second question concerns the criterion that *L. delbrueckii* is a producer of the D(-) enantiomer only. It goes without saying that this criterion per definition excludes the possibility of L(+)-lactic acid manufacturers to use *L. delbrueckii* in their fermentation processes.

Is *Bacillus coagulans* the Culprit?

A highly interesting paper from the Research Laboratory of Arthur Guinness Son & Co. (Dublin) goes back to the problem Pasteur and Delbrück, amongst others, had been investigating in the 1870's and 1880's, that is the acidification of sweet wort (McMurrough and Palmer, 1979). These authors isolated lactic acid producing bacteria from sweet wort fermentations inoculated with the normal wort flora at two different temperatures - 45° and 55°C. Isolated

colonies from these fermentations were again cultured in sweet wort at the temperatures they were isolated. This procedure was repeated until (after three times) one enantiomer of lactic acid was produced exclusively. The next step was the identification of the isolated pure cultures:
The 45° isolates were identified as typical strains of *Lactobacillus delbrueckii*; they were exclusive producers of D(-)-lactic acid, and could ferment glucose, fructose, mannose, maltose and sucrose;
the 55° isolates produced exclusively L(+)-lactic acid, and could ferment ribose, galactose, glucose, fructose and raffinose. The bacterium was identified as *Bacillus coagulans*. An important feature of *B. coagulans* is the ability to form spores, a property not reported for commercial lactic acid producing bacteria. The authors acknowledge, however, that spore production is exceptional, and only takes place under special conditions.
Note: The Danish manufacturer Kemisk Vaerk Køge reported in 1972 when it closed its lactic acid plant that the company had used *Bacillus coagulans*, a "heterofermentative bacterium", producing lactic acid and small amounts of mannitol (private communication).

How Stable Are Lactic Acid Producing Bacteria?

The issue of what kind of lactic acid cultures have been and are used in the lactic acid industry has not yet been resolved. Even the last edition of Bergey's Manual still casts some doubts on whether the correct classification has yet been found. The newest methods based on DNA analyses are a big step forward in determining evolutionary (phylogenetic) relationships, but the results are not yet conclusive. The large differences *Lactobacilli* show in DNA composition are related by Kandler and Weiss to the high phylogenetic age of this genus (Bergey's, 1986). This is another way of saying that evolutionary changes in this genus are a consequence of the adaptation to a large number of different (natural) conditions (See also Cowan, 1962).

It is very likely that (evolutionary) adaptation also occurs at the very special and highly selective conditions of industrial lactic acid manufacture. Many questions arise in this respect, is it for instance possible that genetic material is exchanged between different species of bacteria? Can genetic information be lost or acquired under the selective conditions?

To simplify the problem let us look at the average time of reproduction. For a bacterium this is usually less than one hour. With many plants the reproduction time is a couple of months or a year. Higher animals generate new offspring in periods of time even longer than that.

The number of generations mankind needed to cultivate many of its present crops is of the order of 1,000 to 10,000, that for breeding its domestic animals even less than that. Bacteria, however, multiply about 10,000 times in a period of one year and are therefore apt to change so much in form and functions in the course of that year that some of the original classification criteria easily get lost.

One should also bear in mind that mutational change (and possible adsorption of genetic material from other microorganisms) is highly favored by the extremely large populations occurring in commercial fermentations. The number of bacteria in 1 ml of fermentation broth easily exceeds the total number of humans today living on this earth!

The above deliberations are quite speculative and are not meant to substract much from a remark made by Kandler (Kandler, 1984):

> It is surprising that even firms selling more or less pure L(+)-lactic acid claim to use *L. delbrueckii*, which by definition forms only the D(-) isomer.
>
> \- *Kandler, 1954*

For the time being the conclusion of the writer is that:

1. It is highly unlikely that the bacterial culture historically used in lactic acid production ever has been a pure strain of *Lactobacillus delbrueckii*;
2. It is likely that the bacterial cultures used in industrial lactic acid manufacture depart too much from naturally occurring bacterial species to make the use of an existing species name fully warranted.
3. Until the genetic structure of lactic acid forming bacteria has been unravelled in more detail (for instance by DNA sequencing) classification of these bacteria will remain beset with arbitrary elements.

Notes

[1] Dr. Erwin Kownatzki, assistant in Henneberg's lab.

[2] Most pharmacopoeias specify a density of 1.21, corresponding with a concentration of 90 +/_ 2%. The "French" density of 1.24 can only be reached at calculated concentrations above 100%, such lactic acid concentrations are possible through the forming of high percentages of intermolecular esters. Even in the 1965 edition of Pharmacopée Française (ed. 7) a density of 1.23, corresponding with a concentration of about 100%, is specified. This pharmacopoeia still prefers racemic acid, but a slight

rotation is acceptable.

Holten (Holten, 1971) states that the optical activity of enantiomeric lactic acid is rather small. The molar rotation of L(+)-lactic acid at the wave length of the sodium line (the usual standard) is, at low concentrations, between +1 and +2, at higher concentrations increasing to about two times that value provided no lactoyl-lactic acid is present. The intermolecular esters are levorotatory, however, and, when present, will therefore lower the angle of rotation. As the equilibrium between monomolecular lactic acid and the condensed molecules is attained only very slowly, the actual ratios between the diffent forms depend much on the conditions of preparation. This can result in appreciable differences in optical rotation when an optically active lactic acid is concentrated in slightly different ways.

References

Buchanan, R.E., and Gibbons, N.E., **1974**. in *Bergey's Manual of Determinative Bacteriology*; 8th ed.; Williams and Wilkins: Baltimore, 579.

Briggs, M., and Briggs, C.A.E., **1954**. *Dairy Sc. Abstr.*, **16**, 253.

Cowan, S.T., **1962**. *Microbial Classification*, Twelfth Symposium of the Society for General Microbiology; University Press: Cambridge, 433.

Dammer, O., **1896**. *Handbuch der Chemischen Technologie*; Ferdinand Enke: Stuttgart, Vol. III.

Henneberg, W., **1901**. *Wochenschr. Brauerei*, **18**, 381.

Holten, C.H., **1971**. *Lactic Acid*; Verlag Chemie: Weinheim (FRG).

Kandler, O., **1984**. *Dev. Ind. Microbiol.* **25**, 109.

Kandler, O., and Weiss, N., **1986**. Genus *Lactobacillus* in *Bergey's Manual of Systematic Bacteriology*; 9th ed.; Williams and Wilkins: Baltimore, Vol. 2, 1208.

Lafar, F., **1896**. *Zentralbl. Bakteriol. Parasitenk. Infektionskrankh.*, Abt.2, **2**, 194.

Lafar, F., **1897**. *Technische Mykologie*; Gustav Fischer, Jena, 220.

Leichmann, G., **1896**. *Zentralbl. Bakteriol. Parasitenk. Infektionskrankheiten*, Abt.2, **2**, 281.

Maly, R., **1874**. *Ber. Dtsch. Chem. Ges.*, **7**, 1567.

McMurrough, J., and Palmer, V., **1979**. *J.Inst.Brew.*, **85**, 11.

Nencki, M., and Sieber, N., **1889**. *Sitzungsber. Akad. Wiss. Wien Math. Naturwisss. Kl., Abt. 2B*, **98**, 423.

Pederson, C.S., Peterson, W.H., and Fred, E.B., **1926**. *J.Biol.Chem.*, **68**, 160.

Schardinger, F., **1890**. *Monatsh. Chem.*, **11**, 569.

Sharpe, M.E., **1962**. *Dairy Sci. Abstr.*, **24**, 109.

Tatum, E.L., and Peterson, W.H., **1935**. *Ind.Eng.Chem.*, **27**, 1493.

CHAPTER 6

EFFORTS TO MAKE FOOD GRADE LACTIC ACID

The first producer of lactic acid, Charles E. Avery, was convinced that lactates and lactic acid were well suited for use in food and beverages. The Avery Lactate Company tried to introduce the acid as a replacement for table vinegar, and acid calcium lactate as an ingredient in baking powder. Both ventures failed, mainly for technical reasons related to the lack of purity of the products.

C.H. Boehringer Sohn of Nieder-Ingelheim in Germany also began its commercial lactic acid operations with a trial to introduce acid calcium lactate in a baking powder formula (German patent 87,235 of October 1895). The German manufacturer met no success in this field either and thereupon turned his activities to the promotion of lactic acid for technical applications.

The major uses for lactic acid which had been developed by the end of the century were technical:

- The Avery company had successfully introduced the acid as an auxiliary agent in the preparation of leather;
- Boehringer had established the use of lactic acid and of special lactates in textile dyeing and printing.

Carl Wehmer Finds an Application for Lactic Acid in Alcohol Production

The first commercially viable use of lactic acid in the beverage industry came from Carl Wehmer, the microbiologist who had advised Albert Boehringer at the start up of the lactic acid plant on the use of pure cultures of lactic acid and on industrially useful methods to maintain the productivity of such cultures.

Carl Wehmer was well aware of the beneficial effect of the souring of the wort on the subsequent alcoholic fermentation. But he proposed a solution different from the one advocated by Lafar and Leichmann, that is the use of pure cultures of lactic acid bacteria (Intermezzo 3, 163). In 1898 Wehmer reported that souring of the sweet mash with 1.0 to 2.0% of a commercial solution of lactic acid (50%) would produce a mash just as suitable for alcoholic fermentation as one in which the souring had been achieved by lactic acid fermentation (Wehmer, 1898). On the basis of his experiments he persuaded a Mr. G. Schulz, owner of a distillery in Hanover, to conduct large scale trials in his plant. These trials went on for a period of six months with excellent results. The alcohol yield appeared to be equal to or somewhat better than before and the amount of baker's yeast obtained increased from about 11.5 to 12.5%. Both products, the alcohol as well as the yeast were of excellent quality, Wehmer reported. [The lower yield of the classical process is related to the consumption of sugar when lactic acid is microbiologically made in situ.]

As to the costs and the quality of commercial lactic acid Wehmer had this to say:

> technical grade lactic acid (50%) from Boehringer has a yellowish to yellowish-brown color, is not entirely without odor, but is free from (the detrimental) impurities often present in crude lactic acid. Dark brown colored acid or acid with a butyric smell is not acceptable. The price of the Boehringer lactic acid is about 60–80 Marks per 100 kg, which is one tenth of the price of the chemically pure acid. The low price is the result of improvements in the manufacturing process and of [the scale of production, realized thanks to] the extensive use lactic acid has found in the textile industry.
>
> *- Wehmer, 1898*

The advantage of the use of lactic acid in the alcohol and yeast industry soon became common knowledge in Germany. One of Germany's most popular textbooks of chemical technology devoted in 1900 a small paragraph to the industrial production of lactic acid (by Boehringer) and said about the applications:

> The more or less pure acid assists in the culturing of the yeast in the distilleries, most important is its use as mordant in textile printing.
>
> *- Ost, 1900*

The influence of lactic acid bacteria in the production of bakers' yeast was independently [?] also taken up by the famous Dutch microbiologist Beijerinck at the "Nederlandsche Gist- en Spiritusfabriek" (the Dutch Yeast and Alcohol Factory)[1]. In 1901 Beijerinck published an extensive article about the most important bacteria playing a role in yeast and alcohol manufacture: *Lactobacillus delbrueckii* and *Lactobacillus fermentum*, both having a positive function in these manufactures, and the "wild" *Lactobacillus caucasicus*, which is often responsible for off-productions (Beijerinck, 1901).

Beijerinck, like Wehmer, was convinced that the function of the lactic fermentation on the mash could be taken over by lactic acid. He succeeded in producing several hundreds of kilograms of lactic acid in the laboratory of the yeast factory (for large scale trials), but he did not pursue the matter further as he experienced too many problems in the "production" of lactic acid due to various infections. He concluded that only by using pure cultures of *Lactobacillus fermentum* would production of lactic acid of good quality be possible. [In hindsight, the problems of Beijerinck were probably related to his working temperature of about 40°C for the preparation of lactic acid; at this temperature the danger of infection is much greater than at the temperatures between 50 and 60°C recommended by Wehmer and, for the saccharification of wort, by Delbrück's Institute.]

Wehmer's new application was not in every respect a genuine food application, the function of lactic acid was more that of a technical expedient to sustain the proper course of the alcoholic fermentation. The expedient was included in the fermentation broth, however, and for that reason only indirectly could become a food ingredient. This was not the case when a separation process was required to make the final product, as was the case in the production of alcohol and yeast. In the most important brewery product – beer – lactic acid would, however, become a (minor) constituent of the final product. In the German Reich this was not considered to be an acceptable proposition, according to the "Reinheitsgebot" (the law governing the purity of beer) only barley, yeast, hops and water were permitted ingredients in beer brewing.

Maybe to counter the threat that the brewing industry, for hygienic reasons, would press for permission to add lactic acid to the preparation of brewery wort, the Fermentation Institute in Berlin decided in 1906 to make pure cultures of lactic acid bacteria commercially available to the brewing industry (Glaubitz, 1925).

Unlike Germany the new application of lactic acid in other countries also became accepted in the brewing industry. An interesting example is Japan where the first trials to use lactic acid in sake production were successfully conducted in 1910 (Private communication from Dr. Katsuhara Yasumatsu, 1987). German export figures bear testimony of this development and also of the high quality product the Japanese sake brewers wanted:

The bulk, or all of the Japanese imports were supplied by E. Merck, Darmstadt (source: Merck's archives).

Table 6.1. German lactic acid exports to Japan.

year	volume (t)	German export price (M/100 kg)	ratio of Japanese price over lowest export price
1910	not reported		
1911	24.5	286	6.0
1912	42.4	384	8.7
1913	28.3	276	6.0
#		price in RM/100 kg	
1920*	16.7		
1921*	20.7		
1922*	78.8		
1923	22.3	300	7.1
1924	68.3	302	6.2
1925	127.1	130	2.3
1926	243.1	141	2.3
1927	107.2	180	2.8
1928	146.9	148	2.2
1929	137.6	195	2.3

\# Germany did not publish export statistics over the intermediate years.
* no prices were given for the years 1920, 1921 and 1922.
the 1921 statistics report only from May until the end of the year.
Source: Official German export statistics.

Is Lactic Acid Fit for Human Consumption?

For Charles Avery this question was irrelevant, lactic acid being the acid constituent of buttermilk and "a natural element in animal and vegetable food" its healthfulness could not be doubted.

The C.H. Boehringer Sohn company took a more serious view of the health aspects:
In 1907 it requested Edwin Stanton Faust, Professor of Pharmacology at the University of Würzburg to investigate the possible injurious effects of lactic acid on animals.

In January 1910 Professor Faust published the results of his studies (Faust, 1910):

In the introductory sentences of his paper the author acknowledged the desire of the Boehringer company to establish the suitability of commercial lactic acid to replace citric and tartaric acid in sweets, carbonated soft drinks and fruit juices.

Dr. Faust then reported on the extensive study he had made of the literature:

> dextrorotatory lactic acid [L(+)] is a normal constituent of animal organisms; under pathological conditions sometimes abnormally high levels of lactic acid in animal bodies are found, but these are the effect of the disturbances, not their causes.
>
> - *Faust, 1910*

In the discussion he refers to the Russian scientist Metchnikoff (1845–1926) working at the Institut Pasteur in Paris, who was convinced that lactic acid bacteria as found in Bulgarian yoghurt as well as lactic acid itself had a positive influence on longevity (a view "too heuristic and optimistic", Faust says, to be worth much attention).

In his toxicological experiments Faust introduces a 5 or 10% solution of lactic acid by means of a probe directly into the stomachs of dogs and rabbits. In general the dogs have little or no problems, even with the higher concentration. The rabbits lose their appetite when the 10% solution is given, one of the three rabbits exposed to this concentration dies in the course of the experiments from acidosis. Not an abnormal result, the Professor comments.

The next series of experiments is carried out with "lactic anhydride", which, Faust says, is a normal constituent of concentrated lactic acid[2]. To prevent the hydrolysis of the "lactic anhydride" flour is suspended in the concentrated acid and the suspension is administered again into the stomachs of the animals. The results are fully comparable with those of the lactic acid itself.

The conclusion of the article is:

A human being of 70 kg body weight can digest 100 g of lactic acid per day without detrimental effects. When drinking lemonade acidified with lactic acid to a content of about 1% a daily amount of 10 liter could be consumed without damaging the renal system.

In the last part of his paper Faust refers to experiments with pathogenic microorganisms performed at the University of Strasbourg. These experiments had shown that lactic acid exerts bacteriostatic properties, thereby lending some support to the ideas of Metchnikoff. But, as Faust observes, the favorable effects of lactic acid could also be due to the laxative properties of the sodium lactate into which it is converted in the colon. To preempt a negative inference from this remarks, Faust adds:

the laxative effect of lactate salts is rather small and will certainly not constitute an objection to the intended substitution of tartaric acid by lactic acid in baking powder.

- *Faust*

The official permit to use lactic acid in food products was granted by the "Reichsgesundheitsamt" (the food and health authority of the German Empire) in 1915 (Remenovski, 1929). The difficulties the Reich then encountered in procuring citric and tartaric acid as a consequence of the blockade by the Entente probably had a positive influence on the Reichsgesundheitsamt in granting lactic acid this status.

Opposition to the Use of Lactic Acid in Food and Beverages

Boehringer's efforts to introduce a refined lactic acid in the food and beverage industry was not met with enthusiasm by the producers of other food acids. Most outspoken was a doctor Wilhelm Hoffmann, sworn-in official expert of the Berlin Chamber of Commerce. In an article concerning the manufacture of technical and refined lactic acid he wrote (Hoffmann, 1913):

... recently propaganda has been made for lactic acid as a table acid[3].
Professor [...] Faust had established in physiological experiments with animals that lactic acid could be safely used in food products in concentrations below 5%. Edible lactic acid should replace acetic, tartaric and citric acid.
Lactic acid for food purposes should of course be a chemically pure product, exactly as is demanded of [glacial] acetic acid. ... [Lactic acid] could, however, only be called a fermentative product in the way vinegar is a fermentative product, when the lactic acid bacteria would adapt themselves in such a manner to their own acid that a genuine fermentation process would persist up to a concentration of several percents of the acid, and the final product [so obtained] would possess the typical odor and flavor of a fermentation product.

- *Hoffmann, 1913*

In Hoffmann's view the special purification processes needed and in particular the conversion of the primary product of the fermentation, calcium lactate, with sulfuric acid into lactic acid, forbade the name table product (German: Genussmittel) for the refined product.

The Competitive Situation

In a historical account of the Boehringer company it is stated that one of the reasons to go into the food and beverage market was the increased competition encountered in the technical market (Heusner, 197?).

Henneberg of Delbrück's Institute in an article on lactic acid bacteria gives the impression that a large number of manufacturers exists (Henneberg, 1904):

> ... the species (of bacillus) used in a good deal of the larger lactic acid factories is *Bacillus Delbrücki* (Leichmann) [now designated *Lactobacillus delbrueckii*]. The botanical laboratory of the Institute regularly sends cultures of this species to many of these factories. The writer has experimental proof that *Bacillus lactis acidi* would also be well suited for [industrial] production. In factories which use lactose solutions [whey] only the latter or another lactose converting species can be considered, and not *B. Delbrücki*. It remains a fact that a few of the smaller factories still apply the old recipe with (spoiled) cheese as inoculant.
>
> \- *Henneberg, 1904*

A few years later in a rather extensive article on the manufacturing process of lactic acid (Schäfer, 1907) it is said that "almost two dozen works produce lactic acid, among the three or four larger ones Boehringer is the leading enterprise[4]. The Boehringer output is estimated between 300 and 500 t/a, the total German production at three times this level. The chemically pure quality carries a price about 5 or 6 times higher than the technical grade, Schäfer says.

Besides C.H. Boehringer Sohn at Nieder-Ingelheim and Knab and Lindenhayn at Niederstriegis (Ch. 5, 153) the following companies are recorded as producers of lactic acid in Germany:

- the pharmaceutical company E. Merck at Darmstadt; average production 1907/1909: 33 t/a pharmaceutical acid (1.21 kg/l, 4 t/a calcium lactate and 5 t/a ferrous lactate (Merck's Archives).
- Chemische Fabrik Güstrow Dr. Hillringhaus & Dr. Heilmann at Güstrow (Mecklenburg), an early manufacturer of whey products, which probably used whey as substrate for lactic acid production;
- Chemische Fabrik Flörsheim Dr. H. Nordlinger at Flörsheim near Frankfurt am Main; and
- Chemische Werke vorm. Dr. Heinrich Byk at Oranienburg, about 50 km north of Berlin[5]. After the merger with Guldenwerke and the take-over of

Knab und Lindenhayn in 1921 the company became the second largest producer of lactic acid in Germany.

Interest in producing lactic acid also grew outside Germany. The installation of small lactic acid factories evidently was not too difficult, the fermentation technology resembled that of the distillery process and thanks to the articles of Jacquemin, Claflin and Schäfer much of the required information had become freely available.

In one of the early years of the twentieth century a small lactic acid factory was established at Reims (France) by H. Gouthière (Gouthière, 1909):

The fermentation was run in 8 vats of 3,000 l fermentation, and as raw material acid hydrolyzed starch was used. The fermentation temperature was kept at 35 to 40°C, the fermentation time was very long: 20 days or more. The product, a technical 50% grade, was sold to tanneries and textile dyers.

When the 1914–1918 War began in Europe, the number of lactic acid producers in the USA was at least five. The few details known about most of these companies will be presented later in this Chapter.

Did Boehringer Have a Technological or Economic Advantage?

The scale of production of the Nieder-Ingelheim plant had gradually increased to such levels that the company decided in 1908 on a rather radical extension and renovation of the lactic acid factory. It is very likely that the wish to make lactic acid **for edible uses** was an important motive in this decision.

The small wooden vats were replaced by rectangular concrete, tile-lined chambers of 25,000 l each. (Dreher had already recommended to ferment in a special room with walls of glazed tiles or specially painted concrete, to facilitate cleaning and sterilization by means of steam [Dreher, 1898].) Some years later even fermentation chambers of 40,000 l were installed. These large and well covered vessels made hygienic control and prevention of infections much easier. Handling also became more efficient and the same was probably the case in the purification department. Purification procedures also were improved in 1908 and during World War I when further extensions were made.

The Boehringer chemists (the company employed 2 Ph.D. chemists in 1896, 3 in 1905 and 6 in 1910) also devised a novel and more economic method to concentrate lactic acid, which merits some attention (German Patents 221,112 of 1908; and 222,741 of 1909). On the one hand because it illustrates the technological difficulties in the concentration step, and on the other hand because of the use the method found in later years with at least one U.S. producer.

Problems in the Production of Edible Lactic Acid

Three major problems had to be mastered:

1. The problem of the large amount of impurities originating from the raw material, potato starch. About 20% of the starch was not converted in the hydrolysis step into fermentable sugars, and much of this material remained, together with other impurities introduced in the fermentation process, in the final product. A solution of this problem would have been the use of beet sugar, which is 100% fermentable, but the high costs on this product made such a proposition uneconomical (See Intermezzo 6).
2. Related to the high content of impurities was the strong discoloration taking place when the dilute solution of lactic acid was exposed to the high temperatures needed to evaporate excess water. The application of vacuum evaporators in which the temperature would not exceed 60 or 65°C did not fully prevent the darkening of the product.
3. The third problem arose from the corrosiveness of lactic acid. the reaction of calcium lactate with sulfuric acid to form lactic acid was done in open wooden vats, for the evaporation of the water closed vessels were needed, however. Steel was not a suitable material for such vessels, it had to be coated with lead or glazed (enameled) to make it resistant to lactic acid.

Boehringer possessed and used a copper evaporator, which was also quite suitable. True, copper also dissolves slowly in lactic acid, but it is not susceptible to pitting corrosion and the amount that dissolves is small enough to make removal from the concentrated lactic acid not too difficult (About Boehringer's process see also Exhibit 5,IV, 147).

Purification Through Crystallization of Calcium Lactate

Avery had been the first to purify lactic acid on an industrial scale by means of (repeated) crystallizations of calcium lactate (Ch. 4, 95). He had avoided major corrosion problems by not concentrating lactic acid itself, but by evaporating as much water as possible from the calcium lactate solution and thereupon converting this into lactic acid. Due to the restricted solubility of calcium lactate a higher concentration than about 50% was difficult to achieve in practice. The maximum concentration of acid obtainable from a 50% calcium lactate solution is about 20%. [The writer holds the opinion that the curious concentration of 22% at which much lactic acid was sold until World War II was a "natural" consequence of this first industrial method to concentrate lactic acid.]

More details about the problems encountered in purification through crystallization of calcium lactate, as well as about other purification methods are given in Intermezzo 4.

In the new process developed at Boehringer's the concentration step via calcium lactate was, so to speak, repeated. It had been observed that at the highest possible concentration of calcium lactate the addition of sulfuric acid resulted in a thick porridge of precipitated gypsum which was almost impossible to handle. However, when the hot, concentrated (60%) solution of calcium lactate was diluted with concentrated lactic acid the suspension of gypsum formed in the reaction with sulfuric acid remained workable. In this way a lactic acid concentration of about 50% could be reached without having to concentrate the acid itself. In the second patent, dry calcium lactate was dissolved in concentrated lactic acid, and subsequently in the same way converted into lactic acid. After filtration a solution with a strength of about 80% could be obtained.

The new method, if applied to crystallized calcium lactate, would yield a lactic acid with about a third of the amount of impurities originally present. The lactic acid produced would also be of good smell and color as it had not been exposed to high temperatures in the concentrating step.

The writer assumes that the first edible lactic acid (of a concentration of 50%) put on the market by the Boehringer company shortly before World War I was made by the process outlined above. It remains doubtful, however, whether the same process was applied for the 80% concentration: the plant probably did not have the capacity to make the required large amounts of dry calcium lactate.

Moreover, it was easier to prepare the higher concentration of 80% by concentrating dilute lactic acid in the copper still.

More information on the processes applied by the Boehringer company can be found in the second edition of "Ullmann" (Ullmann, 1931). The subject "Lactic Acid" was treated in this edition by one of Boehringer's chemists, Albert Sander. Sander had begun his career with the company in 1919 in the just established Research Laboratory, the "Wissenschaftliche Abteilung". His contribution to the "Ullmann" also explains how minor contaminations were removed from concentrated edible lactic acid[6].

In the U.S.A the process described in the two Boehringer patents was applied in the 1930's by Sheffield By-Products Co. (Olive, 1936; Burton, 1937). Olive described it as follows:

A TRICK OF THE TRADE

For the production of more concentrated lactic acid a slightly different technique is required. In making the edible 44 per cent grade, technical calcium lactate is dissolved in water and sulphuric acid is added just as in the case of the 22 per cent acid. The calcium sulphate sludge is removed by filtration and then additional calcium lactate is added to the filtered acid, dissolving in its water. With the use of sufficient sulphuric acid for conversion, this lactate, too, is transformed into lactic acid and by this building-up process, a 44 or 50 per cent acid can be formed. This acid is finally purified with activated carbon and filtering .... The 50 per cent waterwhite edible grade is made in exactly the same manner, except that the U.S.P [United States Pharmacopoeia] grade of calcium lactate is used, rather than the technical grade.

- Olive, 1936

N.B.: The Sheffield company used whey, by-product from the dairy industry, as raw material. The amount of foreign, non-fermentable material in whey is so large – organics up to 10% or more – that a special purification step is imperative, even for the technical product. Because Sheffield aimed for the calcium lactate market they had chosen the crystallization of calcium lactate as the means to obtain a pure product.

Introduction of Lactic Acid in Beverages

The first real application of lactic acid for human consumption came in 1911 with the foundation by Boehringer and a Dr. Ernst Büschler of a 50/50 joint company with the name of Chabeso GmbH for the development and sales of lemonades. The year before, in October 1910, Boehringer had registered the trade mark Chabeso (**C.ha.B**oehringer **So**hn) for a number of its (potential) manufactures, including non-alcoholic drinks and fruit juices (Warenzeichenblatt, **17**, 1910, 2141).

The lemonade-extract – Chabeso extract – was to be prepared by Boehringer: one part of the extract contained the flavoring substance, the other part, the "developer-solution", consisted mainly of lactic acid. The bottlers to which Chabeso GmbH sold the extract(s) combined the two, added the required amount of sugar and diluted it with mineral water (or carbonated water) to an acid strength of about 1% (Fürth and Engel, 1930). During World War I (as well as World War II) when lemon flavor was not available, other flavors like apple aroma, were introduced.

Between the two World Wars Chabeso (Figure 6,1) was so successful in the southern part of Germany and in Austria that the name Chabeso became almost synonymous with lemonade (Heusner, 197?).

Fig. 6,1. Bottle of Chabeso lemonade. Courtesy: Deutsches Museum, Munich.

The Boehringer Company Fared Well with Lactic Acid

Not only did Boehringer have the best process, it also had the advantage of the economy of scale achieved in the preceding years through hard work and persistent marketing activities. Herr Boehringer had ploughed back most of the cash flow into his factory, partly in new products in the pharmaceutical field, but also in fairly heavy investments in the lactic acid factory. Moreover he had attracted quite a number of well qualified chemists to support the activities of the company.

Much attention was also given to exports. Table 6,II gives the total German exports. It is, however, impossible to relate how many of these exports are due to individual German producers. The scanty information available points to Boehringer as the most important exporter of technical lactic acid, and Merck of taking care of most of the pharmaceutical exports.

Table 6.II. German export of lactic acid and lactates (in tons).

country year	Great-Britain	France	United States	Austria/ Hungary	Switzer-land	world total
1906	173					547
1907	306	110				835
1908	344	168				972
1909	294	189	119	97	66	1.044
1910	384	252		114	85	1.278
1911	529	345	159	89	118	1.808
1912	521	316	177	99	120	1.780
1913	764	282	204	94	106	2.050
1926*	355	26	415	219**	77	2.221

* the figures of 1926, the first year after the Great War that German lactic acid exports again surpassed 2.000 t, are given as reference.
** Austria, Hungary and Czechoslovakia taken together.
Source: Official German export statistics.

Although financial data of the period are non-existent, evidence of the prosperity of the company can be found in the "Ingelheimer Chronik 1899–1950" (Diehl, 1974):

September 30, 1903. At the close of the wine harvest, which was excellent this year, Herr Fabrikant Albert Boehringer organized a wine party for his laborers.

April 22, 1905. A major extension and renovation of the Boehringer plant is in execution. [The extension concerns the alkaloid extraction plant].

From the Boehringer Journal, June, 1956:

..., 1906. Herr Kommerzienrat[7] Boehringer buys his first car, an Opel-Darracque, from the director and co-owner of the portland-cement works [a neighboring factory].

Again from the Ingelheimer Chronik:

May 11, 1907. The firm of C.H. Boehringer Sohn, Chemical Works, obtains a permit to construct a railroad connection to the Selztal line and to operate this connection with locomotive engines.

July 26, 1910. Jubilee Festivities.

The silver jubilee of C.H. Boehringer Sohn, Niederingelheim, was celebrated. The entire personnel of the firm, laborers and employees, was present to pay tribute to their highly esteemed boss, Albert Boehringer. Headed by the band of Hassemer the impressive procession marched from the cement works to the house of Herr Boehringer. Flames of large torches illuminated the route of the procession. Passing along the festooned and well lit factory buildings the march ended in the garden of Herr Boehringer where proxyholder Giani presented a register with congratulations from all members of the personnel.

...

The next day (Saturday) Herr Boehringer received all personnel together with their wives for a cruise on the Rhine. About 300 people took part ....

The First Social Charter of the Company

When the financial position of the company became firmly established, Albert Boehringer also assumed the social responsibilities attached to a firm of importance. In this respect he showed himself to be ahead or at least on a par with much larger chemical enterprises of imperial Germany. A reproduction in the 1984 Yearly Report of Boehringer Ingelheim gives some information:

(March 13, 1904). Every laborer who has been employed for 5 to 10 years will receive a special allowance of 30 Mark, to be paid annually in April. ... (higher amounts were granted to longer employed personnel).

(April 15, 1910). A laborer who has an unbroken service with the factory of 10 years receives 14 holidays including a holiday allowance of 80 Mark while keeping his normal wage over the holiday period. When he has served longer than 10 years he will be entitled to such a holiday plus allowance every four year. ...

(February ?, 1912). All laborers with an unbroken service period of 20 years or more are entitled to an extra old age pension and payment for disability in excess of the legally required payments. ...

Boehringer Tries to Increase his Influence in City Matters

May 28, 1914. (Again from the Ingelheim Chronik) The provincial committee in Mainz dealt with an interesting legal matter. The city council (of Nieder-Ingelheim) had received a request from Kommerzienrat Boehringer to assign the special seat plus voting rights in the council to him, being the highest-tax-paying inhabitant of the city. The city council as well as the county council have declined this request. The county council had decided that the baroness von Erlanger pays the highest tax and therefore has the right to occupy the seat. Boehringer's request was denied though he pays more taxes when all his different tax payments are added together. The seat in the city council, however, belonges to the proprietor of the largest estate, which is the one of baroness von Erlanger with a value of 937,900 Mark. In Mainz Boehringer's lawyer defended the view that according to the right interpretation of the law his principal being a captain of industry and the highest-tax-payer was entitled to the seat. The same view was held by the city mayor Muntermann. He could not understand why a landowner with 25 laborers should have more interest in community affairs than an industrialist who employed over 200 laborers. He had to admit, however, that the city council had interpreted the law literally. Boehringer's request was also rejected in Mainz.

War between Austria and Serbia

July 26, 1914. (Beginning of World War I) ..., 1916.
Renovation of the lactic acid factory of the firm of C.H. Boehringer at Nieder-Ingelheim. [The plant was enlarged at the same time, see Figure 7,1.]
July 8, 1916. Herr Kommerzienrat Albert Boehringer, who so many times pioneered in welfare work, provided his entire personnel with cost-of-living allowances ....
The Kommerzienrat, in a much appreciated action, has taken care of the families of the employees and laborers, who are serving in the war. The first (group) will be paid the full salary, while the families of the laborers will keep the weekly (financial) support they have received since the war began.

Developments in the United States until World War I

No American manufacturer apparently felt the need to produce edible lactic acid, at least not until other food acids became scarce during World War I. At the turn of the century the Avery Chemical Company was still the sole producer of lactic acid in America. It would not retain this position for long, however. Probably due to the successful introduction of lactic acid in the tannery the use grew quite rapidly, and expectations evidently grew even faster. In 1909 McLauchlan wrote that since the first lactic acid factory had been built by Charles Avery "three other concerns have entered the field ...".

Not very gloriously when we read further on: "The demand for lactic acid is not great enough to warrant the establishment of a plant for its sole manufacture. In America, at least, it is made only by concerns engaged in the manufacture of other chemicals as well" (McLauchlan, 1909).

The Littleton High School student, who has already been quoted in Chapter 4, and who probably had his information from an employee of the Avery plant in town, said in his 1911 sketch:

> the (Avery) company has the largest plant for the manufacture of lactic acid in the U.S. It also manufactures ... chromium lactates, antimony lactate and calcium lactate. ... There is only one other producer which is located at Bradford, PA[8]. There are also several small [sic] factories in Germany.

The year 1911 brought a disastrous fire to the Avery plant at Littleton. A booklet which appeared on the occasion of the 250th anniversary of the town (Littleton, 1964) speaks of complete destruction of the buildings:

> Despite the proverbial enthusiasm of the horse drawn fire engines, brick walls reportedly two to three feet thick and floors so sodden with product wastes that employees had to wear boots at work, the building was completely destroyed...

The Avery company sold the remains: "Using the foundations, ... a new factory was built in 1912, for the manufacture of concrete blocks." In the files of the Littleton Historical Society in the Littleton Public Library Mrs. Nancy Bradbury found an undated handwritten note also reporting on this fire:

> the left side of the building burned at night
> there was no [fire fighting] apparatus to put it out
> it never was rebuilt
> the activities were moved to Tewksbury-Wamesit where the factory exists till today ...
> The white residue of the factory was used by the farmers on their land because of the high lime content. It was given free to them provided they hauled it away themselves. They had to wear boots to do this.
> Later the factory was owned by a man who transformed it into a (white) brick factory (See Figure 4,2, 90).

Although the unknown high school student only talks about one more producer the writer is inclined to attach more weight to the article of McLauchlan which mentions three other concerns. Most likely these other producers of lactic acid were:

Harrison Bros. & Co., Incorporated of Philadelphia (at Grays Ferry Road), the oldest chemical company of the United States and the first producer of sulfuric acid in the country[9]. The interest of the Harrison company in lactic acid allegedly dates back to 1860, when the company began producing it [in pharmaceutical quantities, probably according to Bensch's process][10]. Trade catalogs of the Harrison Bros. company, however, do not mention lactic acid until the company was taken over by Du Pont in 1917. The 1917 Du Pont catalog names as producer of lactic acid Harrison, Inc. and the address: Grays Ferry Road, Philadelphia.

American Alkali & Acid Company of Bradford, PA. This company, at the time the largest producer of oxalic acid in the United States, was owned by Lewis Emery Jr., an important industrialist in the Bradford region[11]. The American Alkali and Acid Company made oxalic acid in the old historic way by treating sawdust with caustic potash. In the first decade of the century this process became obsolete through the invention of a new process in Germany. This caused considerable financial problems to American Alkali & Acid. The company requested and secured help from the federal government in the form of duties on imported material. Shortly after these were imposed the First World War broke out in Europe, which provided shelter for the outdated production till the end of the war.

The writer has not been able to ascertain which was the fourth company McLauchlan indirectly was referring to. It could either have been the **Atlantic Chemical Co.** of Mansfield, MA, or **Woolner & Company** of Peoria, IL. Both companies are mentioned by Haynes as producers of lactic acid during the Great War, but no information has been found on the date they started production of lactic acid.

The basis of one new application of lactic acid in the manufacture of cellulose films was laid in the prewar years by Charles Nelson Waite, one of the first chemists employed by the Avery Lactate Company.

Lactic Acid and Cellulose Films

Although synthetic high polymers were not yet invented, the properties of natural high molecular weight compounds had quite early on in the nineteenth century attracted the interest of chemists. Of particular historical interest is the discovery of nitrocellulose in 1846 by Schönbein, and the subsequent efforts made by many chemists to exploit the superior explosive properties of the product, which correctly should be named cellulose nitrate.

These investigations resulted in 1871 in the invention by Alfred Nobel of blasting gelatin, a gelled solution of nitrocellulose in nitroglycerin.

Solutions of nitrocellulose in camphor poured and dried in the form of films were shown to be suitable for photography in 1873; the material was also used for artificial silk and leather. Similar compositions were put on the market at about the same time to replace horny materials in the production of ornaments, billiard balls, etc.

Efforts to produce less inflammable cellulosic materials for the production of fibers and films gave birth to the viscose industry. In 1892 Cross and Bevan discovered a process for the production of so-called regenerated cellulose fibers: cellulose is first swollen in a solution of sodium hydroxide, then reacted with carbon disulfide to form soluble celllose xanthate. The xanthate when extruded into an acid bath hydrolyzes and the cellulose is recovered. Depending on the way the extrusion of the xanthate solution takes place fibers or films can be formed. The development of an industrially suitable process for the production of a clear and strong sheet of regenerated cellulose was only completed after many years of research by the Swiss chemist J.E. Brandenberger (1908–1912). He established the first manufacturing plant for the new product – Cellophane – in France in 1912.

One of the many American investigators who tried his hands at the development of cellulose films was Charles N. Waite, the man who collaborated in the 1880's with the Avery Lactate Company in the search for improved lactic acid manufacturing processes as well as in applications research. On May 7, 1901, he applied for a patent (U.S. Patent 690,211) on the use of lactic acid in the preparation of cellulosic films. This, he claimed, was achieved by soaking for instance amorphous cellulose in a 5 to 15% solution of lactic acid. Or by adding pure lactic acid to the solution of cellulose nitrate or cellulose acetate. The films or sheets made from any of these materials maintain indefinitely a tough and pliable condition, thanks to the incorporation of lactic acid. [The function of lactic acid in this application is that of a plasticizer.]

The patent rights were assigned to the *Cellulose Products Company* of Wilmington, DE, a company founded in 1899 by Daniel C. Spruance of Philadelphia (Haynes, 1945, Vol. 3, 374). The Cellulose Products Company had bought the American rights to the viscose process from Arthur D. Little. For a number of years it tried, unsuccessfully, to develop a process for synthetic rayon yarns.

The rights to the original British patents on rayon yarn manufacture had at about the same time been acquired by the General Artificial Silk Company, a company owned by Thomas Skelton Harrison, president of Harrison Bros. & Company. The two companies, the Cellulose Products Company and Harrison Bros., with their competing interests in the development of rayon, came together on June 12, 1901, in Wilmington, to found a new company, *The Lactic Company*. Harrison Bros. obtained 50% of the shares in *The Lactic Company*, the Cellulose Products Company got 40%, and the remaining shares went to eight individual persons, one of whom was Daniel C. Spruance (Book of shareholders, Hagley Museum and Library).

The Lactic Company raised additional capital in 1904 and again in 1905, all of the additional shares issued were taken by Harrison Bros. & Co. [This is also the latest year any information about The Lactic Company has been found.]

The coincidence of the 1901 events does not seem to be a chance occurrence and neither was the name of the company chosen fortuitously. The leaders of the two participating companies apparently had visions of large (?) uses for lactic acid in the production of rayon sheet (and yarn?).

The writer has not been able to establish whether Harrison Bros. were already making lactic acid in 1901. Maybe their interest in lactic acid only began with this cooperative venture.

The whereabouts of Charles N. Waite can be traced thanks to his prolific patent activities. Between 1901 and 1910 he moved from Cranford in the north-west of the State of New Jersey to Lansdowne, PA, not far from Wilmington, DE.

An American Lactic Acid Plant

In 1909 a fairly up to date description of the manufacturing process was presented under the above title by McLaughlan (McLaughlan, 1909). His remarks on plant hygiene demonstrate that he was not just an outside observer:

> In all chemical processes depending upon fermentation, it is well understood that a scrupulous cleanliness must be observed; therefore, a lactic acid plant must be arranged so that it can be kept clean and free from all chance of contamination of the fermentation. The floors should be built of cement so that they may be flushed off with water. The fermentation tanks should be placed so that the floor beneath and the walls next them can be easily kept clean. In addition to the inverting [meant is hydrolysis of starch or grain], fermenting [fermenters of cypress wood, 15–30 m^3 in volume], and storage tanks, ten or twelve in all, a steam plant, 40-horse-power engine,

> air compressor, vacuum pan, filter-press, four pumps, a lead-lined acid tank, and iron sulphuric acid storage tanks are required. Such a plant could be built for $ 15,000, and would have a capacity of 100 tons of 22 per cent. lactic acid per month.
>
> *- McLauchlan, 1909*

In McLauchlan's process the acid is not specially purified. The author who evidently possessed intimate knowledge about lactic acid manufacture, is familiar with the method [used in Germany] to purify through crystallization of calcium lactate, which

> gives an acid light in colour, and free from soluble impurities. This operation is not followed in America because the consumer is not particular about the colour of the acid, and further, calcium lactate is a difficult salt to handle. It does not crystallise well, and it is quite soluble even in cold water.

Most of the lactic acid was sold to the leather trade. It is understandable that the tanneries so long accustomed to the smell of fouling bran mashes and dog dung had no problem in accepting the color (and smell) of dark technical lactic acid.

A light colored lactic acid can be made from glucose or starch, McLauchlan writes, but not from grain "or other materials containing coloured matter". The idea to promote lactic acid for edible uses is not mentioned by the author.

Did U.S. Lactic Acid Producers Cooperate in the Market?

The files of the Hagley Museum and Library produced an interesting document: the Book of Shareholders of the *Lactic Process Company*. This company was founded on March 13, 1907 by the Avery Chemical Company, Harrison Bros. & Co., Incorporated and American Acid & Alkali Company. The last named company, situated in Bradford, PA, was probably identical to, or a predecessor of American Alkali & Acid Company of the same city of Bradford, PA, and is also mentioned as one of the lactic acid producing companies of World War I. (Haynes, Vol 2, p. 13, writes that the American Alkali & Acid Company went bankrupt in 1908, due to ruthless price cutting of oxalic acid.) President of the Lactic Process Company was Frederick A. Claflin, father of Alan A. Claflin, and treasurer of the Avery Company.

Neither in the files of the Hagley Museum, nor in other documents relating to the period did the writer come across any additional information about the Lactic Process Company. The inevitable conclusion crops up that the company served a special purpose only, maybe that of providing a forum for cooperation in the market?

Trusts and Antitrust Actions in America

In the decade following the Civil War (1861–1865) the industrial enterprises of the United States had to adapt to new ways of doing business. The country had now become one great market, to which the railroad system had given a large impetus. Merchandise could be moved rapidly, and usually at fairly low costs, from the production plant to almost any place in the reunited Union.

Mass production of commodity goods became the rule, large production capacities were installed and severe price competition followed when capacities grew faster than demand. This, in turn, led to bankruptcies, lay-offs and loss of capital goods, in a word: to depressions.

Businessmen tried to find an answer to these developments by forming business associations, in which the interests of all or most of the competing companies in one field were pooled. Such pools tried to control the market by fixing prices, allocating production and market quota, etc.

The most superior form in which this monopolization of the market was achieved was the trust. In the trust, the first major example of which was the Standard Oil Trust formed by John D. Rockefeller in 1882, the participating companies assigned all of the shares of their companies to the central organization, which was made responsible for all decisions concerning prices, allocation of production, division of market territories, etc.

Though the idea itself to protect the participants from the disasters of regularly occurring business depressions was not unsound, the monopolistic practice showed too many excesses to be acceptable to the population at large: it led to high prices for many commodities (including railroad fares and charges!), to high import tariffs, and to the accumulation of extreme wealth with a few individuals. The monopolistic practices were quite often accompanied or enforced by actions bordering on criminal behavior, or were just that.

Public anxiety and anger over the excesses of the trust system led to political action. In 1890 the U.S. Congress passed the Sherman Antitrust Act, which stipulates in its first section that "every contract, combination in the form of trust or otherwise, or conspiracy in restraint of trade or commerce among the several states, or with foreign nations, is hereby declared to be illegal."

At first the law was hardly effective, the courts were not willing or able to apply the law in the intended way. In 1899 the United States Supreme Court, however, upheld the power of the Federal Government to break up monopolies falling under the definition of the Sherman Law. Federal action against the trusts received a big impetus when Theodore Roosevelt rose to the Presidency in 1901. Under his Administration (1901–1909) a "trust-busting" campaign was set up. Although the successes of this campaign in the beginning only came slowly, the results on the long term were very impressive indeed. Of the trusts and monopolies broken up as a consequence of the actions taken by the Roosevelt Administration two were of the utmost importance for the development of the American chemical industry: Rockefeller's Standard Oil of New Jersey was dissolved in 1911; the du Pont de Nemours & Company, which had a virtual monopoly in explosives production, was split up in the same year.

Under President Woodrow Wilson (1913–1920) the Sherman Law was further strengthened by the passing of acts of Congress which gave the Federal Government more legal power against monopolistic practices. For the enforcement of the antitrust legislation the Federal Trade Commission was then created. The FTC not only was given a task in tracing and combatting monopolistic practices, but it was also endowed with the task of checking unfair methods of competition.

A sense of the feelings of the business community in this period is conveyed in an article by Alan Claflin about the history of the Drysalters Club of New England (Claflin, 1946)[12]. This club had been founded in 1885; Claflin in his function as chairman of the Avery Chemical Company had become a member in 1902.

PURPOSE OF CLUB

In some degree it is fortunate that the original minutes defining the purposes for which the Drysalters Club was organized are lost, because aside from the general one of promoting acquintance and food fellowship today, such purposes as maintaining uniformity of prices, punishing price cutting, restraining competition, and establishing committees to enforce such regulations would be illegal in the extreme. Your historian ... found when he was a fledgling member that such committees as the "Acid Committee", the "Alkali Committee", the "Bichromate Committee", and the "Arbitration Committee", were purely perfunctory honors. This shows that by the early 1900's there was some recognition among business men that the Sherman law and various other anti-monopoly statutes possibly possessed teeth. However, he still remembers the mingled awe and dread with which a decade earlier he regarded a man reputed to be a member of the Acid Committee of the Drysalters Club.

- *Claflin, 1946*

Did Lactic Acid Really Compete with Tartaric and Citric?

Although commercial information about the uses of these three acids for edible purposes is fragmentary only, the relative positions can be defined with sufficient accuracy.

The differences in properties between the three acids are a point to consider when comparing possible substitutions:

Tartaric acid is the tartest in taste, citric acid is more refreshing, while lactic acid has a rather flat mild taste, often accompanied by some off-flavor resulting from the fermentation process.

Under normal conditions of use all three acids are unquestionably safe for human consumption. Acceptance in the market will therefore be strongly determined by price and availability.

For the sake of simplicity the possible replacement of acetic acid in the form of vinegar is not considered. This notwithstanding early efforts by Avery to promote the substitution, and the anxiety expressed by W. Hoffmann about it when Boehringer began with the introduction of edible lactic acid. The threat to the position of vinegar never was very realistic considering the differences in properties and economies of the two products.

TARTARIC ACID

Roughly estimated world production of tartaric acid in the prewar years could have been between 10,000 and 15,000 tons annually (Dammer, 1898, gives for the 1890's 4,800 t; Dehio, 1966, estimates 1930 production at 20,000 t). U.S. import prices are given at 0.48–0.60 \$/kg (Haynes, 1954); German statistics show export prices from 1.80 to 2.44 M/kg, which at a rate of exchange of 4.25 means 0.42–0.57 \$/kg.

CITRIC ACID

Wehmer (Wehmer, 1933) estimates 1913 world production at 3,800 t, and the 1900 price range at 2.20–3.00 M/kg. Ullmann's Encyclopedia (1916) converts Sicilian exports of raw material to citric acid and arrives at an average yearly figure of 3,600 t (1905–1910). U.S. import prices (Haynes) were 0.63–0.81 \$/kg; German export statistics range from 2.66–3.44 M/kg, corresponding to the same price range of 0.63–0.81 \$/kg.

LACTIC ACID

German production was estimated at 1,000–1,500 t 100%, of which about 400 t were exported (Schäfer, 1907). [The export volume given by the official German statistics has been divided by two to take account of the actual concentration of the exported product.] In view of the surge in use of the product on the internal German market, viz. the increase in the number of large producers from one to four, the 1913 production can be set at 2,000–2,500 t. Although statistically nothing is known about the relative productions of the various qualities a sophisticated guess might be:

- technical grade for leather and textile: > 80% of output
- refined and edible grades, mainly for brewing: about 10%
- pharmaceutical grade: 3–4%

U.S. production, according to Haynes, was about half the German export, which put the American output at 400–500 t per year.

Prewar prices (Molinary, 1921):

- lactic acid, 50%, brown : 0.16 $/kg
- lactic acid, 50%, pale : 0.26 $/kg
- lactic acid, 90%, pharmaceutical: 0.90 $/kg

Haynes (Vol. 3, 50) gives:

- technical lactic acid, 44%: 0.09–0.11 $/kg
- edible lactic acid, as 100%: 0.45 $/kg, which at the then usual 50% concentration means: 0.225 $/kg

The only qualities of lactic acid which could possibly compete with tartaric and citric acid were the edible and the pharmaceutical. Table 6,III presents the relative positions of the three acids in volume and price in the years preceding World War I.

Of the two really important food acids tartaric still held the most dominant position, but the relative position of citric had probably much improved since the middle of the nineteenth century.

Neither with respect to production capacity nor with respect to quality did edible lactic acid in these years pose a serious threat to tartaric or citric acid. The pharmaceutical quality only found use in special pharmaceutical preparations and in sake brewing.

Table 6.III. Comparison of the three food acids (about 1910).

acid	world production t/a	price range $/kg*
tartaric acid	about 10,000	0.47 – 0.59
citric acid	3.000 – 4.000	0.63 – 0.82
lactic acid (ph)	about 100	0.70 – 1.00
lactic acid (ed)	about 250	0.45 – 0.52

* lactic acid prices recalculated to 100% concentration.

Notes

[1] Martinus Willem Beijerinck (1851–1931) studied chemical technology at the Polytechnic School of Delft and botany at the University of Leyden. After a short teaching career he was appointed Bacteriologist of the Nederlandsche Gist- en Spiritusfabriek at Delft. In 1895 he obtained the Chair in Biology and Bacteriology at the Delft Polytechnic School. In the mid-nineties Beijerinck developed the technique of the **enrichment culture**. Starting from the knowledge that in many ordinary natural substances and particularly in soil samples a large variety of microorganisms is present, Beijerinck brought a sample of such a substance in a particular well defined culture medium in which only one or a few kinds of microorganisms would grow well. [This is an application of the principle of natural selection defined by Delbrück a few years earlier.]

The method became an effective method for discovering new types of microorganisms. After World War II it was for instance used very successfully in isolating new antibiotics producing fungi.

Beijerinck also coined the name "virus" and gave the first definition of this cell-free "fluid infectious principle". Literature: van Berkel, 1985, Stanier et al., 1976.

[2] Instead of lactic anhydride double molecules of lactic acid – lactoyllactic acid – together with longer chains of intermolecular esters are present. Faust probably used highly concentrated lactic acid, which can contain almost 100% of these lactoyllactic acids (Holten, 1971, Ch. XI).

[3] Boehringer called its edible grade of lactic acid Genussmilchsäure, which literally translated means enjoyable or delightful lactic acid. The translation which is closest to the German meaning of the word is edible lactic acid. The writer has chosen the expression table acid because of Hoffmann's intention to compare edible lactic acid

with table vinegar.

[4] The only example found of a small lactic acid producer is the dairy company Eisenharz at Isny (Ch. 5, 130). It is not unlikely that more dairies converted surplus whey into crude calcium lactate, which was worked up by pharmaceutical and other producers.

[5] Chemische Werke vorm. Dr. Heinrich Byk merged in 1917 with Guldenwerke Chemische Fabrik A.G. of Piesteritz near Wittenberg (Saxony, today G.D.R.) to form Byk-Guldenwerke A.G.

Guldenwerke operated a coal tar distillation plant at Piesteritz, and manufactured chemical intermediates, dyestuffs and tanning auxiliaries. The Byk company was engaged in the production of photographic paper, and of pharmaceuticals and chemical products for many applications (textile, agriculture, tanning, photography). The interest of the company in lactic acid is shown by a number of patents:

- DRP 267,826 (1912) on the manufacture of lactide
- DRP 271,643 (1912) on calcium trilactate for baking powder
- DRP 278,487 (1912) on the preparation of lactic and glycolic esters
- EP 25,182 (1913) on the use of ethyl lactate as lacquer solvent

[6] Sander on removing trace impurities (Ullmann, 1931):

Following the evaporation [or another concentrating method] the edible acid has to be freed from the salts of heavy metals – iron and lead –, ..., and from traces of arsenic and discoloring impurities. Lead and arsenic are precipitated as sulfides with barium sulfide. The barium lactate formed in this reaction combines with the (minor quantity of) gypsum present in the lactic acid (to form insoluble barium sulfate).... Iron is removed with potassium ferrocyanide [which also forms an insoluble compound with copper when present]; in the same process step discoloring impurities are removed with activated carbon. After filtering off the finely dispersed precipitates the clear, almost colorless, technically pure lactic acid [= edible lactic acid] is treated in enameled containers with compressed air until the last traces of dissolved hydrogen sulfide have been blown away. The concentration is adjusted with clear water. The product is shipped in carboys or in paraffin coated wooden barrels.

[7] Kommerzienrat, honorary title of distinguished entrepreneurs. The title was probably conferred to Herr Boehringer a few years later than 1906.

[8] This refers to the lactic acid plant of the American Alkali & Acid Company, see p. 197.

[9] Information from the Hagley Museum and Library at Wilmington, DE.

[10] From a list in "The Grasselli History of Plants and Properties" (Hagley Museum).

[11] Information on Emery and his enterprises has been received from Mrs. Candace Ekdahl of the Carnegie Public Library of Bradford. Additional information originates from Haynes (1945), volumes 2 and 3.

[12] Drysalters is a term used in the nineteenth century for dealers in dry salts, that is concentrated, dried chemicals or combinations of chemicals applied as auxiliaries in the textile industry.

References

Berkel, K. van, **1985**. *In het voetspoor van Stevin*; Boom: Meppel.

Beijerinck, M.W., **1901**. *Arch.Néerl.Sci. Exactes Nat.*,Sér.2, **6**, 212.

Burton, L.V., **1937**. *Food Ind.*, **9**, 634.

Claflin, A.A., **1946**. *Am. Dyest. Rep., Dec. 2*, 1946, 634.

Dammer, O., **1898**. *Handbuch der chemischen Technologie*; Ferdinand Enke: Stuttgart, Vol. 4, 530.

Dehio, W., **1966**. *Die Industrie der Weinsäure, Zitronensäure und Milchsäure*; internal publication of C.H. Boehringer Sohn: Ingelheim.

Diehl, W., **1974**. *Ingelheimer Chronik 1899–1950*; Paul Giese: Offenbach am Main.

Dreher, C., **1898**. *Färber Z.*, **1898**, 277.

Faust, E.S., **1910**. *Chem. Z.*, **34**, 57.

Fürth, and Engel, P., **1930**. *Biochem. Z.*, **229**, 381.

Glaubitz, M., **1925**. *Das Institut für Gärungsgewerbe und Stärkefabrikation zu Berlin.* Festschrift zur Feier des 50 jahriges Bestehens; Parey: Berlin, 152.

Gouthière, H., **1909**. *Bull. Ass. Chim. Sucr. Distill. Fr. Colon.*, **27**, 576.

Haynes, W., **1954**. *American Chemical Industry*; Vols. 1, 2 and 3; D. Van Nostrand: New York. Reprinted by Garland: New York, 1983.

Henneberg, W., **1904**. *Zentralbl. Bakteriol. Paratisenk. Infektionskrankh.*, Abt 2, **11**, 154.

Heusner, A., **197?**. *The history of various processes of C.H. Boehringer Sohn*, internal publication (in German) of C.H. Boehringer Sohn: Ingelheim.

Holten, C.H., **1971**. *Lactic Acid*; Verlag Chemie: Weinberg (FRG).

Hoffmann, W., **1913**. *Z. Spiritusind.*, **36**, 71.

Littleton, **1964**. *Town of Littleton, Massachusetts*, 250th Anniversary.

McLauchlan, W.H., **1909**. *Chem.Trade J.*, July 10, 28.

Molinary, E., **1921**. *Treatise on General and Industrial Organic Chemistry*; translated by Pope, T.H.; Blakiston's Son: Philadelphia, Part 1, 386.

Olive, T.R., **1936**. *Chem. Metall. Eng.*, **43**, 480.

Ost, H., **1900**. *Lehrbuch der chemischen Technologie*; 4th ed.; Kolbeck, F., Ed.; Gebrüder Jänecke: Hanover, 515.

Remenovski, E., **1929**. In *Handbuch der Organischen Warenkunde*; Victor Grafe, Ed.; C.E. Poeschel: Stuttgart, Vol. 3, 408.

Schäfer, G., **1907**. *Chemische Zeitschrift*, **6**, 177 and 189.

Stanier, R.Y., Adelberg, E.A., and Ingraham, J.L., **1976**. *General Microbiology*; 4th ed.; Prentice-Hall: Eaglewood Cliffs, NJ, 17.

Ullmann, F., Ed., **1916**. *Enzyklopädie der technischen Chemie*; 1st ed.; Urban und Schwarzenberg: Berlin, Vol 3, 570.

Ullmann, F., Ed., **1931**. *Enzyklopädie der technischen Chemie*; 2nd ed.; Urban und Schwarzenberg: Berlin, Vol 7, 583.

Wehmer, C., **1898**. *Z. Spiritusind.*, **21**, 342 and 350.

Wehmer, C., **1933**. *Chem. Ztg.*, **57**, 313.

TECHNOLOGICAL PROBLEMS IN MAKING EDIBLE LACTIC ACID

Lactic acid for leather manufacturing could be almost any kind of crude lactic acid containing material, provided heavy metals and in particular iron were absent. This was easily achieved by precipitation of insoluble sulfides or ferrocyanides with sodium sulfide and potassium ferrocyanide respectively.

For textile finishing and dyeing a better quality of lactic acid (and of lactates) was required, but specifications were still not too difficult to meet.

The real problems came when manufacturers attempted to substitute the main food acid tartaric acid or the less important citric acid in food and beverages. Lactic acid has a number of (physical) properties which make purification much more difficult than is the case with the two other food acids. The differences in properties "account" for the different technologies required to manufacture the pure acids from their respective crude raw materials. A summary of the technologies tried to purify lactic acid will help to understand the problems the manufacturers of lactic acid were facing in their efforts to produce an acid comparable in quality to that of the two other acids.

Purification by Means of the Calcium Salt

The industry made all three of the acids from the corresponding calcium salt by the use of procedures going back to the first discoverer of the acids, Carl Wilhelm Scheele.

Calcium tartrate was obtained from the extracts of argols and lees with mineral acid, followed by precipitation of the extracts with lime (Ch. 5, 127).

Calcium citrate resulted from the reaction of lime with citric acid contained in the juices from citric waste and cull lemons. Water was removed from the pressed juices by heating with direct steam until a concentration of about 10% was achieved. Thereupon a boiling hot solution of calcium hydroxide (3–5%) was added slowly to neutralize about 75% of the citric acid in the juice. The remaining acidity was taken away with powdered chalk. The hot suspension was passed over a sieve, which retained the precipitated crystalline calcium citrate. The crystals were washed, dried and packed (Ullmann, 1916).

Calcium lactate was prepared by fermentation from various carbohydrate containing raw materials. At the end of the fermentation the broth was treated to precipitate proteins and other contaminants and thereupon filtrated. Most of the soluble impurities present in the feedstock, as well as in the nutrients used in the fermentation process would, however, still be present in the clear filtrate, a solution of crude calcium lactate.

Precipitation of the Calcium Salts of the Food Acids

With all of the three salts the first step consisted in the precipitation of the calcium salt, either through the addition of milk of lime (with tartaric and citric acid) or through evaporation of excess of water (lactic acid). The purification achieved in this step will be determined by the solubility of the calcium salt and by its crystal form, see Table I4, I:

Table I4.I. Solubilities of the calcium salts of three food acids.

salt	solubility at 20° in g/100g of water	crystal form	water retained by the crystals (%)
Ca tartrate 4 aq	0.017	fine crystals	about 50
Ca citrate 4 aq	0.090	fine crystals	about 50
Ca lactate 5 aq	6.0	aggregates ("cauliflower buds")	about 200

Literature: Ullmann, 4th ed., 1985, headings Weinsäure, Zitronensäure and Milchsäure.
Data for the calcium salt of lactic acid refer to the racemic compound.

When solutions corresponding to salt concentrations of 20% are precipitated or crystallized the results are roughly as shown in Table I4, II.

In other words: after precipitation and one washing of the crystals of calcium tartrate and citrate the amount of contaminants has been reduced to 6 to 7% of the original, while the recovery of the calcium salt is almost 100%. This contrasts very much with the effect on calcium lactate which, after one wash still has retained more than 20% of the original percentage of im-

Table I4.II. Purification effect through crystallization of the calcium salt.

salt	recovery (%)		removal of impurities (%)	
	direct	after one wash	direct	after one wash
Ca tartrate	100	100	88	94
Ca citrate	98	98	87	93
Ca lactate	88	83	57	78

purities, whereas almost 20% of the calcium salt has been lost. The technology demanded to reach these results is another detrimental factor in the efficient production of calcium lactate: calcium tartrate and citrate can be separated from the liquid by centrifuging, with calcium lactate a filter press has to be used to press the water out. Moreover, washing in the filter press is impossible, the cake has to be removed from the press and kneaded with as little water as is practicable before it is filtrated again (the first "wash"). Calcium tartrate and citrate on the contrary can be simply washed in the centrifuge[1].

The difference in properties of the stereoisomers must have formed an extra complication for the early producers of lactic acid. Neither Boehringer nor Merck used pure cultures in the fermentation process, both relied on Delbrück's principle of "natural pure culturing" by using a high fermentation temperature of about 50°C. As the inoculant most of the time came from a previous well running batch, changes in enantiomeric composition could occur easily:

The bacterial culture of *Lactobacillus delbrueckii* principally yields a high content of the D(-) isomer; under the prevailing process conditions with temperatures of 50°C or higher and a high rate of infection, the forming of racemate or even of an excess of the L(+) isomer could be expected, however. (Intermezzo 3, 168) Merck made a study of the optical properties of its product in 1912, and found indeed notable differences. As Krauch wrote in his report about these investigations: only synthetic acid is racemic and inactive, "in every lactic acid fermentation a mixture of unequal part of d- and l- acid will be formed ...."

A further problem in the purification by means of calcium lactate crystallization is caused by the susceptibility to infections. Wet calcium lactate filter cake has to be processed within 24 hours, and all process equipment has to be cleaned thoroughly, to prevent rapid growth of fungi, particularly in the summer season. This problem is given by Boehringer's process chemist Wallach as one of the reasons why his company abandoned purification through crystallization of calcium lactate in the 1930's (Wallach, 1966)[2].

The Purification of the Three Food Acids

The next step in the preparation of all three of the acids is the conversion of the calcium salt with sulfuric acid, followed by the separation of the acid from the calcium sulfate formed in the double conversion. Further purification of tartaric and citric acid is achieved by crystallization. This is not practical with lactic acid which is too soluble in water and of which the melting point is too low, see Table I4, III.

Table I4.III. (Table III, Intermezzo 4)

Solubility and melting point of three food acids

acid	solubility (g/100g of water) 20 °C	40 °C	60 °C	80 °C	melting point (°C)	recovery in one step (%)
tartaric	140	180	220	270	170	50 – 60
citric	145	220	280	360	153	60 – 70
lactic (DL)	completely miscible				18	–
lactic (L+)	770	?	completely miscible		53	?

Literature: Ullmann's Enzyklopädie der technischen Chemie, 4th ed., 1985, tartaric (Vol. 24) and citric acid (Vol. 9); Holten, 1971, lactic acid.

Purification by means of crystallization depends on the condition that the solubility of the product to be purified is strongly temperature dependent, that is relatively high at a high temperature and much lower at a low temperature.

Table I4, III shows that this is the case with tartaric and citric acid, but not with lactic acid.

When purification by crystallization is next to impossible (see on this point, however, Intermezzo 7, 343) purification has to be performed in the liquid state, or by distillation. In the liquid state this was achieved by chemical precipitation of the main impurities and contaminants; Exhibit 5, IV, 147 presents some of the means of doing this. Some impurities are difficult to precipitate in an economically acceptable way, hence other methods have been proposed like extraction with ether, or distillation of a lactate ester.

An essential part of most purification processes and of all finishing operations of lactic acid is removal of excess water to adjust the concentration of the final product. Water removal has to be carried out by evaporation at fairly high temperatures and this poses an additional problem due to the tendency of lactic acid to react with itself (see below).

Another major problem confronting lactic acid manufacturers was corrosion. All acids cause some degree of corrosion of metal equipment, but lactic is one of the strongest corroding acids. This is especially the case at elevated temperatures, which are required to remove excess water by distillation from the dilute solution obtained at the end of the purification process (Ch. 9, 320).

Interesterification; Lactic Acid Disturbs Its Own Purification

Pelouze had already discovered in 1845 that lactic acid, when water was removed from its solution at a temperature of 130°C, formed a syrupy liquid with a composition containing less water than lactic acid itself (Ch. 1, I1, 64).

Written in present day formulas Pelouze observed the reaction: 2 $C_3H_6O_3$ $C_6H_{10}O_5 + H_2O$, or in structural formulas:

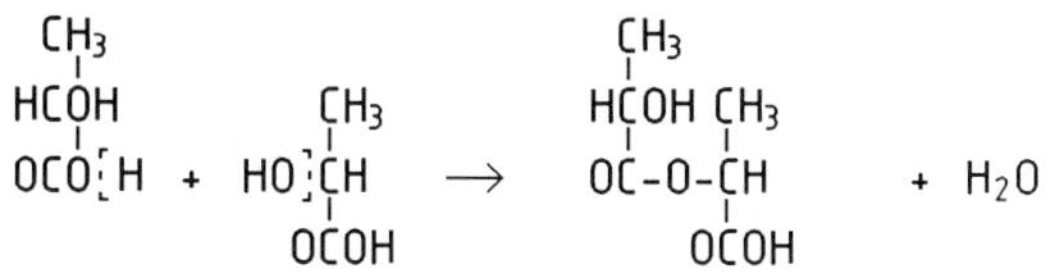

(2 molecules of lactic acid lose one molecule of water to form one molecule of lactoyllactic acid)

As a matter of fact the reaction does not stop at this stage but goes on, forming combinations of 3, 4, etc. molecules of lactic acid called trimers, tetramers, etc. When the reaction proceeds still further polymeric molecules are formed in which large numbers of monomeric molecules (of lactic acid moities) are combined:

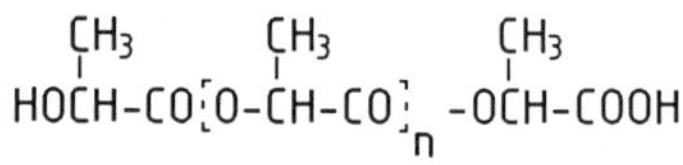

Later work has shown that this reaction always takes place when the lactic acid concentration in the aqueous solution surpasses about 50%. At temperatures below 50° the reaction is very slow, however, it takes months before equilibrium between monomeric lactic acid and higher esterification products is attained. At temperatures closer to the boiling point of water the reaction is more rapid and it becomes a matter of tens of minutes.

The forming of double molecules, etc. of lactic acid at concentrations about 50% is an unpleasant phenomenon, because in many purification processes these intermolecular esters act as contaminants. They always have a negative influence on the yield, and in some cases also obstruct the purification process itself.

What Purification Processes Existed for Lactic Acid?

1. The Bensch process in which purification is effectuated by means of **crystallization of zinc lactate** (Ch. 2, 38). Table I4, IV gives the solubility data. The process is quite elaborate, but gives good results. Merck used the method followed by ether extraction from about 1894 to 1939 to manufacture pharmaceutical lactic acid (Figure I4,1 and Ch. 5, 133).

Table I4.IV. Solubilities of zinc lactate (g/100g water).

solubility	racemic product	stereoisomer
at room temperature	1.7	5
at boiling temperature	17	27
literature	Engelhardt and Maddrell, 1847	private information from CCA biochem, 1987

Fig. 14,1. Zinc lactate centrifuge at Merck, 1937. Courtesy: E. Merck, Darmstadt.

2. **Crystallization of calcium lactate**. As already explained this is a poor purification method for racemic lactate. (In Chs. 6, 181 and 9, 315, it will be shown that Sheffield nevertheless established an acceptable purification process based on crystallization for the manufacture of calcium lactate itself.)

Enantiomeric calcium lactate which exhibits still higher solubilities is even more difficult to purify by crystallization.

3. **Steam distillation**. This method, first proposed by Waite in 1884 (Ch. 4, 96) is highly impractical unless special equipment is used and temperatures above 160°C are applied.

Schäfer (Schäfer, 1907):

The distilled lactic acid has a concentration of 9–10%, to bring it back to the desired commercial concentration water has to be evaporated again for which enameled vacuum pans have to be used, and finally it is treated with blood coal. It usually has a faint burnt smell. The process is not really satisfactory because of high losses, due to decomposition into lactic anhydride [that is interesterification] ..., often up to 20–30% of the starting material.

- Schäfer, 1907

In the 1930's Boehringer developed a more efficient vacuum steam distillation method, see Intermezzo 7.

4. **Distillation by means of a forced stream of gas**. In a patent of Chemische Fabrik Flörsheim, Dr. H. Noerdlinger, of Flörsheim near Frankfurt am Main, Germany, it is claimed that a good purification can be achieved by blowing air or inert gas through a 60% solution of crude lactic acid (German Patent 221,786, 1910). First, at temperatures between 40 and 80°C, most of the water is removed, and, after heating to 120 to 130°C, pure lactic acid distils over as a water clear, syrupy liquid. This liquid solidifies almost completely after cooling to -10°C for a prolonged period of time. The melting point of the crystalline lactic acid so obtained is 18°C. [This is the melting point of racemic lactic acid.] The method is a variant of process 3 in which steam is replaced by (another) gas. The yield of this process supposedly was also rather low (Ullmann, 1931).

5. **Extraction with ether**. Extraction with diethyl ether was a popular method with chemists in the nineteenth century. A mixture of ether and alcohol had been used by Liebig to extract sarcolactic acid from meat (Liebig, 1847). Kolbe was the first to mention the use of ether as a suitable solvent in the purification of lactic acid (Kolbe, 1860).

Schäfer in 1907:

Concentrated lactic acid was placed in a glass vessel with a three fold volume of sulfuric ether [diethyl ether] and shaken for 24 hours. Contaminating material adheres to the walls of the vessel, the clear etherish solution is pressed with air to a second glass vessel, in which it is shaken with very pure decolorizing carbon.

- Schäfer, 1907

Thereupon it is filtrated into a distilling flask and the ether removed by distillation. Sometimes the procedure has to be repeated to obtain a very pure lactic acid. (It will be evident that ether soluble impurities like some amino acids are difficult to separate completely from lactic acid by this method.)

Charles N. Waite patented in 1901 a counter-current extraction method in which amylalcohol was the extracting solvent (German Patent 140,319).

6. **Crystallization of the anilin salt**. Blumenthal and Chain of Berlin obtained in 1904 a patent for a method based on the formation of an insoluble salt with the weak organic base aniline (German Patent 169,992). They added an equimolar quantity of aniline to a 60% solution of lactic acid, which, after

cooling to -5°C gave a precipitate of aniline lactate in a yield of almost 100%. The aniline lactate was filtered off, recrystallized from a small amount of water, and treated with steam to decompose the salt and completely remove the aniline. The lactic acid remaining behind conforms to the German pharmacopoeia (DAB 4).

Krauch tried to imitate the process but did not succeed (Merck's Manufacturing Report 1905/1906).

7. **Distillation of ethyl lactate.** Chemische Fabrik Güstrow, Dr Hillringhaus, of Güstrow, Mecklenburg, Germany, invented a new method for the preparation of ethyl lactate (German Patent 171,835, granted in 1906): Crude dried calcium lactate was admixed with an equivalent quantity of concentrated sulfuric acid and a molar excess of alcohol. The mixture was heated to about 50°C, the liquid ethyl lactate formed in the reaction separated from the gypsum, and purified by vacuum distillation. The pure ethyl lactate was thereupon split again into lactic acid and ethyl alcohol by heating with water at boiling temperature. The alcohol was distilled off, the remaining lactic acid was of pharmaceutical quality.

This process was improved in a patent of Chemische Werke, vorm. Dr Heinrich Byk of Lehnitz, a village near Oranienburg, north of Berlin (German Patent 278,487, 1912). Byk first polymerized lactic acid at temperatures as high as 170–180°C. The resulting polymeric mass was boiled under reflux with excess anhydrous ethylalcohol and a few percent of a mineral acid for a period of up to 20 hours. Alcohol and ethyl lactate were distilled off under vacuum and thereupon purified by fractional distillation.

The **concluding remark on these purification methods** is left to Schäfer (l.c., 1907):

> It is evident that the problem to produce a chemically pure lactic acid in a cheap and reliable manner has not yet been solved. All of the processes (mentioned) are either too expensive or, for other reasons, not suitable for industrial manufacture. As a consequence the pure acid is 5 to 6 times as expensive as the technical grade.
>
> - *Schäfer, 1907*

Notes

[1] The stereoisomers of calcium lactate are still more difficult to purify than the racemic salt, the solubility at 20°is 9 g/100 g of water, which means that removal of impurities is even more inefficient than with racemic calcium lactate, and the losses increase from 20 to about 25%.

[2] The susceptibility of calcium lactate to fungous infection is also a problem of the

finished product. This manifests itself in particular when one or more of the following conditions are met:

a. when the temperature is on the high side,
b. when proteinaceous or other contaminants are present serving as nutrients for infecting microorganisms,
c. when the water content is near or above the equilibrium content of the hydrate.

The water content deserves some special attention:
the racemic product contains 5 molecules of water of crystallization, corresponding to 29.2% of water; enantiomeric calcium lactate, however, binds only 4.5 molecules, corresponding to 27.1%. Most pharmacopoeias specify losses on drying from about 26 to 30%. This means that commercial enantiomeric calcium lactate within this range is more apt to contain free water than the racemic form, making it even more vulnerable to infection (Literature: Holten, 1971). In Holten's book some confusion exists regarding the exact number of molecules of water of crystallization; recent thermodynamic studies at Utrecht University have confirmed the figure of 4.5 already found by Wislicenus in 1873 (van der Kemp and van Genderen, 1988).

References

Engelhardt, H., and Maddrell, R., **1847**. *Ann. Chem.*, **63**, 83.
Holten, C.H., **1971.** *Lactic Acid*; Verlag Chemie: Weinheim, F.R.G.
Kemp, W. van der, and Genderen, A.C.G. van, **1988**. Unpublished results (University of Utrecht, Netherlands).
Kolbe, H., **1860**. *Ann. Chem.*, **113**, 242.
Liebig, J., **1847**. *Ann. Chem.*, **62**, 257.
Schäfer, G., **1907**. *Chemische Zeitschr.*, **6**, 177 and 189.
Ullmann, F., Ed., **1926**. *Enzyklopädie der technischen Chemie*; 1st ed.; Urban & Schwarzenberg: Berlin, Vol. 3, 570.
Ullmann, F., Ed., **1931**. *Enzyklopädie der technischen Chemie*; 1st ed.; Urban & Schwarzenberg: Vol. 7, 583.
Wallach, B., **1966**. *Internal Report on Lactic Acid and Lactate Manufacturing* (in German); C.H. Boehringer Sohn, Ingelheim.

CHAPTER 7

LACTIC ACID AND OTHER FERMENTATION INDUSTRIES IN WORLD WAR I

The Great War (August 1914 – November 1918), as the contempories called it, came quite unexpectedly. Strangely enough it was met with great enthusiasm by the majority of the peoples of the great European nations. Astonishing as this reaction might seem to the present generation, it is emotionally understandable. The economic progress of the era and the increased wealth this had brought to the Western European peoples together with the extension of democratic principles to large parts of the population had created intense nationalistic feelings in the greater nations: Pride in their "own" achievements, anxiety about limits to the developments imposed on their own country by actions of potential adversaries (for instance the division of Africa into colonies) had fostered nationalism, particularly in Germany, Britain and France. Moreover, after four decades of peace in Western Europe reminiscences of the wars of Bismarck and Napoleon III had been reduced to folklore. And those wars had by no means been as cruel and devastating to the population as a whole as, for instance, the Civil War in the United States (1861–1865) or, more recently, the Boer Wars in South Africa.

Quite naturally the soldiers who marched to the frontiers in August 1914 also expected a short war, to them the superiority of the own Nation was almost a guarantee of quick success, and so thought the general staffs of the belligerents. The best prepared undoubtedly was the German Army. The military planners of the German General Staff also only had a war of short duration in mind, the whole German war machinery was aimed at a rapid victory in the West, followed by an impressive campaign in the East to effect the defeat of the Russians not much later.

Germany Was Leading in Chemistry and Chemical Engineering

In the decades before the War Germany had become the leading chemical nation; not only in academic and industrial research, but also in production and application, in textile dyes and drugs and in numerous other chemicals. Almost no industry anywhere in the world could do without the use of at least a few essential chemicals from Germany. Only in the production of heavy chemicals, in particular of inorganics, was Great Britain still leading.

Soon after the outbreak of the War the British declared a total blockade of Germany. This meant that shipping of German chemicals came to a complete standstill almost overnight . Soon, however, Dutch and Italian harbors took over the shipping, and, in the fall of 1914, shipments of chemicals to the U.S.A. reached even higher levels than ever before. The neutral vessels carrying these goods brought back essential raw materials and foodstuffs Germany required for the war operations and to feed its population.

The British then extended the blockade rules, to which the Germans responded by declaring a total submarine war: any merchant vessel, British or neutral, in the waters around the British Isles would be sunk without warning. As a consequence the supply position of many American enterprises deteriorated rapidly in 1915, and prices soared to unknown heights. Severest hit were the U.S. textile and pharmaceutical industries.

The Role of Chemists and Chemistry in the War

What no general or politician had envisaged was the crucial role chemicals were to play in the war. Of course, the inventions of Nobel in developing dynamite for peaceful as well as for military uses had been recognized by the military all over, but the importance of the science of chemistry and the need to mobilize chemists to solve essential procurement problems for the military was recognized only when the soldiers had already dug their trenches along most of Northern France.

Bäumler (Bäumler, 1968), in "A Century of Chemistry", a history of the Hoechst company, sketches the situation in Germany:

> ... neither Government nor army had paid much attention to the significance of adequate industrial capacity. Not that the self-confidence of the army was in any way jolted by this. Walther Rathenau and the highly respected Professor Emil Fischer were sternly rebuffed when, after the start of the war, they drew attention to the bottleneck in the nitrate supplies. This, the High Command declared brusquely, represented intolerable interference in military affairs.
>
> The conferences in the ministry of war revealed clearly how ignorant most military leaders were even in matters of elementary chemistry. When the chiefs of the large chemical companies pointed out on such occasions that Germany would be finished once the supplies of Chile nitrate came to an end, they were asked, in all innocence: 'But what about our high potash deposits?' The fine difference between nitrate and potash had not impressed itself on the brains of the ministry of war.
>
> It was not until the autumn of 1914 that the general staff in Berlin had caught up with its homework in chemistry and economics. But then the pace became military.

The raw materials department of the ministry of war imposed its requirements on the chemical industry.

- *Bäumler, 1968*

The first concern of the High Command was of course the procurement of explosives. With the supply of nitric acid needed to make nitrocellulose and nitroglycerine the staff had all the luck of the world on its side: The fixation of nitrogen from air had been achieved by Professor Fritz Haber only a few years before, in 1908. Haber had presented his process for commercialization to the BASF company at Ludwigshafen, whose director Carl Bosch immediately recognized the merits of the process and the large commercial possibilities of the product ammonia as artifical fertilizer and raw material for the production of nitric acid. Tremendous technological problems had to be solved as a consequence of the extreme reaction conditions of the process (500°C and a pressure of 100 to 200 atmospheres [10,000 to 20,000 kPa]). The first successful pilot plant production took place in 1910, a small commercial plant with a capacity of 7,000 t/a came on stream in 1913.

When the war came along almost the entire construction work force of the chemical industry was enlisted and sent to the front, however. Only a few months later the construction workers and engineers of BASF were ordered to go back to the company to start the building of new installations for the production of ammonia. At the end of the war a whole new chemical industry applying high pressure technology had come into being:

A big factory at Oppau (near Ludwigshafen) with a capacity of 100,000 t/a, and a huge, completely new industrial complex located next to large lignite fields at Leuna (near Merseburg) in Saxonia with a capacity of 250,000 t/a.

The War of 1914–1918, a War of Chemicals

A very readable survey of the chemical aspects of World War I was written in 1928 by the former president of the Avery Chemical Company, Alan A. Claflin (Claflin, 1928). Claflin gives much information about the different types of explosives used, and about the chemistry involved. Of course he also tells the story of the development of the Haber-Bosch process and its crucial importance for the German war efforts. He particularly stresses the fact that the engineers of BASF had to devise completely new steel alloys to resist the pressure at the high temperatures of the reaction. Claflin's general conclusion is that the German High Command made well-organized use of the vast potential of the German chemical industry and its large number of able and well trained chemists.

The Fermentation Industry Supplies Glycerol

Another product of crucial importance in the manufacture of explosives was glycerol. Until 1914 the bulk of glycerol was supplied by the soap industry. The amounts required for the production of explosives far exceeded those of prewar times, however. For its supply of oils and fats Germany had relied in peace time to a large extent on imports from overseas countries, imports which due to the British blockade soon dropped to almost nil. Here again German scientists stepped in to help their country in its war effort. The inventors of a process to make glycerol by fermentation, Connstein and Lüdecke gave the following account of what happened:

> The process was a war baby, born under the stress of the time. For the initiated it was evident that the immense reduction in the supply of Germany with technical fat would, as a consequence of the blockade, in a very short time result in a serious shortage of glycerol. The minor measures the government had taken [in the form of confiscation, etc.] would not be appropriate to supply the "Heeresverwaltung" [the army supply corps] with the minimum amounts of glycerol required to manufacture powder and explosives. Only the discovery of a new source of glycerol could solve this problem, and such a source was found at just the moment the need became felt seriously.
>
> - *Connstein and Lüdecke, quoted from Deite and Kellner, 1923*

Connstein and Lüdecke, who were employed by the "Vereinigte Chemische Werke" of Charlottenburg (Berlin), knew that in the normal alcoholic fermentation minor amounts of succinic acid and glycerol were formed. They reasoned that changing the acidity at which the fermentation was carried out might influence the metabolism of the process and might lead to the genesis of other (by-)products. This proved to be true, under alkaline conditions less carbonic acid (CO_2) was formed and the amount of alcohol also strongly decreased. Instead acetaldehyde and glycerol were obtained in sizeable quantities. Of the substances used to adjust the alkalinity sodium sulfite proved to be the most effective, giving the highest yields of acetaldehyde and glycerol. Connstein and Lüdecke also found that the yield of these compounds was almost proportional to the sulfite concentration applied (Connstein and Lüdecke, 1919).

In the final recipe 40% of sodium sulfite was added, and an amount of yeast of 10%, both calculated on the amount of sugar. The fermentation was carried out at 30°C and took only about 36 hours. Production of yeast did not take place, however, and after a few batches fresh yeast had to be supplied to keep the fermentation going. At the end of the reaction alcohol and acetaldehyde were distilled off, sulfite precipitated, the filtrate separated from the sulfite, concentrated and again filtrated to obtain crude glycerol. The crude glycerol finally was subjected to vacuum distillation to produce glycerol of dynamite quality. In view of all these steps it is not surprising that the overall yield of the process was rather low (Deite and Kellner, 1923): In well run factories the fermentation of 1,000 kg of sugar resulted in

230 – 247 kg of glycerol
373 – 420 kg of carbon dioxide
234 – 300 kg of alcohol, and
50 – 127 kg of acetaldehyde

In the purification and distillation steps about 40% of the glycerol was lost, hence the total yield was only about 140–160 kg of glycerol per 1,000 kg of sugar.

In April 1915 the Vereinigte Chemische Werke applied for patents on the new process, and, within days, reported the invention to the Heeresverwaltung. The army immediately showed great interest and made arrangements for large scale trials. When these trials showed the feasibility of the fermentative production of glycerol a new company, the "Protol GmbH" was founded to organize full scale commercial production[1]. At first the glycerol fermentation step was performed in shut down distilleries, the purification in existing glycerol plants. Soon steps were taken to rationalize production and integrated Protol factories were built. The output of the Protol enterprise was increased step by step as the war went on; at the end of the war a monthly production level of 1,100 t glycerol had been reached. The fermenters built were the largest constructed until that time with maximum contents of 600 m^3.

Deite and Kellner explain that the amounts of sugar used did not affect the food situation too seriously:

> Calculated per head of the population the glycerol production took about 6 kg of sugar per year, which means about 16 to 17% of the average prewar sugar consumption. This use of sugar could not be the main reason of the sugar shortage during the war, which was caused to a much larger degree by the decrease in acreage and the smaller crop yield as a consequence of fertilizer shortage. [Prewar German sugar consumption of about 25 kg per capita dropped to about 17 kg in 1918, of which 6 kg was allocated for glycerol production; only the 1914/1915 consumption was at the level Deite and Kellner indicated.]
>
> *- Deite and Kellner, 1923*

The Kaiser Wilhelm Institute Is Mobilized

A few years before the war the German Emperor Wilhelm II had taken the initiative to found a private learned society devoted to the study of science and medicine and other subjects of particular interest for the German Nation. The background was not only to promote special scientific work of high standard, but also to free some of the best qualified university professors or potential professors from the burdens of teaching. Completely new research institutes were built by the Kaiser Wilhelm Society in rapid progession, and quite a few were already noted for the quality of their staff at the outbreak of the war. After the beginning of the War of the Trenches the Institutes were also called in to help the German war effort.

One of these institutes was the Kaiser Wilhelm Institute for Experimental Therapy at Berlin-Dahlem. The chemical department of this institute was headed by Dr. Carl Alexander Neuberg (1877–1956), who was engaged in the study of the biochemistry of the alcoholic fermentation[2,3].

Neuberg was convinced that one of the intermediate products in the fermentative conversion of sugar into alcohol was acetaldehyde. In 1911 he published the results of his investigations, which showed that when the fermentation with yeast was conducted under alkaline conditions large quantities of acetaldehyde were formed. He also found that the addition of sulfite which under alkaline fermentation conditions reacts almost quantitatively with acetaldehyde, would decrease the conversion into alcohol in the same ratio as the aldehyde was bound. Neuberg considered these experimental results as conclusive proof of the existence of acetaldehyde as an intermediate in the alcoholic fermentation.

In further studies of the fermentation process in the presence of sulfite, Neuberg and Reinfurth discovered that together with acetaldehyde large amounts of glycerol were formed. When the results of the experiments of Connstein and Lübecke became known to the army staff, Neuberg and coworkers were asked to optimize the fermentative production of glycerol.

The Glycerol Committee of the Heeresverwaltung headed by the eminent chemist Emil Fischer wanted to play more than one card in the supply of glycerol, however, and initiated a search for substitutes of glycerol for other applications than the manufacture of explosives. Quite naturally Neuberg's laboratory was called upon again to search for glycerol substitutes.

Sodium and Potassium Lactate as Glycerol Substitutes

Probably in 1915, Neuberg came forward with a proposal to use the alkali salts of lactic acid as substitutes of glycerol in applications in which only the physical properties of glycerol were of importance. To follow Neuberg's own account (Neuberg and Reinfurth, 1920):

> Under normal conditions [solutions of] both salts (that is the sodium and the potassium salt of lactic acid) are viscous, colorless compounds with a salty, not unpleasant taste, good heat resistance and a hygroscopic character; as they only crystallize when very pure, they usually form supersaturated solutions completely miscible with water and alcohol. Being as innocuous and non-irritating as glycerol no objection against their use for technical, cosmetic and medicinal purposes could be raised. The economic importance of the application of lactates has particular merits because in their production, even when based on sugar, a much higher yield can be achieved than in the production of fermentation glycerol. In the fermentation of sugar to lactic acid a yield of 95% or more is possible! Moreover the conversion into salts ... can be effectuated with cheap inorganic compounds such as soda and potash. Naturally other raw materials can be applied instead of sugar, like waste starch, wood sugar, chestnut flour, etc. Consequently the "mobilization" of lactates was also in the interest of the food supply.
>
> *- Neuberg and Reinfurth, 1920*

The particular uses suggested to the Heeresverwaltung were: brake fluids for artillery guns, cooling liquids for machine guns and heat transfer media for field kitchens:

> When our experiments at the chemical department of the Kaiser-Wilhelm-Institute for Experimental Therapy had shown the usefulness of the alkali lactates for the intended applications, the material was presented to the competent military officers, who in turn informed the appropriate department, ... [In due time] ... the War Ministry was requested to replace the now used peace glycerol in the recoil brakes and pneumatic pumps of all guns of the field and infantry artillery by a 45% solution of sodium lactate. Through a number of directives from the War Ministry the introduction [of this solution] for the German guns of the field and infantry artillery and for the trench-mortars was implemented. For these purposes the material has been used until the last war action. It has also been used for the guns of the navy and for the field kitchens. The armies of the States allied with Germany applied the substances in a similar manner. To establish the relation with the familiar glycerol [Glycerin in German] the sodium lactate (solution) received the name Per-Glycerin and potassium lactate Perka-Glycerin.
>
> *- Neuberg and Reinfurth, 1920*

Neuberg and Reinfurth continue with a calculation of the financial advantages of the use of a 45% solution of alkali lactate over a 60% solution of glycerol. The German War Ministry saved about 90 million Marks by the use of the lactates:

The total amount of lactate salts produced in three years for military purposes can be estimated at 9,000 t, corresponding with 3,250 t lactic acid (100%). The production of the material was carried out "in an exemplary manner" by Chemische Fabrik vorm. Goldenberg, Geromont & Co. at Winkel am Rhein, in collaboration with C.H. Boehringer Sohn, Nieder-Ingelheim and Byk, Biebrich (near Wiesbaden). Emil Fischer also arranged for the testing of the lactate salts for pharmaceutical and cosmetic purposes. Interesting results were obtained in dermatology and in the conservation of human organs (for the German War-pathological Collection).

As Sander wrote in the 2nd edition of Ullmann's Encyclopedia (Vol. VI, 1931, 288) the normal quality of technical lactic acid was not suitable for the production of lactate salts, the material charred and formed carbonaceous deposits. The lactic acid, therefore, had first to be purified by means of extraction with diethyl ether (See Intermezzo 4, 206).

German Lactic Acid Production in World War I

At the beginning of the war the supply of raw materials for the production of the other food acids was restricted to imports from Italy only. In May 1915 Italy, although bound by a treaty to Austria-Hungary and Germany, declared war on Austria-Hungary and, as a consequence, also found itself at war with Germany. Trade with the Reich came to a standstill. This included the exports of argols and lees for tartaric acid production and of calcium citrate for the manufacture of citric acid. The food and beverage industry of Germany now turned to lactic acid as a replacement. The Reichsgesundheitamt permitted the use of lactic acid for food purposes in 1915, and a modest start was made in substituting tartaric and citric acid by edible lactic acid.

This situation did not last very long. Only a short time later (1916) when the military recognized the importance of the use of lactate salt solutions in brake fluids and as heat-transfer liquids, the sale of lactic acid for non-military purposes was brought under government control:

> The production of lactic acid was placed under the supervision of the Kriegschemikalien A.G. The production was allocated to the preparation of glycerol substitute and the manufacture of high boiling solvents for the airplane lacquer industry [See Ch. 8, 246]. Waste lactic acid was used in leather manufacture.
>
> *- Ungewitter, 1927*

Lemonade producers were advised to acidify their syrups with vinegar or acetic acid (Luhmann, 1917).

From 1916 onward lactic acid was in short supply. Merck's 1916 Yearly Manufacturing Report gave this appraisal of the situation:

> In the beginning of the year [1916] the demand [of pure lactic acid] was small; in the spring it rose to high levels, however. Due to the shortage of raw materials supply became more difficult towards the end of the year.
>
> - *Krauch, 1917*

In the same year Merck began with small scale production of Perglycerin and Perkaglycerin.

The production of pharmaceutical lactic acid by Merck remained almost at the prewar level in spite of the severance of exports by the British blockade, and the almost complete halt in exports after Japan and the U.S.A. had entered the war in 1917.

During the war Boehringer, as well as Byk-Gulden and Knab & Lindenhayn increased their production capacity considerably to meet the demands of the military. Dr. Gruschwitz, plant manager of E. Merck's lactic acid factory complained in the 1920 Yearly Manufacturing Report about the expansion of the lactic acid factories of these competitors:

Umbau der Milchsäurefabrik der Firma C. H. Boehringer Sohn in Nieder-Ingelheim

Fig. 7,1. Renovation of Boehringer's lactic acid plant in 1916. Courtesy: C.H. Boehringer Sohn, Ingelheim.

Unfortunately factory VIII [Merck's lactic acid plant] is rather obsolete. Furthermore we did not keep pace with the competitors, who expanded their productions during the war and replaced old equipment by large modern installations.

The lactic acid factory of the Boehringer company was extended and modernized in 1916 (Diehl, 1974), a very obvious sign of the importance the government attached to the production of lactic acid (Figure 7,1). All building activities as well as the allocation of investment materials and personnel during the War were strictly controlled by Berlin!

Albert Boehringer, who served in the army, had entrusted the running of the daily affairs of his company to his nephew, Robert Boehringer. When on leave in Nieder-Ingelheim he used the opportunity to meet with his staff to discuss the latest developments. In such a meeting, on August 18, 1917, the quality of the edible and pharmaceutical grades of lactic acid were discussed. "The customers prefer the acid from Knab und Lindenhayn", particularly with respect to the odor. Evidently some lactic acid was still allocated to the food industry. The meeting decided to do something about the quality, and to increase the production with 300 t.

How Did France and Britain React When Imports of German Lactic Acid Stopped?

FRANCE

In the last five years before the war German exports to France had reached an average of almost 300 t/a. These exports fell to a much lower level during the war, but did not stop entirely, see Table 7,I. The average price of the imported material rose to about 2.5 times the prewar price, which was possibly due to the roundabout way the merchandise was shipped (via Switzerland or Holland?). Another factor could be that only the more expensive grades were bought.

A French source about the replacement of German exports:

[The] production [of lactic acid] was taken up in France very recently. It is expected that this manufacture will reach a position to get rid of German imports [once the war is over].

- *Grandmougin, 1917*

Table 7.I. French imports of lactic acid in tons.

	1913	1914	1915	1916
lactic acid	255	156	72	89
iron salts	0.6	0.6	0.1	0.1
other salts	13	7	2	0.2

Source: Grandmougin, 1917.

A factor to consider is the drop out of the small factory of Gouthière at Reims when that city was overrun by the German armies on September 4, 1914. Although Reims was reconquered in the famous Battle of the Marne on September 16, the Germans were able to hold a number of strategic heights to the North of the city. From these positions Reims was heavily bombarded many times, almost until the last days of the war. The splendid Gothic cathedral was converted into a ruin, what happened to Gouthière's plant the writer does not know.

Many industries from the occupied Northern part of France which had manufactured products of military importance were quickly rebuilt on a larger and more modern scale in safer regions of the country (Haber, 1971). One such region was the Rouen region (Haute Normande). Here, at Petit Quevilly, new chemical factories were erected. The lactic acid production Grandmougin had in mind was probably coming from a new plant of the Société Normande de Produits Chimiques at that site.

GREAT BRITAIN

Lactic acid was not made in England on a commercial scale when the war began. The use of lactic acid had reached considerable proportions, however, in particular in the tanning industry. While imports from Germany had only begun in 1906 with a mere 173 t, by 1913 the total volume had already risen to 764 t.

It was the firm of Bowmans (Warrington) Ltd., which took up the wartime challenge. Bowmans was a firm specialized in tanning chemicals. The company had been founded in 1905 by a Glasgow chemist, Robert Bowman, to produce various chemicals for the brewing and tanning industry. The first location of the company was at Warrington, about halfway between Manchester and Liverpool.

In 1915 Bowman left the works at Warrington and took over a building at Moss Bank, Widnes. The site was well chosen, in the neighborhood of large chemical works of what later was to become the Imperial Chemical Industries. It was on the bank of the Mersey and near the harbor of Liverpool. A brochure issued by the company in 1955 to commemorate its 50th anniversary gave a short account of the first trials to produce lactic acid:

> It was [at Widnes] on October 25th, 1915, that Robert Bowman carried out his first attempt to make lactic acid. The account of his method makes interesting reading compared with the processes of today. The ingredients included molasses, 2 cwts [90.72 kg] of yeast, and 3 gallons [13.6 l] 'of well-turned milk'; the milk, it is said, was bought at a nearby farm. With the mixture safely in the vat, Robert Bowman waited. After ten days it is recorded triumphantly there was 'slight working, bubbles rising' – today the same stage is reached in an hour.

Very likely Bowman acted on demand from his customers in the leather trade to supply lactic acid, a product they had previously obtained from Germany, and still earlier maybe from America.

Production of lactic acid was not an easy job the jubilee book acknowledges: "filtering was slow and concentration difficult." Judging from German export figures in the 1920's Bowmans could have succeeded in replacing by then between 300 and 500 t/a of imports.

Acetone-Butanol Fermentation: A British War Baby

At the outbreak of the war the British economy depended to a very large extent on the unrestricted supply of raw materials, foodstuffs and semi-manufactured goods from overseas. A large supplier of many kinds of commodities was, of course, the United States. Shipments from America to the British Isles increased considerably in the first months of the war, causing scarcity and mounting prices in the U.S.

The scarcity of one particular chemical – acetone – would lead to the development of an entirely new type of industrial fermentation. The advanced technology needed in this process would change the perspective of fermentation products for years, and for that reason alone already merits a short treatment. The application of the products of the new fermentation – acetone and butanol – as solvents would, after the war, have a direct influence on the lactic acid industry; another reason to take a closer look at what happened during the war.

Acetone was used as a solvent in the manufacture of gelatinous dynamite, a gel of nitrocellulose in nitroglycerol. Although the solvent was in principle recovered in the course of the manufacturing process, much was lost due to its high volatility. Outside Germany, where the production of synthetic acetone at the outbreak of the war was in its trial phase, the product was only manufactured from wood. Originally it had been extracted from crude "wood spirits", but at the end of the nineteenth century dry distillation of calcium acetate, also a wood chemical, had become the most important manufacturing procedure. (The German industry succeeded during the war in totally converting to synthetic acetone, based on acetylene chemistry.)

The total annual requirements of acetone for the British war efforts soon reached a level of 20,000 t, five times the total production of the American wood industry in 1914. Haynes:

> To this utterly exorbitant demand was later added a brand-new use in the preparation of so-called airplane dope, a variety of sundry cellulose lacquers for coating the canvas covering the wings of the airplanes....
>
> - *Haynes, 1945, Vol. 3, 135*

Of the many methods tried to increase acetone production the one that goes back to Dr Chaim Weizmann (1874–1954) was the most successful. Weizmann was born in Russia and had studied chemistry in Germany and Switzerland. In 1904 he obtained a research position with the also Russian-born I.L. Kondakov, Professor at the University of Manchester. Kondakov was investigating the production of a rubber substitute from butadiene, and Weizmann was recruited to assist in this work.

It was known that rubber was a polymer of isoprene, so research had been initiated to find a method for the synthetic production of isoprene or of a related polymerisable substance (with two double bonds in a conjugated position). Such a substance was butadiene, for which butanol was a suitable precursor. The butadiene route did not directly solve the problem of how to produce rubber at acceptable costs and in large quantities because butanol was only available in very limited quantities.

In 1911 Fernbach of the Institut Pasteur had discovered a fermentation method by which amyl alcohol, normal butanol, and acetone could be obtained by the fermentation of potatoes. The English firm of Strange and Graham Ltd., which had engaged Fernbach for these investigations, also had contacted Weizmann to work on this subject. Weizmann soon separated himself from Strange and Graham and independently took up a very thorough study of microbiology and fermentation processes, particularly aimed at the production of butanol or isoamyl alcohol. In 1912 he isolated a new bacterium, *Clostridium acetobutylicum*, which could ferment grain and, moreover, gave a much better yield of butanol.

In the fall of 1914 the British Admiralty became aware that a severe shortage of acetone was imminent and might endanger the production of the explosive Cordite (smokeless powder, a British brand of dynamite). As a possible source of acetone their attention was called to the fermentation process developed by Fernbach, but efforts to put this process into practice failed. The Admiralty then, in 1916, decided to ask Weizmann to develop his process to an industrial scale. After successful demonstrations Weizmann was requested to organize large scale production. Already in the same year a production level of 1,000 kg of acetone per week was realized. The required large expansion of the production, however, was transferred to Canada in the same year; the British government, confronted with the serious effects the German submarine warfare had on the procurement of raw materials for its industry and food for the population, decided that shipments of the small amounts of acetone from Canada made more sense than that of the much larger volumes of grain to produce it.

After the United States entered the War (April, 1917) two large distilling plants were bought in Terre Haute, Indiana, to assist in the production of ever increasing amounts of acetone.

The Butanol-Acetone Fermentation Process

lit. Gabriel and Crawford, 1930; recently a very complete review of the process and its history has been given by Jones and Woods, 1986.

The butanol-acetone fermentation as developed by Weizmann demanded a more sophisticated approach than the glycerol fermentation, which could better be called a biochemical reaction, performed by large amounts of enzymes supplied to the medium in the form of yeast cells. In the butanol-acetone fermentation the biochemical conversion is carried out by the living cells of *Clostridium acetobutylicum* under carefully controlled and sterile conditions. Fresh inoculant has to be supplied regularly, and the scaling up (from spore-impregnated sand) has to be checked carefully by chemical and bacteriological analyses.

The mash consists of a degerminated corn meal of about 6 to 8% concentration, cooked and sterilized for one to two hours at about 130°C. When the cooking is completed, the mash is rapidly cooled to approximately 36°, then pumped into closed, sterilized fermenters of about 185,000 l capacity (Terre Haute), and inoculated with about 5% of inoculant. Mechanical mixers were not installed, the required agitation was achieved by blowing sterile carbon dioxide through the liquid. Gabriel and Crawford remark on sterilization procedures and the subsequent process.: "All pipe lines, connections, fermenters, etc., are frequently sterilized with live steam and, if necessary, with chemicals". The fermentation itself takes from 2 to 3 days. The products are recovered by fractional distillation.

The fermentation produced the solvents butanol, acetone and ethanol in the ratio of about 6:3:1, at the same time large volumes of hydrogen gas and carbon dioxide evolved. The gases (60% by weight of the fermented substrates) escaped from the process into the air. The yield of acetone on corn flour was only 6%. Nevertheless large productions were achieved: the plant in Canada for instance produced in the last year of the war over 2,500 t of acetone and double that amount of butanol, a small amount of which was converted into methyl ethyl ketone, a solvent which could also be used in making Cordite.

Almost immediately after the surrender of the German army on the western front on 11 November 1918 the plants in Canada and the U.S.A. were closed down because the process was considered uneconomical.

Dr. Weizmann was honored by the British government for his invaluable contribution to the solution of the acetone problem. At his request, Prime Minister Lloyd George promised him full British support for his dearest wish: a Homeland for the Jews in Palestine.

Chemical Problems of the United States

The situation the American chemical industry was facing when the war broke out in Europe, was in quite a few respects the reverse of the German situation:

raw materials in general were or could be made available in sufficient supply, but for the supply of many essential chemicals the country was fully dependent on the German chemical industry. In heavy chemicals for instance a big shortage of potash already became felt in the winter of 1914. [Potash was almost a complete German monopoly thanks to the Stassfurt mines and a shrewd commercial policy.]

The demands of the explosives industries of the Allied Nations France and Great Britain caused a scarcity in the U.S. of nitrates, sulfuric acid and glycerol. The supply of dyestuffs and chemical intermediates decreased to totally insufficient levels in 1915, notwithstanding some arrangements made with the British to overlook certain German imports in return for exports of essential German chemicals to neutral countries like the United States[4]. Quite naturally the scarcity, or fear of scarcity led to speculation in chemicals and starkly rising prices. It also promoted the search for substitutes and the introduction of new technologies.

The shortage of many chemicals resulted in a tremendous effort from the side of the chemical industry to redress the situation: in the course of the war the industry learned with increasing success how to make the sophisticated "German" products themselves. At the end of the war these efforts had resulted in a much more mature chemical industry in the United States, an industry capable of competing with the German industry in most fields of chemistry.

American Production of Organic Acids

Haynes (Vol. 3, Ch. 4) treats the fate of a number of important organic acids when imports from Germany decreased and later came to a complete stop.

Production of **oxalic acid** from saw-dust by the American Alkali & Acid Company of Bradford, PA, the sole U.S. producer had almost been wiped out by the cheaper fully synthetic German product. During the war the old manufacturing method again became applied on a large scale, but one American producer also started to make it synthetically. This one producer, Victor Chemical Works, was the only survivor when German competition returned to the scene in the 1920's.

Raw materials for the production of **tartaric and citric acids** came largely from Italy, smaller amounts from some other Mediterranean countries. Because of shortage in shipping tonnage the supply of raw materials for these productions became difficult, which led to soaring prices and the search for substitutes. Local production of citric acid from cull lemons had already been established in California in 1913. During the war period Californian production was expanded rapidly from 1,240 t in 1914 to 1,830 t in 1917. This helped to some extent in alleviating war shortages of the acid.

The shortages of these acids, however, were not solely due to the difficulties in the supply of the raw materials. Another factor of importance was the increased use of these acids in the production of soft drinks, a consequence of the prohibition of alcohol. During the war, when a shortage of grain threatened food supply, the production and use of alcoholic drinks was forbidden temporarily. This supported the existing strong sentiment among large parts of the population to outlaw the drinking of alcoholic beverages completely. In the course of the war a number of States introduced laws to this effect and in 1919 a National Prohibition Act, the Volstead Act, was passed by the U.S. congress.

Lactic acid profited much from the war; Haynes writes:

> The use of lactic acid expanded rapidly, due in part to the production of the edible grade by these pioneer makers [the Avery Chemical Company] and its use as a substitute for scarce citric acid in the food and soft drink industries.

The American Lactic Acid Industry during World War I

Shortly before 1914 the U.S.A. already counted at least four lactic acid producers (Ch. 6, 187). During the war period Haynes mentions, besides Avery, five other producers:

> American Alkali & Acid Company, Bradford, Pennsylvania;
> Atlantic Chemical Company, Mansfield, Massachusetts;
> Harrison Works (Du Pont), Philadelphia;
> Westchester Chemical Company, Pelham, New York;
> Woolner & Company, Peoria, Illinois.
>
> *- Haynes, 1945, Vol. 3, 50*

Although the **Avery company** continued to export (technical) lactic acid until 1914 (Haynes), German exports of the better qualities to the United States had become a major threat to Avery as well as to other U.S. producers of the acid. These exports had risen from a small amount in 1907 to about 200 t in 1913. German and U.S. statistics (Table 7,II) show the influence of the hostilities in Europe (the volumes are given as such, because the concentrations are not known, most probably these included 50 and 80% light technical grade as well as pharmaceutical 90%):

Table 7.II. American imports of lactic acid.

year	German exports to the U.S.A. (t)	total of U.S. imports (t)	U.S. import price ($/kg)
1912	177	152	0.28
1913	204	72	0.24
1914		125	0.42
1915		65	1.36
1916		8	3.52
1917		0.05	3.62
1918		0.15	
1919		5	
1920	56	61	

Source: German statistics, U.S. statistics (from Haynes).

Although the involvement of **Harrison's** in lactic acid production is beyond doubt, the Trade Catalogues of the company do not reflect this until the take over by Du Pont in 1917. It then lists the following grades (information from the Hagley Museum, Wilmington, DE):

Dark, 22% by weight;
Light Refined, 22% and 44% by weight;
Special Light Refined, 22% and 44% by weight; and
USP (United States Pharmacopoeia)

The writer has not been able to collect any more detailed information about the Atlantic Chemical Company, nor about Woolner & Company. The same is the case with the **Westchester Chemical Company**, but for one newspaper clipping from "The Pelham Sun" of October 21, 1921, found by Mrs. Susan C. Swanson, town historian of Pelham:

Many residents of Pelham Heights have complained of the odors arising from the Westchester Chemical Company's plant ... during the past month.

... the odor was caused by insufficiently supply of water. The manager of the plant assured ... that there is no more lactic acid manufactured at the Pelham plant, but there is a quantity of this acid still in the vats. The lease of the plant terminates on January 1st, 1922. The lease will be renewed if water can be obtained. If not the Westchester Chemical Company will move its ... [copy illegible, probably: apparatus].

Most probably the calcium lactate residing in the vats had been infected with butyric acid producing bacteria, a rather common occurrence in lactic acid production (see Intermezzo 5).

Total U.S. production of the technical grade advanced to 860 t 100%), valued at $ 340,000 in 1917. The U.S. "Census of Manufacturers" reported for 1919 a total production of 2,300 t (Haynes, Vol. 3, 50 and 522). The price of the 44% technical grade rose from a prewar level of 0.09–0.11 $/kg to 0.31–0.35 $/kg in 1917/1918. The new edible quality (50%) introduced by Avery was sold at about 0.77 $/kg (Defren, 1920).

Introduction of Edible Lactic Acid in America

In 1920 George Defren, vice-president of the Avery Chemical Company published a paper on the "Use of Refined Lactic Acid in Food Products" (Defren, 1920).

> Up to the present time the so-called fruit acids, citric and tartaric acids, have been used almost exclusively as acidulants for food products. Within a year, however, refined edible lactic acid has become a factor in competition with the fruit acids mentioned. The only reason that lactic acid has not been extensively used heretofore is due to the fact, that no successful technical method has been devised for the production of this acid on a commercial scale to compete successfully in cost with citric and tartaric acids.

But, Defren says, these technical problems have been solved, and refined lactic acid (50%) "is now selling at [0.77 $/kg] in barrel or larger lots".

The most important application of this refined edible lactic acid was in so-called low-alcohol beers. These beers, containing not more than 2.75% of alcohol, had been introduced in the last year of the war, following the prohibition of stronger types of beer. [Under the Volstead Act the maximum allowable percentage would be reduced further in 1920, to 0.5–1%.]

Lactic acid, at concentrations between 0.02 and 0.1%, suppresses or prevents the characteristic "yeast" or "bread" odor and taste which develops when the alcohol content of regular beer is lowered by distilling off most of the alcohol. A similar effect occurs when low-alcohol beer is made by the "checked fermentation" method: the addition of lactic acid overcomes the influence of the otherwise dominating odor and taste components of the malt wort.

Other uses Defren advocates are in the manufacture of soft drinks, candies and bakery products. He also stresses the preservative action of lactic acid, important for meat and fish products. He furthermore uncovered two of the original ideas of Charles Avery, but without mentioning the name of the founder of his company:

- acid calcium lactate for baking powder ("The production of this acid calcium lactate, while it has already passed the laboratory stage, is confronted with the terrific expense of introducing it to the general public, in competition with the well established baking powders, at present so extensively employed")
- lactic acid vinegar, to be used industrially for pickling and analogous uses.

Both applications have to be viewed in the light of the need to develop substitute products during the war period. Defren ends in a soft key:

> The refined edible lactic acid industry is still in its infancy and much is expected therefrom.

Raw Material for Lactic Acid in the United States

The Avery Company had, in the first phase of its operations, mainly used corn flour as substrate. Later it converted (partially ?) to dextrose. Until the 1920's dextrose was a relatively impure product containing not more than 70–80% of glucose, the remainder being dextrins. Fairly sophisticated methods like purification through crystallization of calcium or zinc lactate had to be applied to prepare edible lactic acid from such raw materials.

Another method is to take a purer carbohydrate as substrate. At the time this could only be sucrose, either in the form of beet sugar or of cane sugar. It is quite well possible that the Avery Chemical Company applied this raw material of which the superior properties were already described by Alan Claflin in his 1897 paper.

An editorial in Chemical Age (Vol. 30, no. 5, 1922) said, however, that the U.S. lactic acid industry is based on inferior grades of corn and the vegetable ivory waste of the button industry.

The use of **vegetable ivory nut meal** had been patented by Sigmond Saxe of New York in 1917 (U.S. Patent 1,249,511). It claimed the use of the meal and waste of the ivory nut, the fruit of the Amazonian ivory palm or tagua palm (*Phytelephas macrocarpa*), as a substrate for lactic acid fermentation[5].

The nut itself, as the name implies, was used to produce substitute products of real ivory, in particular buttons and trinkets. The waste from these processes was ground into a paste, then hydrolyzed with sulfuric acid to convert the polymeric mannan of which the fruit contained about 60% into the simple C6 sugar mannose. After neutralization [and filtration ?] mannose could thereupon be readily fermented to lactic acid. The factory carrying out this process was located in the State of New York (Garrett, 1930), which points to the Westchester Chemical Company (p. 226).

The Effect of the War on Tartaric and Citric Acid

Like many other products, the food acids tartaric and citric acid were also affected by war shortages. The prices of both acids went up considerably as is exemplified by Table 7,III below, which gives yearly averages of the prices, calculated from data given by Haynes:

Table 7.III Yearly averages of list prices of tartaric and citric acid.

year	tartaric acid ($/kg)	citric acid ($/kg)
1911	0.67	0.85
1914	1.32	1.39
1915	0.99	1.32
1916	1.59	1.50
1917	1.74	1.61
1918	1.83	2.03
1919	1.30	2.20
1920	1.57	1.68
1921	0.64	1.01

Source: Haynes.

The rise in prices was, however, not solely due to shortages as a consequence of the raw materials. They were also strongly influenced by the increased demand of the soft drink industry, following the restrictions in and later the almost total prohibition of alcoholic "hard" liquors as well as of beer of normal strength.

The way the Pfizer company responded to these shortages is one of the major historic feats in the history of fermentation technology. This response would also have a dramatic influence on lactic acid making. The successful introduction of fermentation citric acid can be held responsible for a delay in the introduction of edible lactic acid by decades.

Citric Acid by Fermentation

lit. Mines, 1978

In 1913 Pfizer was the largest manufacturer of citric acid in the United States with a production of about 400 t/a from crude Italian calcium citrate. Other producers supplied an additional 840 t to the American market.

Although Pfizer as well as others succeeded in increasing local production from (Californian) waste citrus fruit and from more raw material coming from Italy, the situation grew difficult again in 1917 when the first restrictions on the use of alcoholic beverages came into force.

Earlier in the same year in one of those coincidences which change profoundly the course of historic events, a Dr. Currie from the U.S. Department of Agriculture had called on John Anderson, chief executive officer of the Pfizer company to inform him about a possible route to citric acid by fermentation. James N. Currie (1883– ***) had been in the service of the Department of Agriculture as a dairy scientist since 1912, developing new methods for making cheese. Maybe induced by the war shortages he also became interested in making chemical products by fermentation. Knowing his literature he picked up the experiments of Carl Wehmer with molds to produce oxalic acid by fermentation of sugar. Instead of molds belonging to the genus *Penicillium*, which Wehmer had studied for oxalic as well as citric acid production, Currie tried *Aspergillus niger*, the common black bread mold.

Apparently Currie forgot, or omitted deliberately the addition of a buffering substance. This had a surprising effect: the fermentation produced only small amounts of oxalic acid, but large quantities of citric acid. Quoting from Mines:

> Like a good chemist, Dr. Currie extended his experiments. He selected thirty cultures of his mold and seeded them on sugar solutions. From this group he picked the two or three that gave the highest yield of citric acid. He then experimented further with these top pupils in his class to see if he could again increase their potential. The results of his experiments were published in 1917 in the Journal of Biological

Chemistry under the prosaic if factual title, "The Citric Acid Fermentation of *Aspergillus niger*".

- Mines, 1978, 17

Early in 1917 Currie contacted the major producer of citric acid in the United States, Chas. Pfizer, and asked whether they might be interested in his work. Pfizer grasped the opportunity immediately. On May 15, 1917, Currie joined the company with the special task of making the new fermentation process commercially viable.

However,

the operation, even on a laboratory scale, was appallingly sensitive. There were numerous uncontrollable variables: the quality of the spores, the purity of the cultures, air contamination, temperature and humidity, the nutrient medium and dozens of others. Even the size and the make up of the production apparatus was critical. Larger pans created more failures than smaller pans because of the contamination problem.

Despite problems and failures, the first pilot plant using the new fermentation process went into operation on April 23, 1919....

One of the problems was that of producing large amounts of mold in pure cultures for inoculating the medium. This was blazing a new trail. Fermentation processes up to that time had not been widely used for large scale production of industrial chemicals, and no one knew much about them.

- Mines, 1978, 19

(The last statement is somewhat of an exaggeration; with lactic acid quite a bit of experience existed already, but the demands on sterility were only minor compared with those of the citric acid fermentation. In sterility requirements the acetone – butanol fermentation surely was a good match, but it did not have the additional difficulty of the supply of large amounts of sterile air to the fermentation, about 300 kg oxygen or 1,000 m^3 is required to convert 1,000 kg of sugar into citric acid.)

A team of scientists worked almost tirelessly from 1920 to 1924 to solve all the technical problems, systems had to be designed to provide sterile air and controlled temperatures. An internal report said:

This process was plagued with wide variations in yield and, therefore, costs. Many eight-thousand gallon batches would cover over with a contaminating *Pennicillium* and were discarded. Others developed abnormal mycelia and yielded very little citric acid. However, continued studies finally solved the abnormal growth problems, while increasing aseptic techniques brought *Penicillium* contamination under some degree of control and the process became stable enough so that we were able to get reasonable production yields.

-Mines, 1978, 20

World Lactic Acid Production Capacity Increased by More than 100%

During World War I lactic acid production increased considerably in Germany, in the first place to meet the requirements of the military for glycerol replacers. From a prewar figure of about 1,500 t/a (100%) (Schäfer, 1907), the capacity was enlarged to at least 2,000 t/a, maybe even to 2,500 t/a. Half of this capacity was installed in the Boehringer plant at Nieder-Ingelheim. The second largest producer was the Knab & Lindenhayn plant at Niederstriegis. (Merck produced less than 100 t/a; mainly in the form of pharmaceutical lactic acid and some pharmaceutical lactates.)

American production had grown during the war from probably less than 800 t in prewar years to 2,300 t in 1919. This was still mainly the technical grade, although small amounts of edible lactic reportedly were sold to the brewing industry for the production of low-alcohol beers.

In England and in France (new) production facilities for lactic acid were installed during the war. In both countries production levels probably remained below 500 t/a of technical lactic acid (as 100%). See also Chapter 8.

IN RETROSPECT

It is a sobering and also ironical fact that chemists on both sides of the battle lines received many honors for their successful efforts in winning (for instance Dr. Weizmann) or in prolonging (think of Dr. Neuberg) the war. The achievements of the chemical communities and chemical industries were to a large extent "coresponsible" for the prolongation of the war and the fierceness with which it was fought, causing the deaths of 8,400,000 soldiers.

Notes

[1] Protol is a contraction from *pro*pane*t*ri*ol*, the systematic name of glycerol.

[2] In the years following the isolation of zymase by Büchner (1896) chemists had taken up the study of the influence of enzymes in bacteriological processes in a more general way. This kind of research led to a new science, the science of biochemistry.

[3] Carl Neuberg (1877–1956) was born at Hanover (Germany), studied chemistry at the Universities of Berlin and Würzberg. In 1913 he became Assistant Director of the Kaiser Wilhelm Institute for Experimental Therapy. Under his influence, and from

1923 onwards under his leadership, this institute became one of the foremost centers of biochemical research in the world. Because of his Jewish descent the Hitler regime forced Neuberg to leave Germany in 1939. In the U.S.A. he was appointed Professor at New York University and later at the Brooklyn Polytechnic Institute (Gottschalk, 1956).

[4] The story of how the U.S. chemical industry coped with the shortages in chemicals and, in the end, emerged from the 1914–1918 War as the world's most important chemical nation, is very well accounted for in Haynes "American Chemical Industry", Volumes 2 and 3 (1945–1954).

[5] Details about the use and constitution of the tagua nut and its role in the history of mannitol production can be found in: Haynes, 1946, Whistler and Smart, 1953.

Books Consulted

Bäumler, E., 1968. *A Century of Chemistry*; Econ: Düsseldorf.

Greiling, W., 1938. *Chemie erobert die Welt*; Wilhelm Limpert: Berlin.

Haynes, W., 1945. *American Chemical Industry*; Vols. 2 and 3; D. Van Nostrand: New York. Reprinted by Garland: New York, 1983.

References

Claflin, A.A., **1928**. *Chem. Mark.*, **23**, 353 and 471.

Connstein, W., and Lüdecke, K., **1919**. *Ber. Dtsch. Chem. Ges.*, **52**, 1385.

Deite, C., and Kellner, J., **1923**. *Das Glyzerin*; Julius Springer: Berlin.

Defren, G., **1920**. *Chem. Age (N.Y.)*, **2**, 478.

Gabriel, C.L., and Crawford, F.M., **1930**. *Ind.Eng.Chem.*, **22,** 1163.

Garrett, J.F., **1930.** *Ind. Eng. Chem.*, **22**, 1153.

Gottschalk, A., **1956**. *Nature (London)*, **178**, 722.

Grandmougin, E., **1917**. *L'Essor des Industries Chimiques de France*; H. Dunod et E. Pinat: Paris.

Haber, L.F., **1971**. *The Chemical Industry, 1900–1930*; Clarendon: Oxford.

Haynes, W., **1946**. *This Chemical Age*; Secker and Warburg: London.

Jones, D.T., and Woods, D.R., **1986**. *Microbiol. Rev.*, **50**, 484.

Luhmann, E., **1917**. *Z. Kohlensäure-Industrie*, **23**, 327.

Mines, S., **1978**. *Pfizer ... An Informal History*; Pfizer: New York.

Neuberg, C., and Reinfurth, E., **1920**. *Ber. Dtsch. Chem. Ges.*, **53**, 1783.

Schäfer, G., **1907**. *Chemische Zeitschrift*, **6**, 177 and 189.

Ungewitter, C., **1927**. *Chemisch-Industrielle Wirtschaftspolitik, 1877–1927*; Verein zur Wahrung der Interessen der Chemischen Industrie Deutschlands: Berlin.

Whistler, R.L., and Smart, C.L., **1953**. *Polysaccharide Chemistry*; Academic Press: New York.

BUTYRIC FERMENTATION CAN SPOIL LACTIC FERMENTATION

Lactic fermentation can easily be spoiled by various kinds of infections of which infection by butyric acid forming bacteria is one of the worst.

When present in low concentrations, butyric acid smells of perspiration and of rancid butter, in higher concentrations the smell is more pungent and becomes repulsive. In lactic acid, concentrations of less than one percent suffice to make it unsuitable for edible uses.

Butyric acid was discovered in 1814 by the famous French chemist Marie-Eugène Chevreul (1786–1889) as one of the saponification products of butter.

The preparation of butyric acid by fermentation was for the first time and in detail investigated by Pelouze and Gélis (Pelouze and Gélis, 1844). The best recipe these authors found consisted of prolonged fermentation of (starch) sugar under the influence of fresh cheese, and in the presence of finely powdered chalk. In their process the sugar is first converted into calcium lactate; when the broth containing partially crystallized calcium lactate is left standing for a period of several weeks the lactate will be converted into butyrate. The yield of the process is about 25%, as by-products small amounts of ethanol and acetic acid are formed. Much gas evolves during the fermentation: in the beginning hydrogen and carbon dioxide in a ratio of about 1 to 8, when the reaction proceeds the ratio becomes close to 1 to 1[1].

Pelouze was an adherent of Liebig's fermentation theory in which microorganisms played no role whatsoever. It was not before 1861 that Louis Pasteur discovered the existence of special "infusorial animalcules" which were able to live without oxygen and to act as ferments. These anaerobic ferments [= microorganisms] caused the formation of butyric acid (From Nicolle, 1961).

A description of a method to produce butyric acid on a larger scale was given in 1878 by Georg Thenius (Wagner, 1878). The main features of his method are presented in Exhibit I5,1, together with data from other historical recipes.

EXHIBIT I5,1
PRODUCTION OF BUTYRIC ACID BY FERMENTATION

lit.: Thenius (Wagner, 1878) (T)
Merck (Yearly Manufacturing Reports 1897–1924 (M)
Lefranc (1922, from Playne, 1985) (L)
Jørgensen (1948) (J)
Rehm (1967) (R)

raw material	starch and calcium lactate (T) starch hydrolysate (M) hydrolyzed wood waste (L) cheap carbohydrates (J) malted starches (R)
pretreatment	hydrolysis of starch or cellulose
concentration	8–12% (L) 8–10% (degree Balling) (J)
inorganic nutrients	ammonium and phosphate is required (R)
organic nutrients	blood fibrin (T) yeast, sweet wort or milk (J) a ratio of carbohydrate to nitrogen of 5 to 1 is recommended (R)
neutralization	ground chalk is added at the beginning
sterility of process	anaerobic conditions should be maintained throughout (a carbon dioxide foam layer suffices)
acidity	neutral to slightly acidic; below pH 4.5 production of butyric acid stops (J) and butanol will be formed (R)
fermentation temperature	40–45°C (T) 40°C (L) 35°C (J) 35–40°C (R)
yield	30% (T) 34–38% (L) 40–44% (M) 30–41% (R)
bacterium	*Clostridium butyricum* (J) and (R) [2]
source of bacterium	soil or crushed grain (J) soil (R)

Purification of Butyric Acid

The method of Thenius provides a good example: To the relatively clear and thin solution obtained after filtration of the fermentation broth sodium carbonate is added to convert calcium butyrate into sodium butyrate and insoluble calcium carbonate. After filtration the sodium butyrate solution is concentrated and then treated with sulfuric acid to set free the butyric acid. The acid is separated as an oily layer, converted again into the salt with sodium carbonate, transferred into a closed distillation vessel, treated with sulfuric acid and distilled to obtain pure butyric acid. Thenius reports a yield of 30%. He concludes:

> The total value of lactic acid which can be recuperated in a wheat starch plant ... is not without importance. Twice the value can be gained, however, when lactic acid is converted into butyric acid and this acid is processed to butyric ether [ethyl butyrate], which is a much desired product of trade.
>
> - *Thenius, quoted from Wagner, 1878*

The ether [= ester !] is employed as an essence in the manufacture of rum and brandy.

Between about 1890 and 1926 E. Merck, Darmstadt, produced butyric acid by fermentation (Yearly Manufacturing Reports of the Company). Originally the fermentation was carried out in a dairy plant at Isny in the Allgäu, Germany. Calcium lactate probably was an intermediate step. Crude, dry sodium butyrate was shipped to Darmstadt and there converted to butyric acid. In 1898 the fermentation of butyric acid was transferred to Darmstadt, at the same time as that of lactic acid. For both processes dextrose [more appropriate: starch sugar] replaced the whey of the dairy plant. The manufactures were carried out under the same roof! Not much more is known about the fermentation recipe and conditions than the use of a bacterial culture for the inoculation (Metz, 1983).

Merck's production rose from 14 t in 1900 to 30 t in 1907. A few years later production came to a standstill when cheaper French products, based on bacterial conversion of distillery waste became available (Inventor: Dr. Effront of Brussels, Belgium). After the war Merck renewed production. In 1927 the factory finally was closed and dismantled, now because of competition from cheap synthetic butyric acid. Playne suggests that in those years the Société Lefranc & Cie became an important producer. This company was very active in developing methods to use industrial waste materials in the production of butyric acid. The "synthetic" part of Lefranc's process was the conversion of wood waste with sulfuric acid to fermentable glucose (Lefranc & Cie, British Patent 186,572 of 1923).

Butyric Acid and the Environment

Dr Harald Metz, in an essay on the history of biotechnology at Merck (Metz, 1983), makes some striking remarks concerning the environmental aspects of butyric fermentation and production, remarks which occasionally also were applicable to lactic fermentation, particularly when off-fermentations occurred:

> At the turn of the century easterly winds often could carry the natural not entirely pleasant smell of rancid butter from the old works of Merck through the streets of Darmstadt. This probably went unnoticed as in those days small cities and villages had retained much of their natural scent: many citizens had a pigpen and kept goats and chickens; horses and cattle were housed within the city – dunghills and manure-pits completed the flavor pallette. Smoke from a large number of stoves and furnaces made their own contribution when the wind was right. In earlier years the city must have known still smellier times: Darmstadt had become a modern city since the completion of the sewage canalization in 1895.
>
> \- *Metz, 1983*

No scientific or trade literature relating to nuisance caused by lactic acid fermentation plants has been found. Insiders know that lactic fermentation sometimes goes “wrong” and that such events more often than not are accompanied by the development of unpleasant smells. One of the then largest lactic acid plants in the world for instance had to close down temporarily in the 1950’s because of complaints about excessive butyric acid formation.

One reported case of nuisance caused by lactic acid was accidentally found, however, by Mrs. Susan C. Swanson, town historian of Pelham , Westchester County, NY. At the writer’s request Mrs. Swanson searched local files for data on the Westchester Chemical Company, a company which (most probably) made lactic acid in the 1910’s and early 1920’s from ivory nut waste. She could not confirm the use of ivory nut waste by the company, but instead unearthed a paper clipping from the October 21, 1921, edition of the local “The Pelham Sun” on complaints of the noxious odors arising from the plant. Said the plant manager to Dr Newell, the investigator of the complaint:

“... there is no more lactic acid being manufactured at the Pelham plant, but there is a quantity of the acid still in the vats.” The problems are caused by insufficient supply of water, the manager explains. [Butyric acid fermentation of anaerobic calcium lactate in the vats seems a more likely explanation.]

Notes

[1] A review article by Playne (Playne, 1985) on the production of butyric and propionic acid by fermentation presents the most salient developments in this field since the times of World War I. From the three overall metabolic reactions given in this review the following best fits the observations of Pelouze and Gélis: 1 glucose $\rightarrow$ [2 lactate] $\rightarrow$ 1 butyrate + 2 CO_2 + 2 H_2.
(The theoretical yield based on this reaction is 48.9%.)
[2] *Clostridia* are sporulating bacteria; the spores are very resistant to pasteurizing, even boiling for 30 minutes will not kill all the spores.

References

Jørgensen, A., **1948**. *Micro-Organisms and Fermentations*; (rewritten by Hansen, A.); Charles Griffin: London.

Metz, H., **1923**. Company Magazine: *Merck Informiert*; E. Merck Darmstadt; December 5, 1983.

Nicolle, J., **1961**. *Louis Pasteur, A Master of Scientific Enquiry*; Hutchinson: London.

Pelouze, J.T., and Gélis, **1844**. *Ann.Chim.Pharm.*, [3], **10**, 434.

Playne, M.J., **1985**. *Propionic and Butyric Acids*, Ch. 37, in Vol. 3 of *Comprehensive Biotechnology*; Ed. Moo-Young, M.; Pergamon: Oxford.

Rehm, H.J., **1967**. *Industrielle Mikrobiologie*; Springer: Berlin.

Wagner, R. von, **1878**. *Jahres-Ber. Leist. Chem. Techn.*; Otto Wigand: Leipzig, 746.

CHAPTER 8

THE CHEMICALIZATION OF THE NEW WORLD

ETHYL LACTATE MISSES A CHANGE, AND LACTIC ACID DEVELOPS ONLY SLOWLY IN THE 1920'S

In 1914 occurred two seminal events. One was the outbreak of World War I. The other was the announcement that Ford would pay a $5.00 per day minimum wage – for which he was, of course, denounced as a crackpot or a monster by the entire manufacturing, commercial, financial, and propertied class. I am not sure which of these two events, in the long run, will be remembered as the more significant.

- *John Gunther, 1947*

The story of subsequent events is the story of assembly line manufacture and of the mass motorization of America. In 1908 Henry Ford (1863–1947), then one of the small handicraft producers of automobiles, decided to aim his production at a mass market. His company, the Ford Motor Company, in future should produce only one model in the most efficient way conceivable. The model he chose for this experiment was the famous Model T Ford. The increase in wage of 1914, about a doubling of the average pay to the industrial worker, was accompanied by a reduction of the working day from 9 to 8 hours, and conversion of the factory to a three-shift operation. The basic idea behind the wage increase was that with higher wages more people could afford to buy cars, and this idea Ford applied to his own workers.

In the same year the assembly line experiment was begun, as much out of necessity because of the success of the Model T as well as out of the philosophy that the assembly line would cut production costs and allow for even lower sales prices. The idea worked:
the price of the Model T Ford dropped from $ 950 in 1908 to $ 500 in 1914, by 1927 it could be bought for $ 290; the output of cars rose to unbelievable levels, from about 100,000 in 1910 to about 6,000,000 in 1930 (all car manufacturers together). Ford alone sold 15 million Model T cars between 1908 and 1927, when the Model T Ford finally was taken out of production.

The motorization transformed the American nation so profoundly that almost a new nation was born, a nation on wheels in which nearly every family had its own means of transport and could freely and easily move over vast distances. Suburban living became popular with the well-to-do; picnicking became a popular pastime.

Farming also became mechanized, as is illustrated in Table 8,I by the number of horses replaced.

Table 8.I The mechanization of the American farm.

year	number of horses on farms	steam engines on farms	tractors in use
1910	20.000.000	72.000	10.000
1939	10.000.000	very few	1.500.000

Source: Slicher van Bath (1980).

The great increase in mechanized farm equipment took place in the 1920's. The war had been a prosperous time for the farming community, and when the war was over they found good export markets in the slowly recovering countries of Europe. The minor depression of 1920/1921 hardly affected the urge to mechanize farming. Mechanization led to increased production while, later in the decade, export markets were shrinking.

By the end of the first post-war decade agricultural productivity had become too high to be absorbed by the market. The Great Depression of 1928, to a large extent due to overproduction in general, hit the farming community the hardest. The resulting low prices for agricultural commodities (Table 8,II shows corn prices) would eventually promote the use of (some of) these commodities for large scale production of chemicals.

Table 8.II. Corn prices (in $/bushel).

1906	0.39	1927	0.85
1911	0.68	1927	0.80
1916	1.14	1931	0.32
1921	0.52	1933	0.52
1923	0.81	1935	0.66
1925	0.70	1937	0.52

Source: U.S. Government statistics.

Cars Need Coatings

The assembly line of Ford of the 1910's still bore a major imperfection: the finishing took weeks to complete. On some cars up to 22 coatings were applied, and some of these layers took a full day to dry. What made things worse was the quality of the final finish, at best good enough for one summer only.

The Du Pont company recognized the importance of the paint business for the automobile industry, and, to secure much of this business started to invest heavily in the other large car maker, General Motors. By 1919 Du Pont owned 29% of GM stock and controlled the company.

Two other important steps were taken in the same period:

1. A series of paint and varnish producing companies was bought, culminating in the 1917 take-over of Harrison Bros. & Co. of Philadelphia, a diversified manufacturer of mineral acids, pigments, paints and other products.
2. A group of research chemists of the company was given the task to develop a nitrocellulose lacquer with a higher solids content than possible with the dynamite quality of nitrocellulose (pyroxylin).

In 1923 the Du Pont "Spray Duco Nitrocellulose Lacquer System was put on the market. It revolutionized the automobile finishing operations almost overnight: the application of (colored) coats from now on only took a few hours instead of a few weeks. Not only a big saving in production time and inventory cost was achieved, but the new coatings also proved to be much more durable, lasting often for the entire life of the automobile (Merkle, 1953).

Can Ethyl Lactate Play a Role in Car Finishes?

Dynamite quality nitrocellulose dissolves very well in acetone and methyl ethyl ketone; during the war in particular acetone was found to be an indispensable solvent in the manufacture of smokeless powder. The lower molecular weight type of nitrocellulose suitable for coatings, however, required a different type of solvent. Lacquer solvents should be less volatile, have less smell, and keep good dissolving properties for the nitrocellulose until they have almost completely evaporated.

The introduction of spray nitrocellulose coatings in the automobile industry caused an extreme demand for slow drying high boiling solvents. One of the companies trying to cope with this demand was the U.S. Industrial Chemical Co.

Burnell R. Tunison, responsible for technical development of sales in this company, wrote in 1925 (Tunison, 1925):

The successful utilization of finishes of gums and nitrocellulose for automobile use caused an abnormal demand for higher boiling nitrocellulose solvents. During 1923, the production of butyl alcohol and butyl acetate was small, because of fermentation difficulties. This caused a greater demand for fusel oil and amyl acetate. Fusel oil being a by-product with a limited production, an acute shortage of high-boiling solvents soon resulted. In fact, so great was the demand that the importation of fusel oil and amyl alcohol reported by the Department of Commerce amounted to [1,442 t] in 1923, as compared with only [68 t] in 1922[1].

The lack of adequate supplies of high-boiling solvents and modifiers for nitrocellulose solutions has stimulated a considerable amount of research. That this investigational work was productive is evidenced by the number of solvents now being offered for this purpose.

The fact that in general the gums used for lacquer purposes are soluble in alcohols and not in esters, and that the nitrocellulose is generally soluble in esters and not in alcohols, sometimes makes it difficult, if not impossible, to make homogenous solutions of the required proportions of gums and nitrocellulose. By the use of anhydrous alcohol it is possible to reduce the ester content of lacquer compositions so that the solvent not only dissolves the nitrocellulose but is low enough in ester content that perfect blends with the various gums are possible. Then 'by adding the proper modifiers, non-solvents, any desired result of flow and rate of drying can be obtained.'

Another important solvent the use of which is increasing rapidly in the manufacture of lacquers is ethyl lactate. It is a more powerful solvent for nitrocellulose than any of the ordinary products at present in use. Nitrocellulose solutions in ethyl lactate will stand greater dilution with non-solvents[2] than any other solvent. Ethyl lactate remains an active solvent even after contamination with as much as 25 per cent of water. It gives good flowing properties and prevents 'blushing'.

- Tunison, 1925

The following year Tunison reviewed the situation again (Tunison, 1926):

An ... outstanding development during the year is the pyroxylin lacquer industry. The U.S. Department of Commerce stated that the recent production of lacquers was as follows:

	Number of Establishments	Production in m^3
Jan. – June, 1924	33	5,400
July – December, 1924	45	8,200
Jan. – June, 1925	67	18,500

It is estimated that during the second six months of the year the production exceeded [26,500 m^3]. This enormous increase in the production of lacquers has resulted in a very unusual demand for nitrocellulose and resin solvents. The production of butyl alcohol increased from 2,000 t in 1923 to 6,500 t in 1924[3]. Prices were considerably lower during the year, making possible the substitution for other solvents. Butyl acetate production increased materially, ...

- Tunison, 1926

Another witness of the nitrocellulose lacquer story is presented by Haynes in his History of the American Chemical Industry (1954, Vol 4, 200). This is Bruce K. Brown of Commercial Solvents, who gave the following account:

According to my recollection there was a time in the middle 1920's when ethyl lactate, Cellosolve [mono ethyl ether of ethylene glycol], and normal butyl acetate were the prime competitors for the nitrocellulose market. Butyl acetate came first and I do not remember which was the next, ethyl lactate or Cellosolve, but I think the two came upon the commercial scene at about the same time. For several years J.G. Davidson of Carbide, Don Keyes of U.S. Industrial Alcohol, and I used to tag each other around at American Chemical Society paint and varnish meetings and similar forums to extol the technical merits of our respective solvents. Time proved that Don Keyes was betting a losing horse [ethyl lactate], but, from the standpoint of history, his work and that of his company was quite important in developing the solvent market. U.S. Industrial Alcohol was then a prime business factor in the lighter solvents field, i.e., ethyl acetate, and was determined to have something to fill out its solvent line.

- Brown, 1926, quoted from Haynes, 1954, Vol. 4, 200

Ethyl Lactate, a Potential Lacquer Solvent

U.S. Industrial Chemical Co., a subsidiary of U.S. Industrial Alcohol Co., was founded during the war to produce acetone from ethyl alcohol. The process used could not compete against the fermentation process developed by Weizmann, and production was discontinued, already during the war. The president of the company, Milton C. Whitaker, foresaw, however, increasing opportunities for nitrocellulose type lacquers in the automobile industry and decided to develop production methods for ethyl acetate and butyl acetate, two excellent nitrocellulose solvents (Haynes, Vol. 6). Large scale production of ethyl acetate was started in 1919 by a continuous process worked out by Arthur A. Backus, research chemist of the company. Other solvents developed in the following years by the research department of the company were: methyl acetate, amyl acetate, ether, isopropanol, diethyl oxalate, and ethyl lactate (Haynes, 1954, Vol. 6).

Ethyl lactate had been a laboratory chemical since the days it was discovered by Lepage (Lepage, 1844). Its properties had extensively been investigated, for instance by Wurtz (see Holten, 1971). Commercial interest arose shortly before World War I when it was proposed as a solvent for cellulose derivatives (German Patent 278,487 of Byk, Oranienburg, 1910). During the war some use of ethyl lactate was made in the preparation of airplane dope, a coating applied to the canvas wings of airplanes (Ch. 7, 216). Boehringer also was interested, in 1910 the company had the trademark “Lactonal” registered for its brand of ethyl lactate. The Byk company, then incorporated in Byk-Guldenwerke, registered its ethyl lactate as “Solactol” in 1923.

Neither Byk’s method, nor the one claimed by Güstrow (German Patent 171,835 of 1906), was a very suitable one for large scale production of the ester. When high boiling esters became very promising products for car coating formulations the American chemical industry put much effort into the development of efficient large scale production methods for such products. For ethyl lactate John A. Steffens of U.S. Industrial Alcohol invented a production method (U.S. Patent 1,421,604, 1922).

In this semi-continuous method, reaction water as well as water contained in the reacting substances lactic acid and alcohol, is removed continually in a so-called azeotropic distillation process:

to the reaction mixture a third liquid (usually benzene) is added that is miscible with the alcohol but substantially immiscible with water. The three low boiling liquids, alcohol, water and benzene, distill over in a constant ratio at a temperature much below the boiling point of water. When benzene is added as azeotrope former, the boiling temperature is lowered to about 80°C;

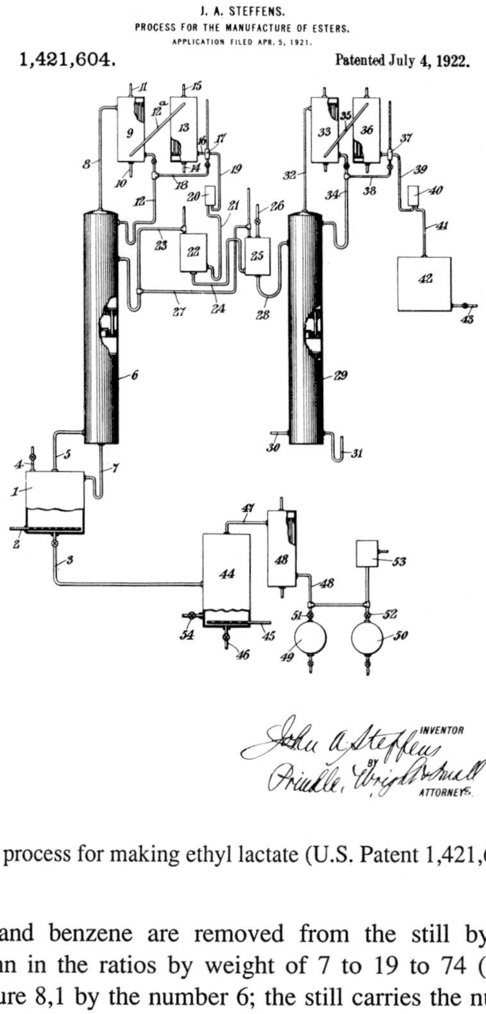

Fig. 8,1. Steffen's process for making ethyl lactate (U.S. Patent 1,421,604).

water, alcohol and benzene are removed from the still by means of a rectifying column in the ratios by weight of 7 to 19 to 74 (the column is indicated in Figure 8,1 by the number 6; the still carries the number 1). The mixture of vapors is condensed, which causes separation into two layers, an upper layer rich in benzene and a lower layer consisting mainly of water and alcohol. In chamber 22 the two layers are separated, benzene is returned via overflow pipe 23 to the rectifying column. To the water-alcohol layer scrubbing water is added to drive the 10% of benzene out which this layer still contains. The benzene again forms the upper layer and is also returned to the rectifying column. The alcohol is concentrated in rectifier 29 and can then be used again. At the end of the reaction the ester, which is formed and left behind in the reactor, still contains traces of benzene and alcohol [as well as condensation products formed by interesterification of lactic acid]. The crude product is passed out to vessel 44 in which it is subjected to vacuum distillation to obtain essentially pure ethyl lactate.

Claim 6, of the Steffens patent summarizes the process:

> The process which comprises esterifying ethyl alcohol and lactic acid in the presence of benzol [benzene], while removing from the reaction mixture vapors of water, the alcohol and the benzol, rectifying said vapors, separating the condensate into two liquids without the addition of water, containing respectively high and low percentages of the benzol, and subjecting the liquid layer containing the low percentage of the benzol to washing to remove said benzol therefrom.
>
> \- *Steffens, 1922*

The Steffens process was an important technological breakthrough; the process has, with minor alterations, remained the most practical and economic process for making ethyl lactate. At the time it was invented one serious problem, however, was inherently attached to the process: the corrosion problem. To drive off the water within a reasonable period of time the still had to be heated to 100–120°C, and almost none of the usual metals could stand the attack of lactic acid at this elevated temperature. It is not known which metal U.S. Industrial Chemical applied but is stands to reason that they used iron clad with lead. This was then the best solution, although it required much maintenance.

The literature search of the period from 1922 to 1940 unfortunately did not result in any direct information on volume produced, nor did it reveal any facts about the pricing of the product. For the further part of the ethyl lactate story we have, therefore, to rely on indirect evidence.

Bruce K. Brown called the U.S.I. product, ethyl lactate, a losing horse. The logical question then arises of what happened to the two other products, butyl acetate and Cellosolve. The fate of butyl acetate is well recorded (Haynes, Volumes 4 and 5), production rose from 360 t in 1923 to almost 10-fold in 1925 (3,400 t) and thereafter more gradually, but still very impressively to the hundredfold level of 35,300 t in 1939. The price of the product decreased dramatically in the same period, from 1.41 $/kg in 1925 to 0.31 in 1931, and to 0.18 in 1939.

About Cellosolve less quantitative information exists. As a solvent it resembles ethyl lactate more than butyl acetate does: like ethyl lactate it has a hydroxyl group which imparts two desirable effects. The presence of this group on the one hand increases the boiling point, and on the other hand enhances the solubility of the gums which are co-dissolved with the nitrocellulose. Table 8,III compares the properties of the two products with those of butyl acetate.

Table 8.III. Some properties of high boiling nitrocellulose solvents.

solvent	boiling point	relative volatility	tolerance for toluene
butyl acetate (85%)	17 – 130	12.5	3.1
ethyl lactate	145 – 155	80	5.5
Cellosolve	132 – 137	43	5.3

Source: Ullmann, 1960.

Ullmann's Encyclopedia (1960) from which these data have been taken, distinguishes between low boiling, medium boiling and high boiling active solvents. "Our" solvents all belong to the high boiling group, and in this group ethyl lactate has the best figures. About the amounts used Ullmann says: about 5% of the solvent mixture should consist of an active high boiling solvent. This puts the maximum amount of such a solvent at about 4% of the volume of lacquer. From known volumes of U.S. cellulose laquer production (Haynes, Vol. 4 and 5) the volumes of high boiling solvents can be estimated; they rose from, maybe, 600 t in 1925 to about 8,000 t in the 1930's.

How did it happen that in the long run Cellosolve (and related products) completely overran ethyl lactate?

More about Cellosolve, and the Beginnings of Petrochemistry

lit. Kirk-Othmer, 1953; Haynes, Vol. 3, 1954.

Cellosolve, the commercial name of ethylene glycol mono ethyl ether [$C_2H_5OCH_2CH_2OH$], was one of the first products of a new industry, the industry of petroleum chemicals, or, for short, the petrochemical industry.

Already during the 1910's it had become apparent that the traditional way of gasoline production in which usually not more than 10 to 15% of the crude oil was converted into gasoline, would not suffice to produce the large volumes required for the rapidly increasing number of automobiles. In 1913 a new process had been introduced called thermal cracking, or cracking for short. In this process the less volatile fractions from the first distillation were returned to a cracking installation in which they were subjected to high temperatures and pressures. The heavy, high boiling fractions were broken down, cracked by this treatment and a large additional amount of gasoline could be distilled off. Thanks to this process the yield of gasoline more than doubled and consequently much less residual product had to be sold at depressed prices. In the new thermal cracking process a special group of gaseous by-products were formed which until then had only been prepared on a laboratory scale, the so-called olefines, lower hydrocarbons with one or more double bonds.

Chemists knew the very reactive properties of the olefines, and already in the early 1910's research was begun to convert these new raw materials to products of higher value than their heating value. The first application was realized by the Standard Oil Company (New Jersey) in the production of isopropyl alcohol and some higher alcohols (1920).

The real birth rights to the industrial development of petrochemicals, go to the Carbide and Carbon Chemicals Company, however. Carbide and Carbon Chemicals, since 1917 belonging to the Union Carbide and Carbon Corporation, had first been active in carbide production and acetylene chemistry. In 1914 the company had started a research program at the Mellon Institute to develop production methods for aliphatic chemicals based on unsaturated hydrocarbons (olefines). A few years later a pilot plant was built to work out suitable processes for the most promising products. A manufacturing plant was built at Charleston, West Virginia in 1924. Carbide and Carbon had planned to begin with isopropanol to be used as antifreeze instead of denatured alcohol, ethylene glycol as a substitute for glycerol, and acetone as a low boiling solvent for nitrocellulose lacquers. While the plant was built prices of isopropanol and acetone started to decline and it was decided to put all emphasis on ethylene glycol production.

Glycol production commenced in 1926 with an output of 2,300 t and twice that amount in 1927. Right from the start ethylene glycol proved to be a winner in the field of antifreezes. By 1939 sales had surpassed 50,000 t, ten times as much as in 1927, the first year of full scale production at Charleston.

In the meantime, the company's chemists tried to develop more derivatives of olefines, and a whole series of new chemicals came from the new plant in the next five years. In Haynes' account:

Cellosolve was one of the first fruits of a new direction which Carbide and Carbon research took at this time: the deliberate synthesis of materials with chemical and physical properties designed to meet specific requirements of users. In the lacquer industry's hunt for improved nitrocellulose solvents, mixtures of alcohol and ether had useful characteristics which were counterbalanced by the disadvantage of exceedingly rapid evaporation. Consequently a chemical was sought that would have the chemical properties of alcohol and ether, combined with a more tractabled rate of evaporation as well as a mild odor. Thus Cellosolve was synthesized and other members of this family followed in logical sequence as the solvent using industry began to define their requirements in more and more specific terms. Carbitol (diethylene glycol monoethyl ether) appeared in 1928, butyl Cellosolve and butyl Carbitol the same year, and methyl Cellosolve in 1929.

- Haynes, 1954, Vol. 4, 198

Why Ethyl Lactate Did Not Make It

The literature does not give a clear cut answer to the question why ethyl lactate did not become an important lacquer solvent. It is rather unlikely that this was related to the properties of the solvent. It is an excellent solvent for nitrocellulose, and a fair one for most of the gums, it has a large tolerance for hydrocarbon diluents and also for water. One drawback was the sensitivity shown by some pigments for residual acidity, this could result in flocculation or another kind of instability. When acid was present in minor amounts in the ester this usually was due to improper manufacturing procedures, however. Sometimes the lacquer manufacturer was at fault when he applied some water in his recipe and exposed the lacquer to too high temperatures over a longer period of timc (Kümmel, 1938).

It is most likely that price aspects had a predominant influence. Although neither the price of ethyl lactate, nor that of Cellosolve has been found in the literature, a certain indication of price developments can be obtained by comparing the prices of lactic acid and ethylene glycol over the period.

Glycol was made by reacting 1.chloro 2.hydroxy ethane ($ClCH_2CH_2OH$), commercial name ethylene chlorohydrin, with water. Cellosolve was made in a similar way by reacting the chlorohydrin with ethanol. The cost of the product Cellosolve would, of course, be closely related to the costs of the two raw materials and, because of the commodity character ethylene glycol achieved within a few years after its introduction, also to the price of glycol. Ethyl lactate was made from a special grade of lactic acid (the so-called plastic grade), which carried a somewhat lower price than the edible grade. List prices of lactic acid published monthly in "Industrial and Engineering

Chemistry" indicate that the price of light colored lactic acid in the 1930's varied between 0.50 and 0.56 $/kg, a level not much lower than the (edible) price in the 1920's. Import prices, including duty indicated a 20 to 30% higher price, but these were probably positively influenced by simultaneous imports of the more expensive pharmaceutical grade from Germany.

Table 8,IV presents the best estimates of the prices of the three products involved.

Table 8.IV. Feedstock prices for Cellosolve and ethyl lactate production ($/kg).

	1927	1933	1935
ethylene glycol	0.66	0.55	0.31
lactic acid (100%)	0.65	0.60	0.55
ethanol (188 proof)	0.10	0.09	0.08

Considering the fact that, theoretically, the lactic acid "content" of ethyl lactate is 76% against a glycol "content" of Cellosolve of 68%, it is evident that Cellosolve already in or before 1933 could be offered at a much lower price than ethyl lactate. Added to the advantage in raw material costs enjoyed by the sole Cellosolve producer came the advantage of an integrated, large scale production from raw material to final product.

According to the 1922 U.S. Census Report on Manufacturing there were two producers of ethyl lactate: the U.S. Industrial Chemical's plant at Curtis Bay, near Baltimore (MD), and Florasynth Laboratories (Inc) at Unionport (NY). It is quite likely that also some imported material from German producers (Boehringer or Byk-Guldenwerke) reached the American market. The Byk company supposedly had begun ethyl lactate production during World War I, a production which, however, probably never surpassed 10 t per month (BIOS report 1652, Solvents & Plasticizers, 1946). About 1935 Byk-Gulden offered ethyl lactate in drums at 0.77 $/kg delivered (British) Indian ports (From a company price list).

The conclusion is evident: ethyl lactate remained a small product despite its promising properties and the intensive marketing efforts of at least one producer (U.S. Industrial Chemical). It is highly unlikely that the U.S. lacquer industry in the period under consideration in any one year used more than a few hundred tons of the product. This conclusion does not automatically imply that the sales of Cellosolve and related petrochemical solvents were in the thousands of tons bracket: for many uses combinations of for instance butyl acetate and butanol were just as suitable and much cheaper.

Production of Lactic Acid Almost Disappears from the American Scene

One requirement for cheap production of ethyl lactate, the availability of sufficient quantities of cheap lactic acid, also was not met. 1919 Production had surpassed 2,000 t (100%), but a few years later the American lactic acid industry found itself to be extremely vulnerable to German competition again. What had happened?

When the war was over all orders for military materials had been cancelled immediately. Production of many goods and chemicals suddenly fell to a fraction of the wartime needs. Lay-offs were the order of the day. The buying-power of the population diminished as a consequence. In short, the American economy was suddenly thrown into a heavy depression. Yet, although the depression only lasted a short time (until 1921), the industry at once seized the opportunity to demand restrictions on foreign imports.

When the Democratic President Woodrow Wilson lost the 1920 elections to the Republican Warren Harding, the industry was happy with a business friendly administration coming into office. Although protection of the American industry had been traditional Republican practice, this time the industry lobby was more successful than ever: customs duties were increased to historically high levels (Fordney-McCumber Act, 1922).

The rates of the duties were fixed only after Congress had collected and investigated relevant material concerning the positions of the manufacturers and users of the various products. Chemical Age covered the plight of the lactic acid industry in detail (Chem. Age, **30**, 1922, 196):

> The dependence of many branches of domestic chemical industry upon the enactment of tariff rates that will permit the continuance of the American industry, is illustrated in the history of lactic acid manufacture, an industry of American origin.
>
> In 1917 there were five American manufacturers of lactic acid with an output of over [860 t] (100% basis), which was increased to over [2,270 t] in 1919. In 1921 German imports amounted to [274 t], which were marketed at a price prohibitive of continued domestic production because of the disordered conditions in Germany.

It is of course remarkable and incomprehensible how an industry selling more than 2,000 t in one year (1919) can be faced with disaster two years later when less than 300 t of foreign product appears on their market. The Wamesit Chemical Co., which had bought the plant of the Avery Chemical Co. at Lowell, MA, gave the Congressional Commission on February 16, 1922, the following written explanation:

> We serve several manufacturers in New England; some are small consumers; others are quite large consumers. Due to our ability to effect immediate and prompt delivery of high-quality acid, to convenience, established business relations, and friendships, we have held the bulk of the business in our immediate vicinity, and these users particularly would suffer if we were forced to discontinue and leave the market to the mercy of foreign manufacturers, who would undoubtedly not be slow to take advantage of the situation and raise prices.
>
> In summary, we can produce both edible and technical lactic acid in the volume and of the qualities required by consumers in the United States, and sell same at reasonable prices consistent with prices which prevailed before the war plus the extra costs of production due to higher costs of labor, coal, raw materials, repairs, etc. We can not compete with the importer of foreign lactic acid unless we have a safe margin of protection in the tariff.
>
> \- *John H. Murphy, 1922 (Chemical Age, l.c.)*

The letter also gives a list of "companies, which have in the past manufactured considerable quantities of lactic acid in the United States ...., have stopped manufacture, and will not resume...":

American Alkali & Acid Co., Bradford, PA
Atlantic Chemical Co., Mansfield, MA
Avery Chemical Co., Lowell, MA
Westchester Chemical Co. Pelham, NY

The tariff suggested by Wamesit which would give adequate, but not "undue protection", is listed in Table 8,V, together with the old tariff and the new one as finally fixed in the Fordney-McCumber Act.

The table shows that the demands of the lactic acid industry were met only halfway. But although the situation had certainly been blown up by Wamesit, it is a fact that for many years only a few U.S. producers stayed in the market. From 1923 onwards United States Census Reports only mention one (1923 and 1924) or two (1925 to 1928) lactic acid producers. One of these producers was the Wamesit company. The last year the Wamesit Chemical Co. is listed in the Lowell Directory is 1926. The company then disappears

Table 8.V. Import duties on lactic acid (in $cts/lb).

concentration or grade	Underwood Act, 1913	proposed by Wamesit	Fordney-McCumber Act, 1922
to 30% c2 1.5		5	2
30 – 55%	1.5	10	4
55% & over	1.5	15	9
edible	minimum 25% ad valoram	25	minimum 25% ad valoram

Source: Haynes.

from the records and no successive company on the same site has been found (private information from Charles Parrott, Lowell Historic Preservation Commission, 1986).

The seriousness of Wamesit's problems is also presented, with quite some exaggeration, in the March 30, 1922 letter of the company (Chemical Age, l.c.):

> The Wamesit Chemical Co. is closing down its plant today having its tanks filled to overflowing with manufactured lactic acid which we have been unable to sell in any great quantities, largely because of German competition. We are willing to make allowance for hard times, but it is the German competition which strangles our industry whether times are good or bad.

In the letter of February 16, 1922 Wamesit had not referred to German competition so explicitly, but the chosen wording pointed in the same direction:

> Upon investigations we find that the exceedingly low cost of labour (about one-twentieth of our cost) and the extremely depreciated value of the currency (which is now over 2,000 per cent below normal value) in foreign countries of manufacture, make it possible for the foreign manufacturers to produce and sell in this country far below our cost of production, and, of course, at a price we cannot compete.
>
> To these facts we contribute our loss of all our export business. We also attach a certified copy of our balance sheet covering our operations during the last six months of 1921, ..., showing a loss to our company of $10,709.96.
>
> *- Robert F. Marden, 1922 (Chemical Age, l.c.)*

The duty imposed under the Ford-McCumber Act amounted to about 60% on the value of the German lactic acid export price to the U.S. For the next two years the protection indeed proved to be quite sufficient (Table 8, VI).

Table 8.VI. German exports of lactic acid and lactates (in t).

year	total of exports	to England	to France	to the U.S.A.	to Japan
1912	1.780	521	316	177	42
1913	2.050	764	282	204	28
1920	893	177	107	56	17
1921 (8 m)	692	108	3	285	21
1922	1.362	373	4	185	79
1923	740	194		34	22
1924	835	112		86	68
1925	1.681	295	28	360	127
1926	2.221	355	26	415	243
1927	2.193	413	22	264	107
1928	2.175	327	26	225	147
1929	2.275	297	62	247	138
1930				196	
1931				162	
1932				186	
1933	1.574			244	
1934	1.345			163	
1935	1.278			11	
1936	891			12	
1937	1.025			22	

References:
1912 – 1929, official German export statistics.
1933 – 1937, figures published by trade journals (Chemical Age and Chemisch Weekblad).
U.S. imports 1930 – 1937, data from Sheehan, 1944 (only concentrations of 55% and higher).

In 1925 imports from Germany started to rise again, still at the export price of about 0.30 $/kg (German statistics). Apparently the American market was growing beyond the capacity of the local producers. This trend continued in 1926, but reversed in 1927. The small boom of a few hundred tons in 1925 and 1926 may reflect the sudden demand for ethyl lactate, which disappeared when ethylene glycol ethers proved to be more economical solvents. (1926 also is the last year that Wamesit Chemical Co. is listed in the Lowell Directory. The company apparently stopped production in that year, maybe also because of the failure of ethyl lactate.)

The U.S. Census Report of 1928 lists as producers:

> Du Pont, Deep Water Point, NJ, (which is opposite Wilmington on the Delaware River), and
> Apex Chemical Co., Elizabethport, NJ (near Newark).

In the same year German export prices to the U.S.A. suddenly increased by over 100% to 0.63 $/kg and remained high in 1929 (0.54 $/kg). The old price was more or less equal to the inland German price for edible lactic, the new price equal to the German price for pharmaceutical lactic acid. It is not clear what happened, but a shortage in the U.S. market seems a more likely reason for the price increase than a complete turn about in the quality demanded. (The volume hardly changed at all.)

The old Harrison plant at Grays Ferry, Philadelphia, since 1917 belonging to the Du Pont company, had by the end of the 1920's become the most important producer of lactic acid in the United States. Company brochures show that technical lactic acid was primarily pushed as a tanning auxiliary (Hagley Museum files, 1988). At a later stage the product was also applied captively in the production of moistureproof cellophane wrappings (Charch and Bateman, 1939).

On December 1, 1928, Du Pont bought the Grasselli Chemical Company of Cleveland, Ohio. The next year the entire business of Du Pont in acids, heavy chemicals, lithopone, dry colors and pigments was transferred to the Grasselli company, now run as a division of Du Pont (Grasselli Company brochure, 1929). The large plant of Harrison's at Grays Ferry was also included in this transfer of activities.

Lactic Acid Production in Germany in the Aftermath of World War I

Was it true that German manufacturers were so aggressive when peace returned to Europe? Did a special reason exist for such aggressiveness, or did they just want to recover lost markets?

An indication of the general state of affairs of the German lactic acid industry in the 1920's can be deduced from Table 8, IV, of German exports and Figure 8,1, showing the annual lactic acid production of E. Merck, Darmstadt.

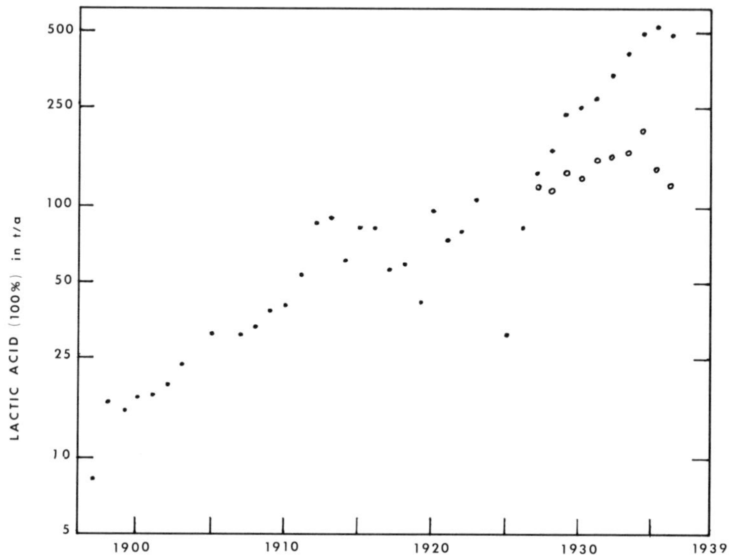

Fig. 8,2. Lactic acid production of E. Merck, Darmstadt.
Output figures have been converted to 100% lactic acid, and totalized. Larger circles refer to pharmaceutical lactic acid.

It is clear from Table 8,VI (p. 256), that German exports only slowly attained their prewar importance. The new producers in England and France as well as the largely expanded American producers did of course not rejoice over the return of the Germans to their market. The German producers, however, were able to deliver better qualities and slowly succeeded in recapturing part of the export market.

Similar efforts were, not surprisingly, made by the whole of the German chemical industry. The Allied Powers had expected such a move and being moreover much interested in the remarkable prewar and war achievements of the German chemical industry had decided (in 1918?) to have a study made of this industry.

In the summer of 1919 a mission of the Association of British Chemical Manufacturers visited the occupied areas along the Rhine. Although accompanied by a military representative the mission had no special powers. For its information it had to depend on personal observation and the courtesy of the German directors of the plants visited. "In the French Zone, the chemist in charge of the process was directed to afford the information given; if he failed to do so, the French officers in the works, who were generally technically qualified men, were usually in a position to give details of the plant and processes." (Association of British Chemical Manufacturers, 1920).

From the section "Fine Organic and Pharmaceutical Products" of the Report of the Mission a few quotations are illuminating:

In the case of fine organic chemicals, the processes examined included those for the manufacture of salicylic acid, phenacetin, antipyrin, salvarsan, lactic acid and chloroform;

The information given to the members of the section was very meagre; they therefore have made it their endeavour, rather than deal with details, to determine the basic principles which must be followed by British manufacturers if they are to enter into successful competition with established German undertakings.

The German success is ascribed to the following factors:

1. Scientific equipment and control
2. Mass production
3. Efficiency and sufficiency of plant
4. Cooperation with other manufacturers and elimination of wasteful competition.

Too much emphasis cannot be laid on the fact that all the large and many of the smaller works are equipped both with laboratories for scientific research, directed to the discovery of new products and for technological inquiry directed to the improvement and control of processes.

The opium house [sic !] of Boehringer has a separate laboratory for the chemists in charge of processes, in addition to a general laboratory for manufacturing processes. [C.H. Boehringer Sohn had taken up the production of cocaine in 1905. Thereafter the Scientific Department had also developed processes for the preparation of pure morphine and codeine (Heusner, 1974).]

It will be of general interest to manufacturers of pharmaceutical products to know that of the staff at Boehringer's 10 per cent of the total male employees are trained chemists. It is clear that to equip their works on these lines British chemical manufacturers will need a large supply of well-trained chemists who are not only able but also anxious to specialise in some particular branch.

- ABMC Mission, 1920

Peaceful conditions did not return to Germany, however. Severe economic problems and much political unrest troubled the country in the post-war years (and in many respects paved the way for the growing influence of Hitler's Nazi Party).

French Troops Occupy the Ruhr

After the defeat of Germany the Allies had imposed large penalties on the country. The total amount of money to be paid in so-called Reparation Payments was fixed at 132,000,000,000 gold marks. This sum was by any standard far in excess of what the country was able to pay.

Reparation payments had to be partly paid in kind. Dyestuffs was an important group of products in which payments in kind were effected. Lactic acid was also included, Belgium for instance received about 300 t of lactic acid as "reparation payment" (official German statistics).

Due to the very poor economics of the country and the huge reparation payments the German government was faced with large budgetary deficits. These were met by the printing machine, more and more money with ever higher nominal values was printed. The value, of course, declined in proportion. In 1922 the value of the mark dropped from 162 to the dollar to 7,000 to the dollar.

In the summer of 1922 the government of the German Republic – the Weimar Republic – not being able to pay the yearly terms of the Reparation Debt without inflicting further great damages to the shaken economy, asked for a moratorium. The French government refused this, and when the Germans actually did not live up to their commitments, occupied Germany's most important industrial area, the Ruhr area in January 1923. To prevent the French from collecting any revenue from the occupied regions the Germans then declared a total strike for the Ruhr. Payments to support the striking population also had to be supplied by the now virtually bankrupt German government. No wonder that inflation reached unbelievable propertions, in July 1923 160,000 marks bought one dollar, in November of the same year the rate was down to 4,200,000,000,000 marks to the dollar!

In the fall of 1923 the Germans gave up their opposition against the French occupation, the Government also reorganized the monetary system. New payment arrangements were made with the Allied Nations in 1924 (the Dawes Plan); the French troops were gradually withdrawn and by the summer of 1925 the situation had more or less returned to "normal".

These years shook the German economy to its foundations. Much private wealth was annihilated by the inflation and business in general was also severely affected. Only in some sectors where it was possible to immediately convert credits into raw materials and other goods, did companies not fare too badly. Such was the case with exporting companies in particular.

Merck Reports Severe Raw Material Problems

The story of Merck tells, however, how extremely difficult the procurement of raw materials during the inflation period was.

During the war Merck had tried all kinds of raw materials for lactic acid production. In the first year of the war raw sugar and potato starch were available in sufficient quantity. In later years, the fermentation department had to resort to any kind of carbohydrate containing product it could lay its hands on (various Manufacturing Reports of those years).

The 1916 Report mentions trials with chestnut meal, shredded sugarbeet, [shredded] whole potatoes and molasses. In 1917 no allocations for potato starch and sugar were granted, which hampered production in an extreme way. The situation did not improve much when the war was over. In 1920 Dr. Gruschwitz reports that the prohibitive price of import sugar again necessitated the use of other raw materials. Chestnut meal was tried again, as well as shredded sugar beet, whey and molasses. Only after much experimenting did the last two products give acceptable results. Gruschwitz had good news, however:

> Accidentally I heard that the competition has used potato waste meal for the fermentation since the beginning of the war. The supplier was the "Trocken-Kartoffel-Verwertungsgesellschaft" (Teka) in Berlin (see Intermezzo 6, 285). This [fact] explains why the competition during the war always had abundant supply of raw material. While we had to use expensive sugar [they] procured potato meal at low prices.... We have immediately ordered from the Teka [as well].
>
> - *Gruschwitz, 1920*

The Manufacturing Reports of 1923 and 1924 again complain about raw material problems and report of many trials done with inferior raw material. Waste potato meal varies much in quality, but is still preferred. In 1927 the reporter is more optimistic; the problem apparently had been reduced to analytical control of the purchased material. The need to accept inferior material had gone.

What Happened to Boehringer?

Allied forces had occupied the Rhineland area in 1918, and stayed there as a kind of guarantee of German "good-behavior". The Ingelheim area belonged to the French zone, it was occupied on December 12, 1918; it would remain under French control until December 1, 1925 when British troops took over the area. Only on June 30, 1930 did the last occupying forces leave the region (Diehl, 1974).

The Boehringer Jubilee Book, published on the occasion of the 100th anniversary, portrays the situation as follows (Boehringer, 1985):

> Great difficulties were experienced when the war was over. The inflation was so fast that the Central Bank had problems to have printed the required currency in time. To cope with this problem private companies were allowed to issue emergency money with restricted validity. This was also done by the firm C.H. Boehringer Sohn [August 1923]. The company was known as reliable and trustworthy, hence the emergency money was accepted readily.
>
> Additional difficulties arose as a consequence of the French occupation. These led to the expulsion of the owner of the company [and a large number of other leading personalities as well as railroad personnel who had participated in the general strike]. Albert Boehringer did not resign, however, but went to Hamburg to realize ... the erection of a second works at a different location.
>
> In 1925 the manufacture of opiates, until then produced in Ingelheim was begun [in Hamburg].
>
> ... the sales department "tartaric and lactic acid" also was headquartered in Hamburg, while the production stayed at Ingelheim.
>
> \- *Boehringer Jubilee Book, 1985*

Albert Boehringer only took up domicile again in Ingelheim in 1930; the sales department "acids" returned in the same year.

Byk-Guldenwerke Did Not Suffer from Occupation

The main works of the Byk-Guldenwerke were situated in the east of Germany and were not directly affected by the occupation of the Rhine area. It is likely that the company used these years to consolidate its position.

In 1921, in an important move supposedly to secure a good raw material position for the production of the new lacquer solvent ethyl lactate, the company acquired Knab & Lindenhayn of Niederstriegis, in size the second lactic acid manufacturing company of Germany. All lactic acid production of the new company was thereupon centralized at the Niederstriegis plant, and ethyl lactate production seems to have been transferred to that Saxony location as well.

In the same year Byk-Guldenwerke moved company headquarters from Piesteritz to a central location in Berlin, and from there began (or restarted) its export activities. In the 1920's Arthur J. Bloemendal was general manager of the company, a very lively polyglot, who energetically pursued the export activities (Ulrich, 1978). Werner Ulrich, who had entered the company as a young sales trainee in 1929, remarked about this period:

> We Germans had a tough time after World War I, the aversion in England and France against us was so strong that we even had to adapt our clothing [to the British and the French habits].
>
> *-Ulrich, 1929*

Of course, Ulrich also had to learn to speak and write English and French meticulously. (Bloemendal controlled the outgoing letters of his export staff daily.)

Byk-Guldenwerke did indeed achieve a position as lactic acid manufacturer about equal to that of Boehringer, both companies claim that they were (around 1930) the largest producer of lactic acid in Europe. The importance of Byk-Guldenwerke is also attested by the fact that it was entrusted with the administration of the "Lactic Acid Convention" of 1926 (This Chapter, 270).

Lactic Acid Production in Great Britain

Before the war British imports from Germany had been close to 800 t/a, a very high figure considering a total volume of use in Germany of not more than 1,000 t/a and a U.S. consumption probably even below that amount. British production apparently was low or non-existent. Some lactic acid was manufactured for the textile trade in the 1890's by the firm of Hickson (Ch. 5, 154), but no information about later lactic production by this firm has been found.

After the war German exports to the United Kingdom never reached the old figures again (See Table 8, VI, 256). Although exact comparison is not possible due to lack of information on the relative amounts of the different grades and concentrations in the import figures, it is evident that much of the prewar German exports had been replaced (by locally produced lactic acid or by substituting acids).

Bowmans (Warrington) Ltd.

During World War I the firm of Bowmans (Warrington) Ltd. had begun to manufacture lactic acid in an attempt to serve its customers in the tanning industry with a chemical they were in dire need of (Ch. 7, 220).

In the 1920's the combined business of lactic acid, tanning oils and other tanning auxiliaries became quite profitable. Bowman, however, used most of the profits to support his other venture, the Cuerdley Chemical Co., which made chromates from imported ore. Reportedly a remarkably clever operation from the chemical point of view, but economically unsound (M.H.M. Arnold, private communication, 1988).

About 1926 the company embarked on a plan to improve or enlarge production facilities. This at least can be concluded from an editorial comment in "Chemical Age" of March 11, 1927:

> A British firm have ordered a plant for the manufacture of lactic acid and the whole of the first consignment of the plant for the earlier process has been delivered and set up as one of the results of the KID act [an act which regulated duty protection].

In the same year this was written, Bowman, the owner of the two companies, died. A consortium of tanners, the main customers of the company, then stepped in and took over the profitable part of the business, that is Bowmans (Warrington) Ltd. S.H.W. Pert was named the new managing director; he would remain at the helm until 1959.

The measures to improve production were continued by the new management: ..."many purer types of lactic acid have been produced [since]", the 1955 Jubilee booklet writes. It seems that some lobbying also was initiated to safeguard the company from German competition. Such is attested by a report on questions raised in the House of Commons in June 1928 regarding the duty on two qualities of lactic acid, viz. pale technical lactic, 50%, and edible lactic acid (Chem. Trade J. (Chem. Eng.), 1928).

The questioner, Mr. Fenby, explained in his request to lower or abolish duties on the purer qualities:

> ... the only lactic acid produced in His Majesty's Dominions is a crude dark cloudy quality; that this quality cannot be satisfactorily used in the textile industry, the artificial silk industry [regenerated cellulose], by manufacturers of light coloured and fancy leathers, and in other important industries; that manufacturers of foodstuffs and soft drinks are bound to use edible quality; ....
>
> *- Fenby, 1928*

The President of the Board of Trade, Sir P. Cunliff-Lister said in reply:

> Lactic acid, which satisfies the requirements of most consumers in this country, has been made here for some years past, and I understand that other kinds are now being produced in this country, and will be produced in substantial and increasing quantities.
> - *Sir P. Cunliff-Lister, 1928*

Considering the loss of German exports to Great Britain of at least 400 t/a the lactic acid plant of Bowmans must have been of that order of magnitude [After World War II the capacity is estimated at about 500 t/a calculated as 100% (Arnold, private communication, 1988). The company then made leather grade material (25, and 44%) and edible (80%).]

Howards & Son, Ltd., Later Named Howards of Ilford, Ltd.

When a group of tanners took over the Bowmans company in 1927 it was joined by one outsider, Howards of Ilford, a manufacturer of pharmaceuticals and industrial chemicals who shortly before had become interested in lactic acid as raw material for solvent production.

The Howards company of Ilford, founded in 1797, had grown in the nineteenth century to become an important manufacturer of inorganic pharmaceutical preparations (salts, etc.). In the 1920's the company became interested in organic chemicals as well, not only for the pharmaceutical trade, but also for industrial uses. Solvents, plasticisers and other technical chemicals were added to the program (Howards, 1947).

One of the solvents the company introduced was ethyl lactate, a venture which had as little success as the somewhat earlier efforts in the U.S.A. to promote ethyl lactate as a solvent for nitrocellulose lacquers. Howards then turned to pharmaceutical lactic acid, which they made by hydrolyzing ethyl lactate.

The Association of British Chemical Manufacturers lists Howards in 1935 as a manufacturer of "lactic acid, of the commercial and the pharmaceutical grade".

In the decade preceding World War II pharmaceutical lactic acid found an interesting new application in Britain in the production of a special quick-release energy drink, Lucozade. The idea came from a pharmacist in the North of England in the late 1920's. From "AIM", a quarterly issued by the Bowmans company from 1961 until 1967, this and the following information is borrowed:

Made from liquid glucose, which is one of the most readily assimilable sources of physical energy, it was originally designed as a drink sufficiently pallatable to seduce the jaded senses of the invalid and, thereby, to inject sufficient strength to make him **want** to get better. The recipe was quickly accepted by the medical profession, and the drink, which was originally called Glucozade, became almost a specific in hospitals, nursing homes and in home sick rooms.

Perhaps of greater social significance is the fact that, over the years, the drink has edged its market beyond the sick bay and is in daily use among many thousands of people who feel the need for physiological "uplift" in order to cope with modern living conditions.

- *AIM, Spring 1963*

As do all lemonade type products, Lucozade contains an acidic component which in this case is lactic acid, "being a very acceptable compound for human consumption" (AIM). Production and sales of Lucozade were

Fig. 8,3. Bottle of Lucozade. Reproduced from the Spring 1963 issue of AIM with permission from Croda-Bowmans, Widnes.

developed in the 1930's by the Beecham company. In the 1960's the the production had risen to "nearly 300 bottles a minute – an indication of the success of the drink" (AIM). Figure 8,3 shows a bottle of Lucozade. This production level then required over 50 t of lactic acid (90%) to acidify.

Howards also made salts of lactic acid (a 1948 brochure of the company mentions calcium lactate, sodium lactate and iron lactate), probably only in minor quantities.

Bowmans Also Make Edible Lactic Acid

In the 1930's Bowmans began with the production of edible lactic and calcium lactate. Their process was akin to that applied by most other producers (Machell, 1959,; Arnold, private communication, 1988):

At the end of the fermentation the broth was made alkaline with lime, boiled up to kill the lactic acid bacteria and achieve flocculation of all insolubles. The clarified calcium lactate solution was concentrated in a triple effect copper evaporator to about 17% (lactic acid content). Under very gentle agitation (one RPM) the solution was slowly cooled and left for about one week to crystallize. The sludgelike crystal mass was separated from the mother liquor by means of a rotary filter. Washing of the filtercake proved to be impossible due to the high solubility of the calcium lactate and the cauliflowerlike crystal form. Calcium lactate was made straight from the scraped off crystal mass, which was melted and heated to remove surplus water. When a water content between 25 and 30% was reached the mass was solidified on a band cooler. The final material, calcium lactate pentahydrate, was only crudely broken up, kibbled, before being sold.

Edible lactic acid was made from the wet calcium lactate coming from the rotary filter by dissolving it in water, treating with the equivalent amount of sulfuric acid, and concentrating to the desired final concentration.

About half of the original amount of lactic acid from the fermentation broth was retained in the filtrate from the rotary filter. This portion was also converted with sulfuric acid to lactic acid. The crude, dark quality so obtained was still suitable for the leather industry, provided traces of iron were removed.

A major problem sometimes arose from the aproximate 50/50 output of technical and edible lactic acid. When sales were not realized in the same proportion large inventories of either product had to be stored, or the company had to resort to dumping of one of the two qualities.

Industrial Cooperation and Cartels in Germany

By the mid-1920's, when the currency was stabilized but the German economy was still in a depressed state, the idea that industry should join forces instead of going on with throat-cutting competition became strongly favored among German industrialists.

The problem of how to cope with the difficult economic situation of the country and the wish to revive the prewar superiority in, for instance, the chemical industry supported the move towards cartel-forming. Another factor of importance was due to the changing character of the chemical industry itself. Before the war a very large part of the production was still made in rather old-fashioned ways, batchwise in wooden, iron or leaded vessels. During the war new methods had been introduced to enlarge production without employing more people. New types of construction materials had been introduced, high pressure technology had been developed for the fixation of nitrogen (Haber-Bosch) in the first step of nitric acid manufacture, which in turn was needed for explosives production. In short: the chemical industry was becoming a capital intensive industry.

The immense financial problems arising from this development had been foreseen by Dr Carl Duisberg, general director of Farbenfabriken Bayer. In 1903 he had laid down his ideas on cooperation between the large German dyestuff works with the aim of cutting down on the very expensive development costs and capital investments by restricting competition and centralizing the purchasing and selling departments of the companies (Bäumler, 1968). The memorandum Duisberg wrote to this effect was sent to the other large dyestuff producers; it led first to the formation of two competing groups of companies (1904), followed in the war by a loose association of all dyestuff manufacturers (1916). After the war, in 1925, Duisberg's dream finally was realized by the founding of IG Farbenindustrie AG which combined amongst others BASF, Bayer, Hoechst, Agfa and Cassella.

The IG, which was organized as a group of companies under one management organization, had much more coherence than usually associated with a trust or cartel. Its formation paved the way for other more loose associations of German enterprises. Unlike the American attitude, the German authorities and public opinion did not dislike the formation of large trusts or cartels. The IG Farbenindustrie was even the object of quite a bit of admiration, it seems. The logic to use scarce financial means by eliminating competition, was evident (Waller, 1928) and in the interest of the nation. Many Germans clearly favored the efforts of the IG to reestablish Germany as the dominant chemical nation in the world (Greiling, 1938).

Of course, much of the deliberations favoring the formation of cartels were valid for other capital intensive industrial activities as well. In the 1920's a large number of fairly loose cartels were formed, also primarily with the aim to restrict (too much) competition. Saving on capital expenses and concentrating on foreign markets were other objectives, depending on the nature of the business. Usually such cartels were confined to one or a small group of related products. As a consequence the number of cartels became very large, in the chemical industry alone 47 cartels were unearthed in an enquiry of the German Ministry of the Interior (Enquete Ausschuss, 1930, reported by Haber, 1971).

For food [sic] acids cartels existed for tartaric, oxalic and formic acid (Waller, 1928):

> Production and sales of tartaric acid were controlled by the 'Vereinigte Weinsäurefabriken GmbH' at Wiesbaden, which company had also made arrangements with the Italian industry [Italy was the main supplier of raw material].

As German manufacture of citric acid had been terminated in 1924 no arrangement existed for this product. German production would be resumed in 1932 on the rather small scale of a few hundred tons per year (Dehio, 1966).

Wagenführ (Wagenführ, 1931) in addition mentions the "Essigsäure GmbH" with four members, founded in 1923. The argument given for the founding of this cartel provides a nice illustration of German opinion of the time:

> [The cartel has been founded] because of events on the market for technical and food grade acetic acid, in particular the price speculations from traders and the consequences thereof. The foundation is in the interest of the consumer, the legitimate trade, and is necessitated to sustain the well organized manufacturer and the continued employement of the personnel. The measures taken by foreign countries against German export also require a coordinated action by interior branches of related industries (quotation by Wagenführ from the 1930 investigation by the Ministry of Internal Affairs, the Enquete Ausschuss).

The German Lactic Acid Industry Fixed Sales Prices

Wagenführ presents a list of about 90 products (or product groups) for which certain price and other agreements are reported. About half of these agreements relate to pharmaceutical substances, the other half to chemical products. One of the products listed is lactic acid.

The lactic acid agreement has been retained in the archives of E. Merck, Darmstadt; partners in the agreement (called "Konvention") were

1. C.H. Boehringer Sohn, Chemische Fabrik, Hamburg,
2. Byk-Guldenwerke Chemische Fabrik Aktiengesellschaft, Berlin,
3. Chemische Fabrik auf Actien (vorm. E. Schering), Berlin,
4. E. Merck, Darmstadt.

The lactic acid "Konvention" was an agreement on prices and sales conditions only, no provisions were included to allocate fixed sales and production volumes to the partners. The terms of the Konvention applied to all sales, inland as well as export sales, with the exception of the United States of North-America [sic], Canada and England. The Konvention was concluded for a period of five years.

A few of the inland ex works prices for deliveries of at least 5,000 kg, inclusive packing (150 kg containers), fixed per January 4, 1926, were:

technical lactic acid 50% 0.67 RM/kg
technical lactic acid 80% 1.08 RM/kg
edible lactic acid 80% 1.26 RM/kg

French Producers Join the "Konvention"

Some renewed German competition had also been felt in France, in particular in 1925 and succeeding years (Table 8, VI, 256). Official French statistics present only figures of the total imports from all countries, but in the case of lactic acid these imports can be assumed to come from Germany only (Table 8, VII).

The volume of German sales to France in the second half of the 1920's was not more than about 10% of the prewar sales volume. Then, however, in 1929, a sudden increase to 62 t was registered at a much lower price than the years before. The French statistics show the effect even more clearly because

Table 8.VII. French imports of lactic acid.

year	industrial l.a. (t)	pure l.a. (t)	total (t)	price industrial l.a. (RM/kg)*	price pure l.a. (RM/kg)*
1928	28	8	36	1.46	2.57
1929	56	10	66	0.86	2.17
1930	61	5	66	0.86	3.08
1931	11	2	13	1.17	3.53
1932	5	1	6	1.07	4.70
1933	5	1	6	1.03	2.50
1934	4	1	5	1.07	2.40

* rate of exchange taken 1 RM = 6 French Fr.
Source: Official French statistics.

of the separate listing of industrial grade lactic acid and the pure product (pharmaceutical lactic acid).

In 1931 something of importance again happened to the French imports: the price of imported (German?) product rose by 26% and the volume dropped to less than 20% of that of the year before. It was not a coincidental matter that the new import price was very close to the price of the German Konvention.

French production of technical lactic acid had been considerably increased during World War I (Ch. 7, 218). In 1924 ten companies were listed as sellers of lactic acid, probably not more than three or four of these companies were real producers. According to Dehio (Dehio, 1966) only two producing companies survived into the 1930's[4]:

1. The Société Normande, since the mid-1920's a subsidiary company of the Société Chimiques des Usines du Rhône, and
2. H. Gouthière of Reims.

In 1928 the Société Chimiques des Usines du Rhône merged with another large chemical group, the Établissements Poulenc Frères[5]. The new company was named Société des Usines Chimiques Rhône-Poulenc, Rhône-Poulenc for short. Supposedly the new group did have two lactic acid plants, one in Petit Quevilly belonging to the Société Normande, and a second one in Saint Fons near Lyon, originally built by the Société Française des Glycerines.

The Société Normande manufactured the bulk of the French production of about 800 t/a (Chem. Trade J., 1930). It is not clear whether this statement includes the production of the Saint Fons factory. Commercially all lactic acid made in the Rhône-Poulenc organization in the 1930's was sold through the Société Normande (Dehio, 1966).

Based on the statistical evidence French producers probably joined the German Konvention sometime at the end of the year 1930 or the beginning of 1931. Strangely enough the Rhône-Poulenc group was represented twice in the extended Franco-German lactic acid convention. This can be concluded from the agreement concluded in 1933 with the then recently founded Dutch producer Schiedamsche Melkzuurfabriek. This contract was signed by the four original German producers and three French companies. The French manufacturers were:

Société Normande de Produits Chimiques, Paris,
Société des Usines Chimiques Rhône-Poulenc, Paris, and
H. Gouthière & Cie, Reims.

Concluding Remarks About the State of the Lactic Acid Industry Around 1930

Total world production of lactic acid in 1930 probably did not exceed the capacity of German and American producers of the last year of World War I. Idle capacity of German producers was to some extent responsible for increased competition in the 1920's. New producers in France and Great Britain also made the market tighter. American output of lactic acid fell during this decade, it seems that the industry only served the leather manufacturers with a low quality acid.

The production of edible lactic acid, although hampered by poor economic conditions, made some progress in Germany. In England steps were also taken to manufacture edible lactic acid, pharmaceutical lactic acid was introduced in the preparation of a special type of soft drink.

Notes

[1] Pounds and gallons in Tunison's papers have been converted into tons and m^3 , respectively.
[2] By non-solvents toluene and benzene are meant, these coal-tar based products were much cheaper than good lacquer solvents like butyl acetate and ethyl acetate.
[3] As a matter of fact production levels kept on increasing by percentages between 20 and 40%, except in the depression years. But even then growth figures of about 10% were achieved (Haynes, vol. 4, 1954).
[4] Only limited information regarding the development of lactic acid manufacture in France has been found. The largest manufacturer, Rhône-Poulenc, did not save prewar archives relating to lactic acid (private information).
[5] The tartaric acid producer "Fabriques de Produits Chimiques de Thann et de Mulhouse" also was taken up in the Rhône-Poulenc group of companies.

References

Association of British Chemical Manufactures, **1920**. *The Chemical Industries of German Rhineland*; ABCM publication: London.

Bäumler, E., **1968**. *A Century of Chemistry*; Econ: Düsseldorf, 66 and 74.

Boehringer, **1985**. *100 Jahre Boehringer Ingelheim*; Boehringer Ingelheim; Weisbecker: Frankfurt am Main.

Charch, W.H., and Bateman, D.E., **1934**. *U.S. Patent 2,159,007.*

Chem. Age (London), 1927.

Chem. Trade J. (Chem. Eng.), 1928.

Chem. Trade J. (Chem. Eng.), 1930, Jan. 7, 57.

Dehio, W., **1966**. *Die Industrie der Weinsäure, Zitronensäure und Milchsäure*; internal publication C.H. Boehringer Sohn: Ingelheim.

Diehl, W., **1974**. *Ingelheimer Chronik, 1899–1950*; Paul Giese: Offenbach am Main.

Greiling, W., **1938**. *Chemie erobert die Welt*; Wilhelm Limpert: Berlin.

Gunther, J., **1947**. *Inside U.S.A.*; Hamish Hamilton: London.

Haber, L.F., **1971**. *The Chemical Industry, 1900–1930*; Clarendon: Oxford.

Haynes, W., **1954**. *American Chemical Industry*; D. Van Nostrand: New York. Reprinted by Garland, New York, 1983; Vols 3, 4 and 6.

Holten, C.H., **1971**. *Lactic Acid*; Verlag Chemie: Weinheim, F.R.G.

Howards, **1947**. *Howards, 1797–1947*; Howards & Sons: London.

Kirk-Othmer, **1953**. *Encyclopedia of Chemical Technology*; Interscience: New York, Vol. 10, 184.

Kümmel, P., **1938**. *Farbe + Lack*, **1938**, 449.

Lepage, S.H., **1844**. *J. Chim. Méd. Pharm. Toxicol.* Sér.[2], **10**, 8.

Machell, G., **1959**. *Ind. Chem.*, June 1959, 282.

Merkle, C.R.E., **1953**. *Harrison Brothers and Du Pont in Philadelphia*; presented for "The City Historic Society of Philadelphia", April 21, 1953 (reprint received from the Hagley Museum).

Sheehan, M.J., **1944**. *Lactic Acid (Synopsis of Information)*; United States Department

of Commerce, Bureau of Foreign and Domestic Commerce; Report 2–11096.
Slicher van Bath, B.H. **1980**, in Fischer, E.J., Ed., *Geschiedenis van de Techniek*; Martinus Nijhoff: Den Haag.
Tunison, B.R., **1925**. *Chem.Metall.Eng.*, **32**, 93.
Tunison, B.R., **1926**. *Chem.Metall.Eng.*, **33**, 25.
Ullmann, **1960**, *Enzyklopedie der technischen Chemie*; 3rd ed.; Verlag Chemie: Weinheim; Vol. 11, 299.
Ulrich, W., **1978** *"So war es damals", im Bykoskop*; Internal publication of Byk-Gulden Pharmazeutika; Konstanz.
Wagenführ, H., **1931**. *Kartelle in Deutschland*; Krische & Co: Nürnberg.
Waller, P., **1928**. *Probleme der deutschen chemischen Industrie*; H. Meyer: Halberstadt.

THE RAW MATERIAL CHOICE: SUGAR OR STARCH

From the first beginnings of lactic acid production the choice of raw material has been one of crucial importance: for the quality of the final product as well as for the economics of the operation.

The first producers knew already that from the technological point of view sucrose was the best raw material available, either in the form of so-called raw sugar, or, preferably, in the form of refined or centrifugal sugar (Claflin, 1897, Schäfer, 1907). Before dealing with the supply position in more detail some attention should first be given to the chemistry of starch and sugar.

The Chemistry of Starch and Sugar

Sugar, of which the chemical name is sucrose or saccharose, is a disaccharide, composed of equal parts of fructose and glucose (Figure I6,1).

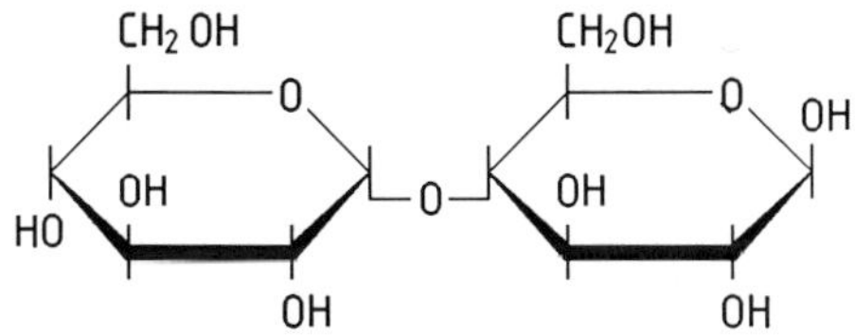

Fig. I6,1. Sucrose.

Starch, a polysaccharide, is a mixture of two glucose polymers. Potato starch and corn starch contain about 20% amylose, which has a straight chain, and 80% amylopectin which is branched (about one out of every 25 glucose units carries a side chain). The scientific name of amylose is poly-α-glucose, the α indicates the way in which the successive glucose molecules of the chain are bound to each other (Figure I6,2).

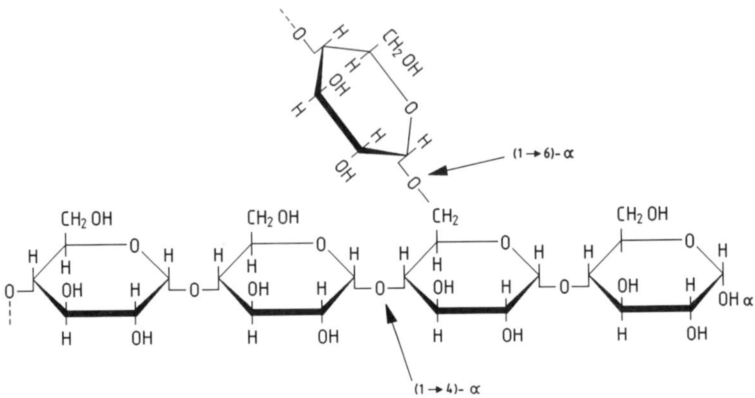

Fig. I6,2. Starch.

Starch Sugar

Starch is easily broken down into the primary constituent glucose (Figure I6,3). The chemist who detected this in 1811 was Gottlieb Sigismund Konstantin Kirchhoff then Professor in Petersburg, Russia (now Leningrad, U.S.S.R.). He boiled starch for some time with dilute sulfuric acid, purified and concentrated the reaction product, and found that he had obtained a sweet substance. Kirchhoff called this substance starch sugar; he believed it to be identical to cane sugar.

Primitive factories for the new sugar were erected in Germany within the next few years, the British blockade of French dominated continental Europe favored this industry as much as the beet sugar industry which also had its beginnings in the Napoleonic era. The new factories were not successful, and after the fall of Napoleon interest in starch sugar ebbed away. When after 1830 beet sugar began to replace cane sugar it was found that the bitter tasting molasses from beet sugar production could not replace sugar syrups based on cane sugar molasses. Then commercial interest in starch sugar and starch syrup then received a new impetus.

Starch or Glucose Syrup

By 1830 the main component of starch sugar was recognized as being identical with the sugar (grape sugar) that could be extracted from grapes, and different from cane or beet sugar (Berzelius, 1839). It was also known that incomplete acid hydrolysis of starch yielded a syrupy product with good digestive properties and suitable to replace colonial syrup in bakery products, and in other applications in the household.

The product was called starch syrup, or glucose syrup, or glucose for short. The commercial product was a viscous, amber colored fluid of about 80% concentration. The degree of conversion of the starch became indicated by the dextrose content, abbreviated as DE. (Dextrose is another name for the chemical compound glucose or grape sugar.) As the determination of the DE is based on a reduction reaction, higher reducing sugars also contribute to the DE number (pure maltose, see below, for instance has a DE of 50). Commercial starch syrups range in DE from 20 to 60.

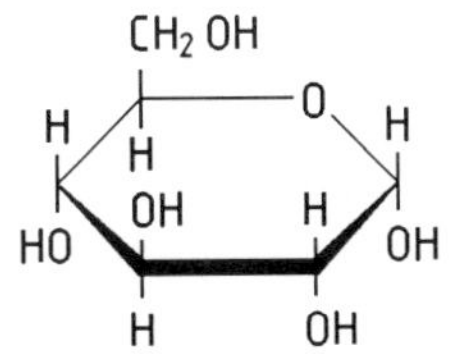

Fig. I6,3. Glucose.

Malt Sugar or Maltose

Starch can also be hydrolyzed with enzymes. This was already known to Berzelius, the famous Swedish chemist who called the action of acids as well as that of enzymes on starch catalytic processes (Berzelius, l.c.). The sugar formed under the influence of diastase, a mixture of enzymes isolated from malt, is not identical with glucose, however, but is a dissaccharide, a molecule composed of two α-glucosidically linked glucose molecules (Figure I6,4). This sugar was named maltose, and the enzymatically produced syrup maltose syrup, or malt sugar. Sometimes the commercial product which contained at best about 60% of maltose on dry substance, was also called maltose.

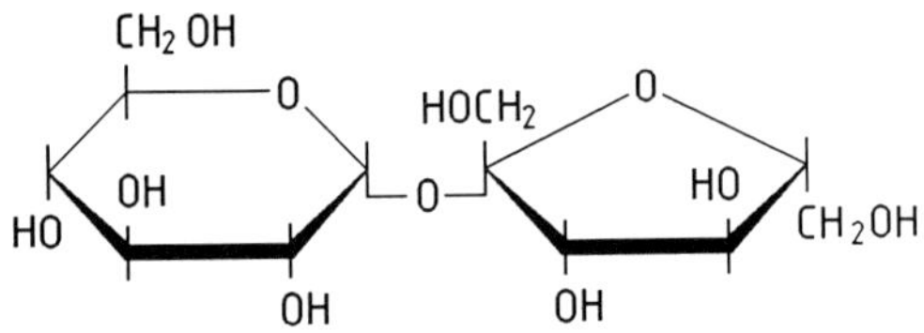

Fig. I6,4. Maltose.

A few production figures of 1890 (Dammer, 1896) reflect the relative importance of the various sugars on the German market:

beet sugar	400 plants, total output 1,200,000 t
potato starch	553 plants, total output 100,000 t (Enquote Ausschuss, 1930)
starch sugar	20 plants, total output 10,775 t
starch syrup	34 plants, total output 27,807 t
maltose syrup	2 plants, total output unknown

Dextrins

In acid hydrolysis as well as in enzymatic hydrolysis a substantial part (usually about 20%) of the starch is not fully converted to lower sugars. This non-fermentable fraction was called the dextrin fraction from a name earlier given to degraded starch which had been treated in the dry state with heat or acid.

Confusion in Nomenclature

For the uninitiated the vocabulary in use to denote the various commercial forms of glucose and starch syrups is highly confusing. Fortunately German and American names are rather similar, however. In this book the trade names are followed as closely as is practical:

The name glucose will be used whereever the properties of the pure chemical are at issue. The scientific name, also including information on the stereoisomeric properties, is D(+)-glucose.

The plus in the above formula means that a solution of glucose turns the angle of rotation of polarized light to the right, this is the reason that pure glucose is also called dextrose. We will reserve the name dextrose for chemically pure glucose in the crystalline form. The most important form in which pure crystalline glucose nowadays is traded is the hydrate, which we will call dextrose hydrate.

Glucose syrup, also called starch syrup, is the commercial syrupy liquid, of which the relative glucose concentration is given in DE.

Some enzymatically prepared syrups contain high percentages of maltose and very little glucose, their sugar content is nevertheless also indicated by a DE number. To obtain the maltose concentration of such a syrup the DE has to be multiplied by two. Syrups of this kind, when recognized as such, will be called maltose syrups.

Before the industrial production of pure dextrose hydrate made its headway, a rather crude dry product obtained by acid hydrolysis of starch was also called dextrose. We will call this product starch sugar.

The Raw Material Situation in Germany

In the German lactic acid industry potato starch was the almost universally used feedstock. It was much cheaper than sugar, and relatively easy to work with in the fermentation process. Only the purification was more complicated when production of edible or chemically pure lactic acid was considered. Until about 1910 the main applications of lactic acid were of a technical nature and the need for refined or chemically pure material was not really felt.

An exception was pharmaceutical lactic acid, which had been produced on an industrial scale since the mid-1890's by the firm of E. Merck in Darmstadt. Initially this firm used whey as feedstock, but in 1898 they changed over to starch sugar. Starch sugar at the time could only be obtained in a comparatively crude form, with a glucose content between 72 and 75%. From Merck's Yearly Manufacturing Reports:

1898/1899. Starch sugar at 0.22 M/kg is a suitable raw material
1899/1900. Technical starch sugar is 5% cheaper
1900/1901. The yield (calculated on the glucose content of the starch sugar) is 84%.
1904/1905. Instead of starch sugar white potato starch has been introduced as raw material. The conversion to glucose has been effectuated with sulfuric acid, which has the advantage of automatically producing sterile solutions.

In the two next decades conversion of potato starch with malt, the process used by Merck's competitors (Boehringer, and Knab und Lindenhayn), was contemplated many times and rejected because no improvements were seen. During World War I and some years afterwards inferior starches and even less suitable carbohydrate containing raw materials had to be used, resulting in inferior grades of lactic acid. Inferior potato starch (dried meal of whole potatoes) gave rise to colored lactic acid, and this was probably the reason that starch conversion with malt was given a serious try again:

1921. A sixth part of the required malt is suspended in water, the [total amount of] starch added under strong agitation, after which the suspension is heated to 70° C. The mixture soon attains a low viscosity, is thereupon cooled to 55° C and the remainder of the dry or green malt added. The saccharification is carried out at 55 to 56° under agitation until the iodine reaction of starch has disappeared. Since the introduction of the malt process we have been able to produce ever clearer acids. Decolorization nowadays is only performed with active carbon and hydrogen peroxide. We are also able to manufacture the 50% edible grade directly, omitting the ether extraction.

Why Sugar Was Not Used in Germany

The simple answer to this question is that refined sugar was too expensive for the fermentation industries, potato starch was much cheaper. This was not only the case in Germany, but in most other European countries as well. In view of the predominance of Germany in lactic acid production up to World War II the German situation will be treated in more detail.

Beet sugar production in Germany had its primitive beginnings in the Napoleonic era. When Napoleon proclaimed the Continental System, an embargo on the importation of British goods, the British retaliated with a counterblockade of all European countries which obeyed the French emperor. This caused a severe shortage of sugar. It was a fortunate coincidence that production of sugar from a certain variety of beets had been discovered a few decades earlier and some pioneer production had been carried out in the Kingdom of Prussia since 1802 (Deerr, 1950).

Industrial production of beet sugar seemed to be economically feasible and became strongly fostered by the imperial authorities. Production was set up in France as well as in Germany, with poor results, however. After Napoleon's defeat manufacture of beet sugar came practically to a standstill.

In the 1830's production of beet sugar again became of some industrial importance, first in France and Russia, not much later also in the German countries (Baxa and Bruhns, 1967). When the governments observed that the growing of sugar beets and the manufacture of sucrose could become valuable economic activities, they began to impose taxation on production. Levies on beet sugar were also strongly advocated by the importers of cane sugar – colonial sugar – which product, being a luxury item, in most states carried rather high duties.

In time the tax revenues on sugar production would provide many a European government with a relatively large source of income. It is known for instance that 10% of the income of the new German Empire, founded at Versailles in 1871, came from the sugar tax (Petzel, 1926).

By 1880 German sugar production began to exceed the needs of its own population. Export now became an important activity of the German sugar industry. Exporters received a refund related to the inland sugar tax. As the tax refund was given per actual kg exported, and taxation was levied on a fixed (low) factory yield, refunds per kg usually were higher than the tax. Table I6,I presents data on German sugar production, exports, prices and per capita use.

The substantial influence of the taxation on the inland price is very clearly demonstrated by this table. Another prime factor influencing the cost of sugar was the foundation, in 1901, of a sugar cartel in which all sugar producers and refiners took part. One of the first actions of the cartel was a fixation of a minimum sales price before taxes of 0.375 M/kg for centrifugal sugar.

The effects of a high inland price combined with a refund on exported sugar higher than the inland sugar tax were twofold: Sugar consumption in Germany was kept at a low level, the per capita consumption stayed at less than half that of the British or American consumer, and industrial use of sugar was, of course, rendered completely uneconomical.

As a consequence sugar exports increased to unprecedented levels. German exports in the first decade of the century were among the highest in the world, amounting to about 20% of the total international sugar trade.

A 1908 comment on these price policies (Vogt, 1908) – which were, with some modifications, also applied in other continental European countries – will still sound very familiar to present day students of the sugar policy of the European Community:

> The combined efforts of the cartels and of the bounty system were that within the bounty fed countries the people were burdened by the high price of sugar; without, people were enjoying the lower price of sugar because of the returns received from government by European producers on each pound exported.
>
> \- *Vogel, 1908*

Table I6.I. Sugar: production, export, price factors, and use in Germany.

year	production (1000 t) ref. 1	export (1000 t) ref. 2	cost-price (M/kg) ref. 3	sugar-tax (M/kg) ref. 4	export refund (M/kg) ref. 5	inland price (M/kg) ref. 6	use per capita ref. 7
1840	14			0.045			2
1850	53			0.12			3
1860	126		0.74	0.15		0.94	4
1870	186		0.68	0.20	0.23	1.11	6
1880	573	295	0.62	0.20	0.21	0.81	7
1890	1.320	520	0.32	0.20	0.22	0.55	10
1900	1.979	1.150	0.20	0.20	0.225	0.56	14
1910	2.590	1.130	0.20	0.14		0.42	21
1920	1.085	4	2.77	0.14		6.02	19
1925	1.593			0.21		0.52	
1930	2.529	430	0.21	0.105		0.60	26
1940	2.207	10		0.21		0.70	34
1950	1.600	360					32
1960	(2.400)	240	0.80	0.10		1.07	33
1970	(2.400)		0.78		0.65	0.78	(35)

General remark: the figures relate to the Wilhelminian empire until 1910, to the Weimar republic and Third Reich from 1920 until 1940, later figures to the Federal Republic (West Germany) only.

ref. 1: Dammer, 1896; Ullmann, 1969
ref. 2: Ullmann, l.c.; Teichmann, 1955
ref. 3: Figures are only indicative, 1960 and 1970 figures are based on government and E.E.C. fixed beet sugar prices.
Petzel, 1926
Teichmann, l.c., until 1910 cost price of raw sugar, the cost of refining was then estimated at M 0.07/kg.
Baxa and Bruhns, 1967 (raw sugar)
ref. 4: Teichmann, l.c. the 1840 and 1850 figures relate to 1841 and 1851, respectively. In addition to the sugar tax an excise tax was levied, about 0.9% until 1931 when it was doubled to 1.8%. By 1960 the level had been raised to 5%.
ref. 5: Teichmann, l.c.; 1970 E.E.C. export refund
ref. 6: Teichmann, l.c.; Baxa and Bruhns, l.c.; Günther et al., 1962; (1970) E.E.C. intervention price
ref. 7: Baxa and Bruhns, l.c.; (...) trade figures; various sources.

Between the two world wars the sugar cartel succeeded in keeping the German inland price almost constant. The government again helped by adjusting the sugar tax on inland production and a high levy on imports.

After World War II the combined efforts of the cooperating sugar companies, agrarian interests and the government of the Federal Republic of Germany succeeded in a continuation of the same policy still aimed at giving the farmer a reasonable compensation for his work and the industry a fair reward for its manufacturing costs. World market effects were hardly taken into consideration, the inland market was protected by high import duties, the consumer had to pay a high but more or less constant price, and excess production was exported with refunds to make the product competitive on the world market.

When the European Economic Community formulated its agricultural policy in the years 1967 and 1968 a system of intervention prices and export refunds was adopted which followed closely the lines the Germans (and the peoples of most other continental European countries) had been familiar with since the late 1890's (Ch. 12, 436).

Potato Starch and Starch Sugar for the German Fermentation Industry

German starch production figures are rather scarce, and, moreover, conflicting. The presentation in Table I6,II, from which starkly deviating figures have been omitted, is only meant to give an impression of the German market for this carbohydrate and to allow comparison of the economics of the use of sugar with those of starch or starch sugar.

In the nineteenth century production of potato starch had been set up as an adjunct to farming. The eastern part of pre-World-War-I Germany counted a great number of large rural estates, many of which were engaged in semi-industrial activities, more particularly in the conversion of potatoes into alcoholic spirits, or into starch.

In 1890 Dammer (Dammer, 1896) writes that 2,656 farms east of the Elbe River were engaged in alcohol production. The majority of the 553 potato starch plants of the country were also run by farmers in the eastern part of Germany. From an industrial point of view these rural plants were small.

The average alcohol producer processed about 370 t/a of potatoes to obtain 400 to 500 hl of distilled alcohol. The starch producers were a little bigger, they processed between 600 and 1,500 t of potato in a season of about 150 days to obtain between 100 and 300 t of starch.

Table I6.II. Starch and starch sugar: production and prices in Germany.

year	potato starch production (1.000 t) ref. 1	starch sugar production (1.000 t) ref. 2	potato starch price (M/kg) ref. 3	starch sugar price (M/kg) ref. 4
1880	45	5	(0.39)	
1890		11	(0.25)	
1900	115 (?)	9	0.19	0.23
1910	174	10	0.20	0.24
1922	22	3		
1925	103		0.29	0.38
1940	190	15('39)		
1950				
1960				
1970	70		0.41	

General remark: the figures relate to the Wilhelminian empire until 1910, to the Weimar republic and Third Reich from 1920 until 1940, later figures to the Federal Republic only.

ref. 1: wet (50% H_2O) and dry starch (20% H_2O) taken together, recalculated to dry.
Bode, 1965
Enquete Ausschuss, 1930
Meyer, 1948
Aehnelt, 1951
Kartoffelbau, 1972

ref. 2: see ref. 1

ref. 3: Wágner, 1876 mentions about 0.39 M/kg
Dammer, 1896: in 1884 about 0.25 M/kg
Ost, 1919
Enquete Ausschuss l.c.
1970: Kartoffelwirtschaft, 1973

ref. 4: Enquete Ausschuss l.c.
Schmidt, 1943

The yearly production of potato starch varied enormously under the influence of climatic conditions. Another factor of importance formed the market conditions, because import duties were often too low to keep more efficient foreign producers out of the market.

Potato Starch in World War I

When World War I began in August 1914 the German authorities immediately took steps to control the use of flour in bread production. Germany was a large importer of wheat and had a good surplus of potatoes, hence one of the ideas that came forward was to use potato starch or meal to replace imported wheat and rye in bread baking (Skalweit, 1927).

Already on August 7, 1914, the Ministry of Internal Affairs decided to establish a central agency for the production of dry whole potato meal (The Teka = Trockenkartoffel – Verwertungsgesellschaft). An important reason for the organization of the production of meal of cooked and dried whole potatoes was the limited durability of raw potatoes and the fear that substantial amounts of the expected large 1914 crop might be lost through decay and wasteful use. Production of dry potato meal became an important industrial activity in the first war years (Table I6,III).

Table I6.III. German production of potato meal and starch during World War I.

year	1914/15 (t)	1915/16 (t)	1916/17 (t)	1917/18 (t)
starch	73.000	112.500	1.000	91.700 (starch, drum dried flour and flakes together)
drum dried flour	43.000	50.500	7.300	
flakes	80.700	14.400	–	

Source: Skalweit, 1927.

The fermentation industry was able to buy dry potato meal from the Teka during the war years and for some years afterwards. The price was fixed at 0.11 M/kg on the starch content, much below the normal starch price. The quality, however, was often very poor (Merck, Internal Manufacturing Reports, 1920, and later years).

Raw Material Use in Germany during the Interbellum

After World War I a substantial part of the eastern production areas were ceded to the newly established republic of Poland. Another cause of decline in potato starch production was the very depressed state of affairs in Germany. For a number of years the people were virtually on the brink of famine, and, naturally, potatoes were then used primarily for nourishment.

While the potato starch industry was slowly recovering, an event took place which would result in a large change in its competitive position: In 1923 Deutsche Maizena Gesellschaft A.G., a newly found daughter company of Corn Products Refining Company (New York) erected a large and very modern corn starch plant in Barby on the Elbe (today G.D.R.) (Enquete Ausschuss, 1930). Within a year fierce competition between the producers of potato starch glucose (syrup) and corn starch glucose ensued, which was solved in the customarily European manner of the time: by negotiating and concluding a kind of cartel agreement (see also Aehnelt, 1951).

As of January 1, 1926, a convention on prices, delivery and payment conditions valid for all German glucose producers came into force. As a matter of fact even the introduction of production quota was contemplated to limit the use of corn starch in glucose syrup manufacture to 49%. In reality the percentage use of corn starch often was much higher because of high potato and potato starch prices.

For the fermentation industries the whole episode contributed to a stabilization of the price of their most important raw materials, starch. Imports from Holland also helped to keep German starch prices from 1928 onward at about 0.30 RM/kg[1].

Cartel conditions for potato starch itself did not really exist until early 1929 when all producers of dry starch joined a newly formed cartel, in which producers of wet potato starch and those of other starches did not partake, however (Enquete Ausschuss, 1930).

Walter Dehio, Boehringer's Sales Manager of the "Acid Sales" Department, wrote:

> Since the stabilization of the currency (Reichsmark) in 1924 we have been able to buy second grade potato starch with a starch content of about 70% for prices of 0.20

RM/kg or less. This price very much favored a profitable production of lactic acid.

- *Dehio, 1966*

Boehringer occasionally also used tapioca starch, crushed rice and molasses. When the world market sugar price reached extremely low prices in the beginning of the 1930's sugar became the most favored substrate. After Hitler came to power in 1933 autarky became a leading principle in the economic policy of Germany. For agricultural products this implied restrictions on imports and redirection of existing German agricultural produce to fill up ensuing shortages.

Let us again listen to Dehio:

In negotiations with the Economic Association of the German Sugar Industry we had succeeded in 1932 to obtain tax-free sugar (at RM 0.20/kg) for the production of export lactic acid. This was a very favorable price, which we did not enjoy for a long time, however. The economic measures of the Hitler-regime soon lead to an increase in sugar consumption, and excess sugar at export prices for chemicals production was no longer made available.

The raw material supply of Ingelheim aggravated further with the introduction of the Four-Year-Plan in 1936, which resulted in a more calculated utilization of German agricultural products for nutritional purposes, and restricted availability and higher prices for other uses. The amount of second grade potato starch of the total production dwindled, and we were soon obliged to buy better and more expensive qualities, ...

After difficult and prolonged lobbying in the Ministry of Finance he [Julius Liebrecht, son-in-law of Albert Boehringer] succeeded in obtaining abolishment of the sugar tax of RM 0.21/kg for the production of technical lactic acid, for inland use as well as for exports. During a certain period we were then able to buy tax-free sugar at 0.20 RM/kg. Customs officers controlled the sugar intake in the lactic acid factory. Extension of the use of tax-free sugar for the production of edible lactic acid was refused, however. Processing of fully taxed sugar at RM 0.65/kg made the production of edible lactic acid completely uneconomical. [Moreover] the prices of industrial products were frozen and kept under the control of the Price "Commissar". Increases were only granted in exceptional circumstances.

In the last pre-war years the procurement of raw material for the lactic acid factory was severed further. Reasonably priced potato starch was extremely difficult to track down, because nearly all available potato starch was needed for food products. This was due to severe restrictions on the import of food products in favor of military supplies.

In those days the Association of the Chemical Industry ... in Berlin, tried very hard to alleviate our raw materials problem. In the "Third Reich" the economic section of the Association (headed by Dr. Claus Ungewitter) had been transformed into an instrument of economic control by the government. Because of the relative importance of large scale chemical syntheses for the autarkic economy, later for the war economy, the I.G. Farbenindustrie A.G. had attained a dominant position. Dr. Ungewitter and collaborators did their utmost, however, to become familiar with the requirements of the medium and small sized firms and provided valuable assistance on many occasions.

The procurement of raw materials for the lactic acid production became very difficult [in the first years of World War II], until in 1941 the responsible authorities in Berlin decided to make German sugar available for the total production. For edible uses we had to pay the fully taxed price of RM 0.65 per kg. In view of the war requirements we had to live with the losses this high price entailed. More important was the continuation of the production and the retention of key personnel. Tax free sugar for technical lactic acid did cost RM 0.44/kg.

- Dehio, 1966

Developments in the United States

In contrast to Germany the United States has always been a net importer of sucrose. For obvious reasons this implied a sugar policy with a more open eye for the world market.

Local production was also of considerable interest, however. The South, in particular the State of Louisiana, had been an important producer of cane sugar since the middle of the eighteenth century. In the first decade of the twentieth century the yearly production of this State amounted to about 300,000 t, 12% of total U.S. consumption (Petzel, 1926).

American beet sugar was successfully introduced in California in the second half of the nineteenth century. By 1908 sugar beet was grown in other western states as well; the annual production of beet sugar had by then reached about 400,000 t – 17% of total U.S. consumption (Vogt, 1908).

Until 1891 the U.S. government had levied a duty on imported sugar. This duty had become an important source of income, Dalton (Dalton, 1937) estimated that of the total government receipt in the 1880's 13% came from the sugar duty (income and corporate taxes were still non-existent !). Following a short period in which sugar could be purchased duty-free on the international market, tariffs were reimposed in 1894, and were never abolished again. The level of duty varied over the years, however. Some of the more important changes are shown in Table I6,IV.

A major change in sugar policy came with Roosevelt's New Deal. In 1933 Congress approved the Agricultural Adjustment Act (AAA). Under this act the production of a number of agricultural commodities were curtailed by a combination of measures aimed at controlling [restricting] the output of

Table I6.IV. American production and consumption of sucrose.

year	production (1.000 t) ref. 1	consumption (1.000 t) ref. 2	world-market price ($/kg) ref. 3	import duty ($/kg) ref. 4	import duty on Cuban sugar ($/kg) ref. 5	inland price ($/kg) ref. 6	use per capita (kg) ref.7
1890	200	1.800	0.08	0.05	0.05	0.14	29
1900	600	2.500	0.07	0.04	0.04	0.11	30
1920	1.000	4.000	0.24	0.02	0.02	(0.40)	38
1930	1.200	6.200	0.03	0.05	0.04	0.10	45
1940	1.700	6.300	0.04	0.04	0.02	0.10	47
1950	1.900	7.400	0.11	n.f.	n.f.	0.17	45
1960	2.600	8.000	0.06	n.f.	n.f.	0.19	44
1970	5.600	9.500	0.09	n.f.	n.f.	0.18	44
1980	5.500	9.200	0.18	n.f.	n.f.	0.66	37

ref. 1: rounded off figures from various sources, U.S. mainland plus Hawai and Puerto Rico

ref. 2: estimates, based amongst others on:
Debus, 1976
Internal report HVA, 1975
Chem. Markt. Rep., Aug. 2, 1984
(the figures on consumption are not always consistent with the capita figures)

ref. 3: centrifugal sugar prices
Spencer, 1945
Ullmann, 1969
Debus, 1976

ref. 4: Spencer, l.c.
n.f. = not found

ref. 5: Spencer, l.c.
The preferentially low duty on Cuban sugar ended in 1960

ref. 6: refined sugar from U.S. government statistics the 1920 price between brackets was a peak price;
in 1974 the U.S. price rose to 0.66 $/kg, the same level was reached in 1980 (Dale, 1987)

ref. 7: Kolodny, 1976
Verbanic, 1986

those commodities. Farmers were offered government contracts to lie fallow part of their land in return for a fixed minimum price for the crop grown on the remaining acreage.

Sugar, not being an American surplus crop, initially fell outside the scope of the AAA. Following the stock market crash of 1929 world market sugar prices dropped to extremely low levels, the lowest price of 0.00125 \$/kg was reached in 1933. The country hardest hit by this dramatic development was Cuba, a close ally and in many respects a protectorate of the United States. Disorder and riots in Cuba led to fears that the political situation of Cuba could become very unstable. This argument plus the dwindling of U.S. exports to Cuba prompted the Roosevelt Administration to take steps to alleviate the economic plight of the island. In the early spring of 1934 Roosevelt proposed to bring the production of sugar also under the Agricultural Adjustment Act. The idea was to apply a quota-price system not only to local U.S. producers, but also to producers in United States dependent overseas countries and territories (Hawai, Puerto Rico, Virgin Islands and the Philippines). In addition the Secretary of Agriculture was to fix a quota for Cuba, which together with restricted production of the United States itself and a lowering of the import duty on Cuban raw sugar to about 50% of the standard duty, would give Cuba a more than fair chance of recapturing its position in the U.S. sugar market.

In the following decades Cuban exports to the U.S.A. rose to such levels that, after World War II, they often covered more than 40% of the American consumption. An all time maximum of 3,580 t was reached in 1947 when the Philippines had not yet recovered from the destructions of their plantations and factories in World War II.

After the overthrow of the Batista dictatorship by Fidel Castro in January 1959 the relations between the U.S.A. and Cuba came under strain. Already in the same year the Castro government nationalized the American sugar interests on the island and in retaliation the Eisenhower Administration cut off Cuban sugar exports to the United States (1960). Within a short time the American obligations were taken over by the Soviet Government and the Peoples Republic of China.

The sugar supply for the American market was transferred to the Philippines and some Latin-American countries. In the following years import quota were gradually lowered, however, which provided more opportunites for U.S. sugar producers. As it turned out, this was, in particular in the interest of the producers of a new sugar, based on corn starch: High Fructose Corn Syrup (hfcs).

American Sugar Prices Were Too High for Lactic Acid Manufacture

Although the advantages of sugar in the purification process were recognized, the use of sucrose never became of any importance in U.S. lactic acid production. The reason was an economic one: sugar prices were almost always substantially higher than the prices of other fermentable carbohydrates.

Avery inititally used corn flour, later admixed with (crude) glucose. Claflin in his New York lecture of 1897 said that he used grape sugar (= starch sugar) together with 10 to 15% of cane sugar. Maybe he did not observe the better conversion of cane sugar and the higher purity lactic acid obtainable with sucrose. At that time sucrose probably was cheaper than starch sugar, certainly when the active content of fermentable matter was taken into account.

This situation changed in the first decade of the twentieth century as a consequence of a fierce battle between two groups of corn starch processing companies, both striving at monopolistic control of the American market. This fight was won by the New York Glucose Company, which had put up a very modern glucose plant with Standard Oil capital. In 1906 the other company, the Corn Products Company, was forced to merge with the New York Glucose Company, and a new company was formed, the Corn Products Refining Company which then controlled 75% of U.S. production (Bode, 1975).

A number of other modern plants were set up within the next ten years. The single most important product of these starch plants was glucose syrup, which thanks to the modern way of production and the economy of scale achieved now could be offered below the cost of sugar.

The Economics of Corn Starch Production

Although production of other starches preceded that of corn starch, in the long run it appeared that the economics unquestionably favored the production of corn starch[2].

a. Corn is an almost ideal crop for the (Midwestern) farmer; it can easily be grown in large quantities. The farmer can use a lot of it himself as animal feed and so convert it into pork, beef or poultry meat. This gives him some leverage against market influences: when corn prices are low he can use more of it to produce meat, and the other way around when corn prices are more favorable. [Some farmers did and do use 80% or more of their crop to grow hogs.] The use of about 90% of corn for all kinds of animal fodder reflects the importance of this built-in stabilizing factor.

b. The processing and storage of corn can be done efficiently and cheaply.

In the beginning of the century not more than about 2% of the corn harvest was used for the manufacture of corn starch. After the introduction in the late 1920's of a process for making pure dextrose this percentage increased gradually to about 3% in 1939. Since the 1960's the level has risen to more than 4% of the also much higher corn output (Table I6,V).

Table I6.V. Development of the American starch industry.

year	corn product. (1.000 t) ref. 1	starch product. (1.000 t) ref. 2	dextrose product. (1.000 t) ref. 3	hfcs product. (1.000 t) ref. 4
1901	51.000	400		
1911	74.000	(500)		
1921	88.000			
1931	77.000	(800)	60	
1939	77.000	(1.100)	120	
1952	84.000	(1.500)	330	
1964	111.000	2.800	450	
1970	108.000	3.000	500	
1975			520	500
1979		4.000		800
1985	251.000	(5.700)		5.000

ref. 1: U.S. statistics
ref. 2: U.S. statistics of starch production; figures between brackets are corrected to include captive consumption in the manufacture of hydrolysates.
1985: estimate, not including hfcs
ref. 3: Peckham, 1979
Kooi, 1965
various trade sources
ref. 4: Chemical Marketing Reporter

Corn starch production itself has developed to a very efficient process in which not only a superior quality of starch is obtained, but also complete recovery of all by-products has been achieved: corn oil, a valuable oil rich in unsaturated fatty acids; gluten and fibrous material, valuable for feed purposes. The concentrated steeping liquid has found an important outlet as nutrient in fermentation processes.

The importance of corn for feed production has also made it valuable in international commerce; in recent years more than 20% of the U.S. production has been exported. All these factors combined have made corn starch one of the major industrial products based on agricultural "surplusses" (in modern politically oriented slang: based on renewable resources).

Sugar is Gradually Replaced by Dextrose

The replacement of sucrose by glucose syrup in foodstuffs had become of some importance from 1910 onwards. World War I provided a strong impetus when German U-boats took heavy tolls of the Allied shipping volume and overseas sugar imports had to be restricted to make more vessels available for the transport of militarily useful material and personnel. By the end of the war about 5% of the sugar consumption had been replaced by glucose syrup.

A large influence on the development of glucose was due to the temperance movement. Historically it was a strange coincidence of events that the Prohibition Movement had such a success during the Great War:

In 1917 the production of alcoholic liquors with the exception of beer and wine was forbidden to save grain for the use in food. Not much later the alcohol content of beer was reduced to a maximum of 2.75% by weight. In 1919 Congress passed the Volstead Act which gave the federal authorities means to enforce prohibition over the whole area of the United States. The Volstead or National Prohibition Act fixed the maximum percentage of alcohol in beer at the still lower percentage of 0.5 (From The Universal Standard Encyclopedia, 1955).

One of the effects of the prohibition of alcoholic drinks was the emergence of illicit alcohol manufacture. The making of "home brew" and "moonshine" by means of readily fermentable carbohydrates like sucrose and glucose became so widespread that the consumption of these sweeteners rose by more than 10%, maybe even 20% (500,00 to 1,000,000 t annually). An extra increase in corn grinding capacity had to be installed to meet some of this extra demand, although most of it was supplied by the sugar trade.

Cheap Chemically Pure Dextrose is Invented by Newkirk

Unrelated to this extra demand for fermentable sugars was a development which also had its beginnings in World War I, that is from the search then initiated to develop a manufacturing process for chemically pure glucose.

Historically starch sugar, or dextrose as it was commonly called, was a crude product containing not more than about 80% of the pure substance glucose. It was prepared by the acid hydrolysis of starch; in the U.S.A. corn starch was the preferred raw material, hence the use of the name corn sugar in that country. The acid hydrolysis was generally carried out with sulfuric acid, often in autoclaves at temperatures above the boiling point. Even so not more than 80 to 90% of the starch could be converted. When glucose concentrations become too high it will react with itself under the forming of glucose polymers, dextrinlike, non-fermentable products. Other inconvenient by-products in the form of coloring matter originate from the action of sulfuric acid on the starch hydrolysates and proteinaceous material at elevated temperatures.

After the reaction the acid was neutralized with chalk (calcium carbonate). Precipitated gypsum was removed by filtration. In the concentration process most of the discoloration was removed by a treatment with animal charcoal.

The resulting clear and only slightly colored solution was concentrated to about 80% and then left to solidify. This was the final product, sold as dextrose, grape sugar or corn sugar.

The most logical way to get rid of the 10 to 20% of impurities (higher sugars and dextrins) would have been to purify the product by crystallization in the same way as crude cane and beet sugar were crystallized to obtain white crystalline centrifugal sucrose.

Generations of practitioners had tried to invent a practical crystallization process for glucose, but had met with only limited success. By 1900 in Germany as well as in America laborious processes had been worked out to make 92 to 95% pure anhydrous dextrose. Commercially these products had some success in the pharmaceutical field, but pricewise they could not compete with ordinary sugar.

During World War I when sugar had become a relatively scarce commodity in the United States, the Bureau of Standards of the U.S. Department of Commerce had again initiated research on the crystallization of dextrose[3]. In a lucky strike the Bureau enlisted in 1918 the sugar chemist William Bentley Newkirk[4] to assist with the work on dextrose crystallization. The thorough knowledge of sugar crystallization processes and the scientific ability of the Bureau's new chemist were soon recognized by Corn Products Refining Company's head of dextrose research, who stood in close touch with the research people of the government laboratory. Corn Products decided to engage Newkirk to develop a dextrose crystallization process for the company instead of for the government. In 1920 Newkirk accepted a research position at the Edgewater (NJ) Research laboratory of Corn Products Refining Company.

Until Newkirk, all work on dextrose crystallization had been directed at the crystallization of anhydrous dextrose. Newkirk decided, however, that it might well be worthwhile to take a closer look at the crystallization of the hydrate. Within a few months he was able to grow dextrose crystals of a uniform size which by centrifugation could easily be separated from the mother liquor. The following part of the story is borrowed from Peckham:

> By the early part of 1921 Newkirk was confident that his inventions could successfully be worked not only in the laboratory but also on a commercial scale in the factory. Accordingly, joined by Frost (head of the research) and Charles Ebert, the Edgewater superintendent, he went to New York to try to persuade corporate management to build ... a plant containing sixteen special crystallizers, enough capacity to produce by his process 50,000 pounds per day of dexrose hydrate. His arguments convinced Bedford and the other executives, and as a result he had in operation by the end of the year a pilot for this process in Building 47 at Argo. This facility, which in 1922 turned out [3,600 t] of C.P. dextrose monohydrate, allowed Newkirk to conduct experiments that in time led to further improvements in his process, ...
>
> By 1923 Newkirk had established the optimum conditions for efficient crystallization of dextrose, and another plant was being erected at Argo to manufacture this new product at full commercial scale.
>
> *- Peckham, 1979*

Production of the new sugar rose rapidly in the next years:
from 9,000 t in 1923 and 18,000 t in 1924 to over 150,000 t in 1940.

Within a few years four companies were making dextrose hydrate: besides Corn Products Refining Company these were American Maize-Products Company under a licence agreement with CPRC, and Clinton Corn Syrup Refining Company and Penick & Ford, who both claimed to have developed processes outside the scope of Newkirk's patents (Peckham, 1979). The new producers probably had capacities of a magnitude of 5,000 to 10,000 t/a, while CPRC in the early 1930's expanded its capacity to over 100,000 t/a.

Markets for Dextrose Hydrate

Dextrose hydrate quickly found markets in the baking industry and for a multiplicity of other food industries. Much of it allegedly was sold for illegal purposes, in particular material not fully conforming to trade specifications. Peckham (l.c.): "one government official estimated that only 40% of the corn sugar purchased in 1929 went for lawful business uses."

When Prohibition ended, the pace with which corn sugar was accepted did not diminish. By keeping the price at least 10% below the sucrose price starch companies were able to capture a gradually increasing part of the sugar market.

Industrial uses were also found, amongst which Peckham mentions two fermentation products: citric acid and riboflavin (vitamin B_2). It should also be mentioned here that the first experimental lactic acid fermentations of second grade dextrose were already made in 1928 by American Maize-Products (Inskeep, et al, 1952).

Overseas activities were initiated as well:
In Germany a production unit for the new starch sugar was set up some years later by Deutsche MaizenaWerke A.G., with the aid of knowhow and capital from Corn Products Refining Company.

Notes

[1] Potato starch prices in the Netherlands dropped from about 0.31 RM/kg in 1928 to 0.12 RM/kg in 1932. The Dutch government then decided to give a price guarantee to the farmers and the industry: potato starch producers were to receive at least DFl. 7,65 per 100 kg for their starch, which amounted to about 0.13 RM/kg. This guaranteed price was gradually increased until in 1939 the free market price rose to the higher level of 0.18 RM/kg (Information from Avebe, Veendam, the Netherlands, 1988).

[2] The first wheat starch plant was put into operation in 1807, the first rice starch plant in 1815, production of potato starch followed in 1820. Only in 1842 was the first corn starch plant installed when a feasible process for the extraction of the starch kernels from the grain had been found (Bode, 1975).

[3] The story of the invention of the dextrose hydrate crystallization process has been described in detail in the dissertation of Brian W. Peckham (1979). Thanks go to the author for providing the information which was not available in the Netherlands.

[4] William Bentley Newkirk (1888–1954) studied chemistry at the University of Southern California in Los Angeles. From 1910 through 1918 he worked for Californian fertilizer and sugar companies (Peckham, 1979).

References

Universal Standard Encyclopedia, 1955. Unicorn: New York, Vol. 19, 6854.

Aehnelt, W.R., **1951**. *Stärke, Stärkesirup, Stärkezucker*; Theodor Steinkopff: Dresden.

Baxa, J., and Bruhns, G., **1947**. *Zucker im Leben der Völker*, Dr. Albert Bartens: Berlin.

Berzelius, J.J., **1839**. *Leerboek der Scheikunde*; 3rd. ed.; translated by Mulder, G.J.; P.H. van den Heuvel: Leyden.

Bode, H.E., **1975**. *History of the Corn Starch Industry*; in Whistler, R.L., and Paschall, E.F., *Starch: Chemistry and Technology*; Vol. 1; Academic Press: New York.

Claflin, A.A., **1897**. *J. Soc. Chem. Ind. London (NY)*, **16**, 516.

Dale, B.E., **1987**. *Trends Biotechn.*; **5**, 287.

Dalton, J.E., **1937**. *Sugar: A Case Study of Government Control*; MacMillan: New York.

Dammer, O., **1896**. *Handbuch der Chemischen Technologie*; Ferdinand Enke: Stuttgart.; Vol. 3.

Debus, L., **1976**. *Voraussetzungen für die Wirksamkeit internationaler Rohstoffabkommen, abgeleitet aus den Erfahrungen mit den Weltzuckerabkommen*; Alfred Strothe: Hanover.

Deerr, N., **1950**. *The History of Sugar*; Chapman and Hall: London.

Enquete Ausschuss, **1930**. *Die Verwertung der deutschen Kartoffelernten*; Vol. 2; E.S. Mittler und Sohn: Berlin.

Günther, H.E., Schwarz, W., and Hartmann, E., **1962**. *Die Zuckerwirtschaft in EWG und EFTA*; August Lutzeyer: Baden-Baden.

Inskeep, G.C., Taylor, G.G., and Breitzke, W.C., **1952**. *Ind. Eng. Chem.*, **44**, 1955.

Kartoffelbau, **1972**. (Market Report), Vol. **23** [8], 206.

Kartoffelwirtschaft, **1973** (Trade Report), Vol. **26**, [5], 42.

Kolodny, S., **1976**. *Sugar Azucar*, July 1976, 20.

Kooi, E.R., **1965**. In *Kirk-Othmer's Encyclopedia of Chemical Technology*, 2nd ed.; John Wiley and Sons: New York, Vol. 6, 919.

Kooi, E.R., and Armbruster, F.C., **1965**. *Production and Use of Dextrose*; in Whistler, R.L., and Paschall, E.F.; *Starch: Chemistry and Technology*; Vol. II; Academic Press: New York.

Meyer, H., **1948**. *Dtsch. Lebensm. Rundsch.*; **44**, 114.

Ost, H., **1919**. *Lehrbuch der chemischen Technologie*; 10th ed.; Dr. Max Jänecke: Leipzig.

Peckham, B.W., **1979**. *Economics and Invention: A Technological Study of the Corn Refining Industry of the United States*; University of Wisconsin-Madison, Thesis.

Petzel, W., **1926**. *Die Entwicklung und Aussichten der Weltproduktion von Rohrzucker*; Selbstverlag: Kniephof-Jarchlin (Germany).

Schäfer, G., **1907**. *Chemische Zeitschrift*, **6**, 177 and 189.

Schmidt, A., **1943**. *Die industrielle Chemie in ihrer Bedeutung im Weltbild*; Walter de Gruyter & Co.: Berlin.

Skalweit, A., **1927**. *Die deutsche Kriegsernährungwirtschaft*; Deutsche Verlagsanstalt: Stuttgart.

Spencer & Meade, **1945**. *Cane Sugar Handbook*; Spencer & Meade: London.

Teichmann, U., **1955**. *Die Politik der Agrarpreisstützung*; Bund-Verlag: Köln-Deutz.
Ullmann, **1969**. *Ullmann's Enzyklopädie der technischen Chemie*; 3rd ed.; Verlag Chemie: Weinheim, Vol. 19.
Verbanic, C., **1986**. *Chem. Mark.*; May, 23.
Vogt, P.L., **1908**. *The Sugar Refining Industry in the United States*; University of Philadelphia: Philadelphia.
Wágner, L. von, **1876**. *Stärkefabrikation*: Vol. 19. Braunschweig.

CHAPTER 9

FOOD ACIDS GREW DURING THE GREAT DEPRESSION

CITRIC BECOMES THE SOFT DRINK ACID OF CHOICE

In the 1920's the American economy had grown at an unprecedented pace bringing prosperity to large parts of the population. By the end of the decade more than 20 million families out of a population of 125,000,000 owned a car. A new way of shopping for food and daily necessities was becoming popular thanks to the introduction of large retail shops, which was fostered very much by the family car and the wide-spread installation of refrigerators.

Then came the ill-fated stock market crash of October 1929, which shook America's economic and social system to its foundations and caused economic chaos, closures of manufacturing plants, failures of banks and large scale unemployement all over the world. The American people, that is to say the more affluent, did not easily give up the new way of life, however. The purchase of new cars was postponed, and so were expenditures for housing and appliances. But the people did not stop shopping at the new retail stores, supermarkets, nor did they stop driving. The new way of life which mass motorization had made possible continued. The freedom to move about, to live in the suburbs instead of in the innercities, to take the family out at the weekend, this freedom was restricted financially, but not curtailed completely.

In these years the production of packaged foods for sale through supermarkets brought about a profound change in food formulation. Other methods of preservation had to be found to prolong shelf life, convenience foods were developed which could be stored in the home refrigerator. A particular item of interest for our story forms the soft drink industry: The total usage of soft drinks grew in the period from 1929 to 1939 at an average yearly rate of 7%, which per capita meant about 6%. Over this decade the average wage increased with only about 1.5%/a, however; and, as the price index decreased with an average of 1.7%/a, the growth in the standard of living was not much more than about 3% per year (Porter, 1976).

Although the popularity of soft drinks in this period was still influenced positively by the prohibition of the use of alcoholic beverages, the new way of life made possible by the automobile, the refrigerator and the supermarket exerted a much greater influence.

Table 9.1. Growth of acidulant use in the United States 1929 – 1969.

	1929 (t)	ref.	1939 (t)	ref.	1969 (t)	ref.
in food and beverages:						
tartaric acid	3.000	2	4.000	2	500	1
citric acid	700	3	3.300	e	38.000	1
lactic acid	300	4	600	5	1.400	1
total organic acids	4.000		8.000		40.000	
consumption of soft drinks:	[35.4]	6			[229.6]	6
equivalent to food acid	4.000 – 4.800	e			40.000 – 48.000	e
tartaric acid in beverages	1.500 – 2.000	e			a few 100	e
citric acid in beverages	200	3			28.000 – 30.000	1
lactic acid in beverages	less than 20	e			a few 100	e
phosphoric acid in beverages	2.000 – 2.500	7			11.000 – 14.000	1

References:
1. Extrapolated figures from SRI reports "Citric Acid", 1969 and 1982
2. Estimates based on production and consumption figures of Haynes, Vols. 3 and 4 (1954)
3. Mines (1978)
4. Based on German exports to the USA
5. US Statistics
6. Number of 12 oz. containers of carbonated soft drink per capita, National Soft Drink Association, 1978 Sales Survey
7. Estimated at 50—60% of total acid use for soft drinks

e. Rough estimates by the author.

Table 9,I presents some data relating to the growth of food acids in the soft drink industry from 1929 until 1969. In this period the share of citric acid rose from about 5% to 70%. Interpolation of the data gives a 1939 use of citric acid of 2,300–2,500 t, 30% of the soft drink market.

Pfizer Makes Citric Acid a Large Volume Commodity

In Chapter 7 the Pfizer citric acid story as written by Samuel Mines (Mines, 1978) was retold until the beginning of the 1920's. The new decade brought the breakthrough: By July 1923 the research team had produced 45,000 kg of calcium citrate. In August 1923 production began on a semicommercial scale.

It was a round-the-clock operation, never shutting down, fifty-two weeks a year. The demand for citric seemed to expand without limit.

In 1926 the plant was expanded. ... In 1927 the purchase of calcium citrate from outside sources was terminated. The plant had to be expanded again in 1929. Currie resigned in 1933, ... [he] had brought the process through from laboratory research to mass production, and continued as a Pfizer consultant for five years more.

The success of the Pfizer process for making citric acid by fermentation helped break the Italian monopoly. Imports of citrate of lime from Italy had been about [1350 t] a year. As they began to shrink, the Italian government retaliated in 1927 by clamping an embargo on exports. It was too late; Pfizer was already in independent full-scale production.

There were economic repercussions with a domino effect. The United States became free of foreign domination in citric supplies, with Pfizer as the leading producer. The price of citric acid dropped steadily from [2.75$/kg] in 1919 to [0.44$/kg] in 1939.

In 1929, citric acid usage was divided as follows:

Medicinal	68%
Technical	2%
Soft Drinks	9%
Food	15%
Candy	6%

Citric acid production rose from [1,750 t] ... in 1928 to [6,190 t] ... in 1938.[1]

- Mines, 1978, 21

The Pfizer story presents a vivid account of the immense problems the company had to master to learn how to make citric acid by fermentation. One wonders how lactic acid would have fared had any one company persisted so long and at comparable costs (about $4 million) to work out a process to make pure lactic acid for a price the food and beverage industry were prepared to pay.

Lactic Acid Producers Do Not Meet the Challenge

The question is not mere speculation. The lactic acid manufacturers did not grasp the "opportunity" because they were in an entirely different position: With citric acid the fermentation is much more difficult, the purification, however, much easier. Hence, once the fermentation problems have been solved, the manufacture of the pure acid does not pose additional problems, as it had been mastered long ago.

Lactic acid fermentation, on the other hand, is fairly easy, but purification is complicated and expensive. (In Intermezzo 4 the difference in purification of the three food acids is portrayed.)

A further point of importance concerns the number of producers: Pfizer was the sole (U.S.) manufacturer of citric acid by fermentation; lactic acid production was taken up by three more producers in the 1930's, as will be shown later on in this Chapter. Average 1939 production of the four lactic acid manufacturers was 350 t (100%), while Pfizer made over 6,000 t of citric in that year.

The future of lactic acid would depend on the development of markets for the so-called edible quality. Rising German exports of this quality to the United States in the late 1920's already indicated this trend away from technical lactic acid. The tremendous success of citric acid in the food and beverage industries supplies further "proof" of this development. Edible lactic, though far from chemically pure, was well suited for many food products, particularly for those products that were made by a related fermentation process (dairy products, bread, beer, distilled liquors, olives, pickles and sauerkraut, to name the most important) (Eisenman, 1930; Garrett, 1930). In these products the typical fermentation flavor which commercial lactic acid more often than not characterized, was not objectionable. For soft drinks the fermentation flavor usually was a disadvantage, here tartaric and citric acid were preferred unless a very pure lactic could be offered.

Large differences existed in the 1930's between the developments in the U.S.A. and those in Europe. In the U.S.A. carbonated beverages became increasingly popular, while in Europe where refrigerators were still luxury items, lemonades continued to be the household soft drink of choice.

In particular in carbonated drinks the taste of citric acid appeared to be the most refreshing. (The cola containing carbonated drinks are acidified with phosphoric acid, they have been left out of the discussion because eventual replacement of this acid by an organic food acid is not possible for taste reasons. The importance of cola drinks can hardly be overestimated, however. Their market share usually was in the 50% bracket or higher.)

In lemonades, economics was probably the most important consideration when new formulations were sought. Lactic acid supposedly had a small technological advantage in making the syrup less susceptible to the action of fungi (Ullmann, 3rd ed., Vol. 1, 1953, 299).

Price Developments

The combined effects of competition and strong increase in demand on the prices of the three organic food acids are demonstrated in Table 9,2.

The figures clearly show the gradually decreasing price differential between tartaric and citric acid, from 33 cents a kg in 1930 to 4 cents in 1937 and a reversal of the difference in 1938, when citric for the first time sold at a lower price than tartaric acid. Edible lactic acid, which in 1929 was competitive with citric acid in acidity (but inferior in purity), then was about 60% more expensive[2]. The conclusion is evident: in the decade between 1929 and 1939 citric acid not only replaced tartaric acid as the most important organic food acid, but also confined lactic to special market niches for which the other two acids were less suitable.

The decrease in the citric acid price in this period of world wide depression was partly due to the economy of scale achieved by Pfizer: from 1928 to 1939 the production rose by 250%!

Pfizer was also helped somewhat by the declining world market sugar price, which reached an all-time low level of 0.0125 $/kg in 1933. Due to the import duty on Cuban raw sugar of $ 0.04/kg the American retail price (of white sugar) stayed at the relatively high level of about $ 0.10/kg, however.

A much cheaper source of sugar were molasses, the mother liquor or residue of the sugar crystallization. In the United States blackstrap molasses, that is the molasses from which the maximum amount of sugar has been extracted in the crystallization process, was the most economical raw material. It contained about 50% of fermentable sugar and was used on a very large scale in the production of ethyl and butyl alcohol. The duty on molasses was very small, Cuban material was charged with 0.005 $/kg unless "imported to be commercially used for the extraction of sugar or for human consumption" (Tariff Act of 1930, quoted by Spencer, 1945).

No wonder that the fermentation specialists of Pfizer also intensively studied the use of this feed stock. In the words of Mines (1978):

> The long struggle to improve the [fermentation] process made a dramatic breakthrough in 1934. Jasper Kane, ..., developed a method of using molasses instead of the much more expensive white sugar. The new method saved millions in raw material and production costs, was faster, gave higher yields, and was less susceptible to *Penicillium* contamination.
>
> *- Mines, 1978*

Table 9.II. Prices of organic food acids, 1928 – 1939.

year	tartaric acid ($/kg) ref. 1	citric acid ($/kg) ref. 1	lactic acid (as such) ($/kg) ref. 2	ref. 3	lactic acid (100%) ($/kg) ref. 4
1928	0.79	0.99	1.08	1.05	
1929	0.84	1.01	1.07	0.92	
1930	0.79	1.12	0.84		
1931	0.62	0.84	0.74		
1932	0.51	0.68	0.76		
1933	0.48	0.64	0.80		
1934	0.55	0.64	0.93		
1935	0.53	0.62	2.98		
1936	0.51	0.57	0.95		
1937	0.51	0.55	0.92		0.49
1938	0.57	0.51	1.38		0.47
1939	0.64	0.46	1.27		0.47

ref. 1: Haynes, Vols. 4 and 5, price lists of important representative chemicals.
ref. 2: prices of imported lactic acid.
1928 and 1929 lactic acid, over 55% concentration (U.S. statistics);
1930 – 1939 Sheehan (1944);
1935 and later figures relate to very small volumes only!
import duty (9ct/lb) added
ref. 3: German export statistics; rate of exchange RM 4.08 = 1$;
import duty added
ref. 4: Sheehan (1944)

This event, the introduction of molasses feed stock in the fermentation of citric acid in 1934 became the turning point in the history of the use of food acids, from this year onward the future became dominated by citric acid. In America Pfizer, the only producer of citric acid by fermentation, took the lead, the company expanded production almost every year, a development which did not slow down until many years after World War II.

In technology, Europe did not stay far behind, however, but the market grew at a much slower pace. The first European companies which mastered the production of citric acid by fermentation were "Citrique Belge S.A." at Tienen, Belgium (1928), and "Montanindustrie vorm. Joh. David Starck A.G." at Kasniau (near Plzen), Czechslovakia. About 1930 the British firm of John & E. Sturge became the third European producer (Dehio, 1966; U.S. Census Report, 1928). Dehio reports that the firm of Starck already in the 1920's made citric acid from molasses!

American Production of Edible Lactic Acid

The unfortunate fate of ethyl lactate reported in Chapter 8 had forced Wamesit Chemical Co., the successor of the Avery Chemical Company, to close its factory at Lowell (1926). The only remaining producer[3] now was Du Pont, the largest chemical company of the United States and since the 1920's in a process of vigorous expansion. A large step in that process was the 1928 acquisition of the important manufacturer of mineral acids and a large number of other chemicals, the Grasselli Chemical Company of Cleveland, Ohio (Ch. 8, 257). The next year the Grassselli (division) management was given the responsibility for lactic acid, one of the products of the old Harrison plant at Grays Ferry. A list of investments (in the Hagley Museum) shows that the new management considered lactic acid a worthy addition to its line of chemicals:

> 1931. Installation of a plant for edible lactic acid by [di]isopropyl ether extraction (Unit no. 1)
> Nov. 1933. Isopropyl Ether Unit No. 2.
> June 1934. U.S.P. [pharmaceutical lactic acid conforming to the United States Pharmacopoeia].

A New Method to Manufacture Edible Lactic Acid

The investments referred to above relate to a new method for preparing edible lactic acid. The inventor, Grasselli chemist Joseph Adam Jenemann, applied for a patent on the process in 1931, which was granted two years later (Jenemann, 1933).

Technically the method is a counter-current extraction method in which the (low) solubility of lactic acid in diisopropyl ether is used to extract it in a number of successive extractions from the water phase. The extraction with ether was an old idea, going back to Liebig and Kolbe, new was the idea to apply a counter-current method (see also Intermezzo 4, 206).

In Jenemann's description (compare Figure 8,1):

> I have taken a commercial lactic acid obtained through usual treatments of fermented liquors and concentrated it to contain 44% lactic acid by weight. This was run into the top of column 2, containing a series of 6 partitions. At the same time I ran 16 volumes of isopropyl ether for each volume of concentrated lactic acid into the bottom of the same column. The aqueous liquor discharged from the bottom of this column contained substantially all the impurities of the original aqueous solution, together with about 20% of the total lactic acid charged. This acid contains some impurities in such concentrations that they can be removed by well known methods and the acid thereby made available for certain technical uses where a colored, impure acid is suitable.
>
> The isopropyl ether solution discharged at the top of the column contained about 2% by weight of lactic acid. It was substantially free from inorganic salts, residual sugars and the coloring matters present in the original solution used.
>
> This solution was then run into the bottom of column 4, where it met through its various pools a stream of water introduced through the top in the ratio of 1 volume of water for each 25 volumes of isopropyl ether solution. The aqueous solution obtained at the bottom of column 4 contained 35% lactic acid. It had a slight amber color and after boiling off the small amounts of isopropyl ether dissolved it was suitable for all edible purposes.
>
> *- Jenemann, 1933*

Although large amounts of diisopropyl ether were used in the extraction, losses were comparatively low because of the circulation system applied (via reservoir 3). As long as the demand for technical lactic acid for the tanning industry did account for a large outlet of lactic acid the coproduction of about 25% of dark crude acid did not constitute a major drawback.

Figures on the actual production of Du Pont are not available. When the plant was closed in 1963 total production capacity was estimated at less than

April 25, 1933. J. A. JENEMANN **1,906,068**

LACTIC ACID

Filed March 2, 1931 2 Sheets-Sheet 2

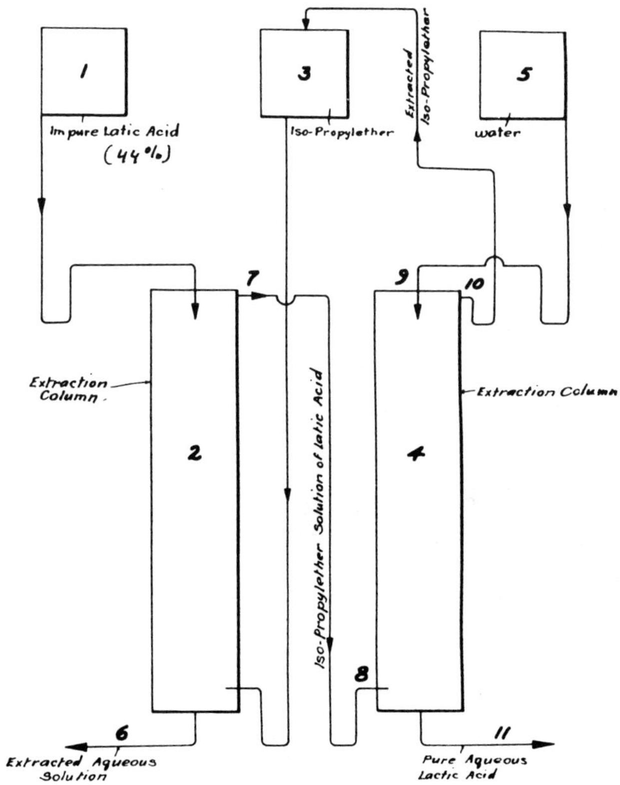

Fig. 2

JOSEPH A. JENEMANN
INVENTOR

BY

ATTORNEY

Fig. 9,1. Jenemann's counter-current extraction process for lactic acid (U.S. Patent 1,906,068).

1,000 t/a (J. Newton, private communication, 1986). Total U.S. use in the tanning industry was around 850 t/a (Sheehan, 1944), sufficient to absorb all crude by-products of the Grays Ferry works.

Pharmaceutical (USP) lactic acid, the purest product available was sold in even smaller quantities. As far as is known until 1934 this quality was only made in Germany by Merck and Boehringer, and in England by Howards of Ilford. Merck produced about 120 t/a; for Boehringer Wallach reports an output of 110–160 t/a. In 1935 total output of these two companies declined to 190 t, and a year later to 120 t, as a consequence of "the loss of the American market" (Wallach, 1966).

This loss is most probably due to the start-up of USP production by Du Pont. Du Pont's production capacity of this grade supposedly did not exceed 100 to 200 t/a, and the product therefore remained too expensive for food applications or chemical conversion (L.F. Marnett, private communication, 1986).

New Raw Materials

In the 1920's new raw materials for enzymatic and chemical modification received much attention. The success of the butanol-acetone fermentation spurred this development. Another reason was the decline in price many agricultural products experienced even before the Great Depression came along.

There also was a widespread belief that crude oil reserves would be depleted within a few decades. Outside the field of solvents and antifreeze, chemicals based on oil were low in volume and expectations were low as well. The future supply of raw materials for the chemical industry was therefore thought to depend primarily on coal and on agricultural raw materials.

Figure 9,2 taken from Kirk and Othmer's Encyclopedia of Chemical Technology (Kirk-Othmer, 1953) shows in a very illustrative way the relatively unimportant position of petrochemical production in the interbellum years. (The figure also shows how World War II changed the situation dramatically.)

A successful example of the introduction of an agricultural raw material in the industry is provided by the development of the soybean processing industry. It was started by the A.E. Staley Manufacturing Company, or more correctly by the founder and chief executive Augustus Eugene Staley himself,

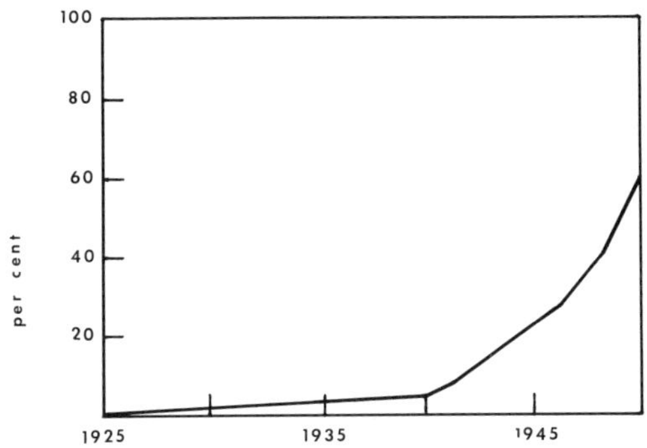

Fig. 9,2. Percentage of total organic chemicals produced from petroleum and natural gas. Source: Kirk-Othmer, 1953. Reproduced with permission from the publisher.

in 1922 with the processing of about 200 t of beans. Five years later already over 14,000 t were converted into a growing number of food products as well as raw materials for the chemical industry. (From "The Kernel & The Bean", Forrestal, 1982; a lively account of the history of starch and soybean processing at the Staley company.)

In the same decade the corn starch industry made crucial advances, particularly in the manufacture of much purer qualities of glucose (See Intermezzo 6, 294). The dairy industry, supported by the U.S. Government also began to pay attention to the revaluation of its by-products (This Chapter, 312).

All of these various activities aimed at the large scale industrial use of agricultural raw materials found a strong advocate in the person of William J. Hale and his Chemurgy Movement. When the Great Depression following the stock market crash of October 1929 brought utter distress to the farming community, Hale's movement began to receive widespread political support[4].

The Chemurgy Movement

Hale observed that the depression was felt much more heavily in agriculture than in industry. This he explained quite simply as a consequence of the introduction of mechanization to farming: in the old days the farm formed to a large extent a self supporting cycle of activities, most of the food and fodder produced was returned to man and beast which kept them alive and provided them with energy for the cycle of tasks they had on the farm. Surplus was sold or traded, manure was returned to the land, all waste found some practical use. Farm mechanization had brought a complete overturn; production had increased through the use of better equipment and fertilizers, and also because no fodder production was needed anymore to feed the horses. The larger surplus had to be sold to outside users, however.

Hale does not propose a return to the golden ages of the past, he knows that this is impossible, and being a real American he is looking forward and expects the solution from new technologies.

In his book "The Farm Chemurgic" (Hale, 1934, 142) Hale coins the word chemurgy to denote the manufacture of chemicals by nature as well as the modifications of these "chemicals" by man:

> ... the growing of cotton is but the manufacture of alpha-cellulose by nature under man's supervision; a true chemurgical operation.
>
> Chemurgy brings out in relief the correct interpretation of agriculture. No longer a pursuit to supply man with food and raiment, but a pursuit that shall bring into existence a vast array of chemicals to fit a myriad of ends. It presents the most fascinating of pictures and the most awe-inspiring of nature's wonders. The tearing asunder of composite matter, heretofore known and used generally under such prehistoric names as 'corn', 'wheat', 'potatoes', 'straw' and 'wood' and the like, and the allocation of the several homogeneous components thereof under their proper and clearly defined chemical characteristics, opens up a new world to man. It is the alphabet of a knew knowledge; a knowledge that nature is now to incalculate in man.
>
> *- Hale, 1934*

An important aspect of Hale's ideas is the proposed erection of "Agricenters", centers for the processing of farm products to industrial end products or raw materials for other process industries. The activities of these centers would include the complete splitting up of farm products into useful components, including the processing of wastes into valuable substances.

Stable and low prices of agricultural products form, of course, a precondition for the implementation of Hale's ideas. Such a combination does not give the farmer the best deal, but it might protect him from the worst. When Hale wrote his book prices certainly were much too low to allow the farmer a decent living. Widespread was the feeling that something more had to be done than just let the forces of the free market determine farm income.

In 1932 a new president – Franklin Delano Roosevelt – was elected who had promised the voters to fight the depression. In the first 100 days of his Administration a large number of new laws were passed by Congress to redress the economic system. (The whole of these and subsequent measures became known as the New Deal.) One of his measures was an agricultural recovery program, laid down in the Agricultural Adjustment Act of May 1933. Under this act an Administration was created with authority to make payments to producers of seven farm commodities – wheat, cotton, corn, hogs, rice, tobacco, and milk – in return for reduction of their output. Although the program did not help as much as expected – a return to the purchasing power of the farmers to prewar levels had been intended – the stabilization of prices at a higher level was a positive effect.

Political pressure coming from the Chemurgic Movement and other agrarian interests was rather influential in another action taken by the U.S. Government in 1939: in the establishment of four research centers by the United States Department of Agriculture (Haynes, Vol. 5, Ch. 16; see also McMillen, 1939). These centers which became known as USDA Regional Research Centers were established to find new (industrial) uses for existing surplus crops:

- The Southern Regional Research Laboratory at New Orleans was to study cotton, sweet potatoes, and peanuts;
- The Northern Regional Research Laboratory at Peoria corn, wheat, and farm wastes;
- The Eastern Regional Research Laboratory at Philadelphia apples, vegetables, potatoes, tobacco, and milk;
- The Western Regional Research Laboratory at San Francisco alfalfa, fruits, vegetables, wheat and potatoes.

Each laboratory was appropriated a yearly budget of $1,000,000. Extensive research programs were soon set up, in some cases these programs incorporated the work of existing research agencies. Within a few years the first results from these programs became patented and published.

One of the projects the Eastern Regional Laboratory (ERL) had taken up was the utilization of whey, which not surprisingly included the conversion of whey into lactic acid and the further chemical conversion of lactic acid.

By now we are about ten years ahead of the lactic acid story. To pick up the thread again, attention will be focussed on the Bureau of the Dairy Industry of the United States Department of Agriculture, one of the governmental laboratories which preceded the establishment of the USDA

Regional Research Centers. The Bureau fo the Dairy Industry operated in the 1920's a [small ?] laboratory working on topics of interest to the dairy industry. This laboratory was then carrying out rather fundamental biological studies relating to cheese making and milk pasteurization (North, 1939).

From the Microbiology of Cheese Making to Lactic Acid Fermentation

Dr. L.A. Rogers, director of Research Laboratories of the Bureau, introduced the use of pure cultures in the American dairy industry. He was one of the first to apply the new concept of pH measurement in the study of fermentation processes. He used these measurements for instance to follow the growth of lactic acid bacterial populations (Rogers and Whittier, 1928). In 1931 Roger's laboratory published a paper on a **continuous fermentation method**, using whey as raw material (Whittier and Rogers, 1931). The drawback of the normal batch-type fermentation in the eyes of the authors is that

> during the time required for cleaning, sterilizing, and emptying the equipment and for the growth of the inoculated organism to its maximum effective numbers, very little actual fermentation takes place. If these time-consuming factors could be eliminated and the actual fermentation process carried out continuously, the ratio of equipment cost to production could be considerably reduced[5].
>
> \- *Whittier and Rogers, 1931*

Whittier and Rogers used as raw material sweet whey pretreated with hydrated lime to prevent undesired (butyric) fermentations. Otherwise, they did not operate their system under completely sterile conditions. The fermentations were run at 44° C and a rather low pH between 5.0 and 5.8, lime suspension was intermittently added to keep the pH in this range. As the authors say, the acidity is on the high side, impeding the growth of *Lactobacillus bulgaricus* or *L. casei*). The low pH is desired, however, to suppress competing organisms, and a mycoderm[6] is therefore added to accelerate the action of the *Lactobacillus*. [No explanation of the effect of the mycoderm is given: maybe *Lactobacilli* become attached to it, which results in a higher concentration of bacteria, and hence in higher productivity per unit of volume.]

Laboratory fermentations in 115 l glass vessels lasting two weeks were performed with good results (90% yield). Larger scale trials in 2,000 l vessels at a creamery were also successful.

Enters National Dairy

The work of Rogers and associates was followed closely by National Dairy Products Corp., a merger of ice cream businesses and dairy companies headed by Thomas H. McInnerney[7]. In 1929 McInnerney "reasoned that economic uses could be developed for dairy by-products through research, and that such research would implement the purpose upon which National Dairy was founded in 1923 – to further the dairy industry by combining talents and resources fo experienced dairymen" (Haynes, 1949, Vol. 6, 297). In the same year National Dairy Research Laboratories was founded, a dairy research laboratory was established in Baltimore. (Part of) the plant of Sheffield Farms Company at Hobart, NY, was renovated or enlarged to manufacture the by-products originating from the work of this research laboratory. The first products of the Sheffield plant which then carried the name Sheffield By-Products Company, were casein for paper coating, crude and refined lactose, and albumen (Chemical Trade Journal, 1938).

The National Dairy's Laboratories also studied the manufacture of lactic acid by fermentation of whey. Dr. S.M. Weisberg modified the Whittier and Rogers continuous process to a batch type process, which he thought to be more practical, as "side reactions were found to be more easily avoided". Pilot plant investigations were then undertaken in the Hobart plant by Stoddard Stevens; the large-scale operation was realized in 1936 at the Norwich, NY, plant of the company under the personal supervision of plant manager F.L. Chappell (Olive, 1936).

Initially the plant made technical and USP grades of lactic acid, calcium and sodium lactate, with a total capacity of about 1,000 kg/day (100%). Because of the corrosive properties of lactic acid the company had installed 316 stainless steel evaporators. From the very beginning calcium lactate was the most important product of the Norwich fermentation plant.

During World War II a new plant was installed to comply with extra demand from the military [supposedly for sodium lactate; compare Ch. 7, 214]. The fermentation capacity always stayed at a modest level of about 500 t of lactose equivalent per year, however.

Until about 1952 Sheffield remained a small producer of technical and edible lactic acid. The production of calcium lactates (many grades were made by the company) gradually reached such levels that production of lactic acid became of less interest. In 1952 it was abandoned altogether. A former employee said this about the decision to the author: "Markets for edible lactic were knocked out by citric acid."

For some time Sheffield also made small quantities of USP lactic acid for which double crystallized calcium lactate (USP quality) served as feed stock (For production method see Ch. 6, 180).

All lactic acid grades were delivered in new oak barrels. Edible lactic acid was sold to one of the ice cream plants of National Dairy to give "a desirable tartness to sherbets" (Burton, 1937). Another customer was the pickled olive business which used it to correct the acidity of olive brines. In this way losses due to growth of bacteria could be avoided (Ch. 12, 384).

Sheffield's methods of fermenting and purification to obtain calcium lactate are presented in Exhibits 9,I and 9,II.

EXHIBIT 9,I
SHEFFIELD'S FERMENTATION PROCEDURE
lit.: Olive, 1936; private information, 1986

raw material	sweet whey from casein production
concentration	5.5% (lactose)
sterilization	not required
inorganic nutrient	supplied by the whey
organic nutrient	also from the whey
neutralization	with lime slurry
fermentation temperature	43°C
fermentation vessels	five wooden vats of 25 m^3, net content 19.3 m^3 each
sterility of process	the vats are cleaned with sodium hypochlorite
mixing	the fermenters are equipped with stirrers
acidity	maximum acidity 1.3 per cent (pH 5.0 – 5.8)
fermentation time	48 hours
control of end	?
yield	about 90% (USDA, Bureau of Dairy Industry)
type of bacterium	*Lactobacillus bulgaricus*
cultivation method	scaling up every day from a pure lab culture under sterile conditions; scaling up factor about 10
amount of inoculant	about 20%
source of bacterium	from American Type Culture Collections (every 6 months)
stereoisomery	probably more or less racemic
control of purity	microscopically

EXHIBIT 9,II
SHEFFIELD'S PURIFICATION PROCEDURE FOR CALCIUM LACTATE
lit.: Olive, 1938; Burton, 1937; private information, 1986

treatment of broth	excess lime is added at the end of the fermentation, the albumen coagulated at 88°C (together with the biomass)
decantation	when the supernatant is clear, the settlings are passed over a filter press and combined with the decanted supernatant
decolorization	with activated carbon, followed by filtration
concentrating	the calcium lactate solution is concentrated in a double effect stainless steel 316 evaporator to 10° Bé, and further in a single effect copper or stainless steel pan to 15° Bé (28–30% ?); crystallization at 20–22°C in water cooled 1,100 l pans (with stainless steel lining)
recovery	by means of a Hepworth centrifuge, the crystals of calcium lactate 5 aq. are washed with cold water. Originally the crystallization was run twice to obtain USP calcium lactate, later with better process control one crystallization step sufficed.
drying	on dryer trays in a tunnel dryer using filtered air under accurate temperature control
packing	in 200 lb barrels
yield	about 75% based on crude calcium lactate

Starch Manufacturers Introduce Dextrose as Raw Material

In the years Rogers of the Bureau of Dairy Industry was conducting his fermentation experiments with whey, two American corn starch producers were also studying the production of lactic acid (and calcium lactate) by fermentation. The interest of these companies, American Maize-Products Company and Clinton Corn Syrup Refining Company, was the outcome of their activities in the production of high purity corn sugar. The history of the invention of a process for making a very pure crystalline sugar (corn sugar or dextrose) from starch is given in Intermezzo 6, which deals with the relative merits of sugar and starch as raw material for lactic acid fermentation.

Around 1930 the word fermentation had no immediate connection with making lactic acid, a very unfamiliar product. It was the time of the Prohibition Period and when people talked about fermentation, they meant illegal production of alcohol. To many people the new starch sugar – corn sugar –

was just another raw material for alcohol fermentation.

It is estimated that annually more than 200,000 t corn sugar of various grades was sold illegally. Peckham (Peckham, 1979) reports an estimate of about 60% of total corn sugar sales in 1929, being sold illegally (see also Intermezzo 6, 293). It is not surprising that exact figures about this illegal trade have never been published. Leading personnel of A.E. Staley Manufacturing Company do remember, however, that their company sold about 20,000 t/a corn chip sugar to broker's warehouses in a few eastern states, allegedly for resale to the tanning industry. (Staley's product was a crude glucose, obtained by acid hydrolysis of corn starch, the concentrated solution of which solidifies upon cooling, and was sold as such.) As Forrestal was told by Staley veterans

> Bootleggers were our best customers for chip sugar for making moonshine. It always made us nervous when they came because they weren't supposed to be in the plant. They couldn't have bought the corn chip sugar directly from us, anyhow. But they liked to come in and see it made, knowing they'd find a way, somehow to get their hands on it for their bootleg whiskey and gin.
>
> \- *Forrestal, 1982*

Repeal of the Eighteenth Amendment

In early 1933 the U.S. congress took action to repeal the Prohibition Law. First beer was allowed to be sold legally, in the summer followed by wine (maximum alcohol concentration 3.2%), by the end of the year the national ban on hard liquor was also removed. Immediately there followed a rush to start brewing and distillery operations (Haynes, Vol. 5). The corn starch industry, confronted with a loss in output of about 200,000 t/a of corn starch sugar, then became very much interested in recuperating at least part of that output by installing or promoting other uses of corn sugar.

One new use for corn sugar was thought to be the production of lactic acid. Defren of the Avery company had advocated the use of lactic acid in beer brewing already in 1920 (Defren, 1920) and with much brewing activity now just around the corner, manufacture of lactic acid might become a profitable affair.

Around 1930 Boehringer and Byk-Guldenwerke, important German lactic acid manufacturers, began to spread commercial literature claiming the advantages of the addition of lactic acid to remove temporary hardness from the process water used in the brewing industry (Lüers, 1929; Koudelka, 1929) and the improvements in the alcoholic fermentation through pH adjustment of the mash with lactic acid (Windisch et al., 1928).

With an envisaged production capacity of the U.S. distilling industry of 400,000,000 proof gallons of alcohol per year (Haynes, Vol. 5) and additions of at most 0.2 l lactic acid (100%) per m^3 of process water a total future use of lactic acid by the U.S. distilling industry of, say, 10,000 t/a could well be within reach. It is not known whether the two starch companies had such ideas about potential lactic acid sales in the brewing and distillery industry, maybe the growth of acidulant use in general had convinced them of the good prospects for lactic acid.

American Maize[8] began research on lactic acid fermentation in 1928. A process survey of American Maize's lactic acid production in "Industrial and Engineering Chemistry" suggests that the lactic acid business was seen as an important outlet for a by-product of dextrose production (Inskeep et al., 1952). During the drying operation an appreciable portion of the dextrose hydrate formed pills which were rejected by the screen in the classifier. Instead of reprocesssing these pills it was thought cheaper to use them in some kind of fermentation process [Which until 1933 was the illegal production of alcohol.]

> The original plan was to produce calcium lactate rather than lactic acid as an end product. The first equipment was of a decidedly makeshift nature: [11,000 l] fermentors were secured from a defunct brewery.
>
> In 1930 the idea of using the sodium salt of lactic acid as an antifreeze was suggested. The company investigated this but decided the idea was impractical.
>
> The production of sodium lactate was therefore discontinued until late in the 1930's, when other uses became apparent. At that time it was used in the textile and paper industries as a sizing or a substitute for glycerol.
>
> Since 1932 American Maize has been actively engaged in the production of calcium lactate and both technical and edible grades of lactic acid [Plant site: Hammond, IN].
>
> *- Inskeep, Taylor and Breitzke, 1952*

The United States Tariff Commission lists American Maize for the first time as lactic acid producer in 1933.

Clinton Corn Syrup Refining Company[9] began investigating the possibility to produce calcium lactate about 1930. A commercially viable process was not developed, however. Attention was then directed at a process for making edible lactic acid. One of the reasons for lactic acid production was the same as given for American Maize: utilization of by-product of the dextrose hydrate manufacture (J. Newton, private communication, 1986).

The Clinton company historically was a technologically oriented company, very interested in diversification in the field of starch products and starch derivatives. Dr. John M. Newton, who supplied the author with much detailed information, joined the research staff in 1941; about 10 people were employed in R&D at that time, under the direction of Dr. Geo T. Peckham Jr. The technical staff of the company always included one professionally trained bacteriologist; a company-trained fermentologist was responsible for the handling of the bacterial cultures and the fermentation plant.
Exhibits 9,III and 9,IV show the fermentation and purification procedures of the two starch companies. See also Figures 9,3 to 9,6.

EXHIBIT 9,III
FERMENTATION PROCEDURES OF AMERICAN MAIZE AND CLINTON
lit. on American Maize: Inskeep et al., 1955 (AM);
on Clinton: Peckham, 1944; Newton, private communication, 1986 (Cl).

raw material	dextrose (pills, second grade product) starch hydrolysates
source	corn wet-milling plant
concentration	15% (AM); 12.5% (Cl)
sterilization	not required (AM) pasteurization at 90° (Cl)
inorganic nutrients	0.25% diammonium phosphate (AM) inorganic salts (Cl)
organic nutrients	0.375% malt sprouts (AM) corn steep liquor (Cl)
neutralization	excess calcium carbonate is added; in 1943 and 1944 AM used hydrated lime which was added automatically through the use of a pH controller, electrode fouling and voltage variations made the method impractical. excess of ground lime stone (Cl).
fermentation temperature	49°C (AM) 50–55°C (Cl)
fermentation vessels	wooden vats, 8 of 25 m^3 and 2 of 110 m^3 (AM) 10 wooden vats, 23–30 m^3 (Cl)
sterility of process	no special precautions washing, filling with water, boiling (AM)

mixing	the small fermentors with two cross arm timbers on a wooden vertical shaft; the large fermentors with stainless steel side-entering propellers (AM) continuous, slow agitation (Cl)
fermentation pH	5.8–6.0 (AM) 5.0–5.6, sometimes hydrated lime was added to pH 6.0 (Cl)
fermentation time	4–6 days, sometimes longer
control of end	by determining reducing sugars: to maximum 0.1% for technical acid: 0.4% (Cl)
yield	85–90% (Cl)
type of bacterium	*Lactobacillus delbrueckii* (AM) *Lactobacillus* isolated from corn steep liquor (Cl)
cultivation method	regularly (?) from pure cultures, scaling up factor about 12 per step (AM) the pure strain is kept on agar (5°C), scale up factor about 10 (Cl) Clinton often performed cross-inoculations between batches after one or two days
amount of inoculant	8% (AM); 10% (Cl)
source of bacterium	U.S. Department of Agriculture (AM)
stereoisomery	racemic mixture (AM); about racemic, (sometimes ?) high in L+ (Moore et al., 1941) (Cl)
control of purity	microscopically

EXHIBIT 9,IV

PURIFICATION PROCEDURES OF AMERICAN MAIZE AND CLINTON

lit. on American Maize: Inskeep et al., 1955 (AM)
on Clinton: Peckham, 1944; Newton, private communication, 1986 (Cl)

treatment of broth	extra lime is added in a separate settling tank (AM); the broth is heated to 82°C (AM), or to the boiling point (Cl)
decantation	yes
alkaline filtration	of the sludge from the decantation operation, AM used gypsum from the subsequent acid filtration as filter aid
acidification	with a slight excess of sulfuric acid
acid filtration	over a filter press (AM); over a stainless steel rotary vacuum filter (Cl)
purification	with sodium sulfide to remove heavy metals (AM); with ferrocyanide to precipitate iron (Cl)

decolorization	with activated carbon (a number of times), followed by filtration
concentrating	AM: calcium lactate first to 32% in a single effect evaporator, lactic acid in a single or double effect 316 stainless steel evaporator. Cl: a continuous evaporator (316 stainless steel); scaling is prevented by high circulation over the heat exchange tubes. Finishing in a glass lined evaporating pan.
packaging	in wax coated oak barrels coopered at the plant from externally bought staves, or in carboys.

Fig. 9,3. Transfer of microorganism (American Maize). Reprinted from Inskeep et al., 1952. Published by the American Chemical Society.

Corrosion problems

All of the three new producers (Sheffield, American Maize and Clinton) reported major problems with corrosion, the worst of which occurred at the heat transfer surfaces of the evaporator. Peckham:

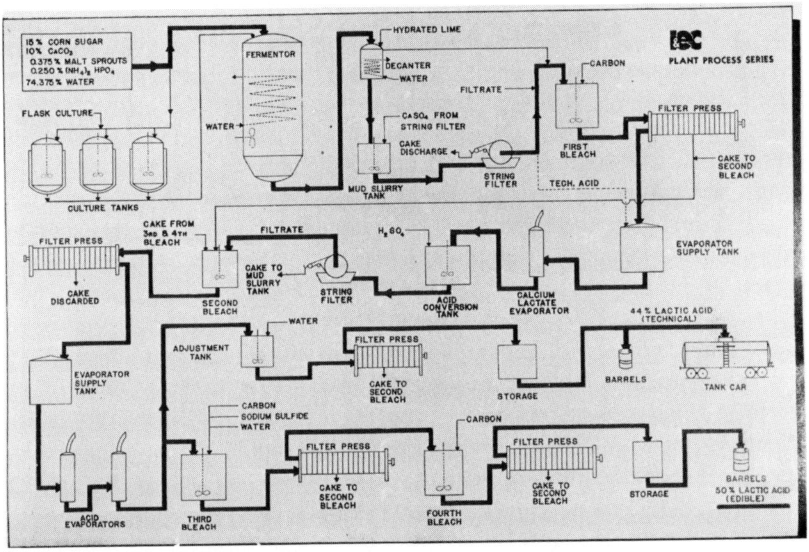

Fig. 9,4. Flowsheet of the process of American Maize- Products Co. Reprinted from Inskeep et al., 1952. Published by the American Chemical Society.

Fig. 9,5. Continuous string filter removes precipitated calcium sulfate (American Maize). Reprinted from Inskeep et al., 1952. Published by the American Chemical Society.

Fig. 9,6. Filterpress (Schiedamsche Melkzuur Fabriek). Courtesy: CCA biochem, Gorinchem.

Experience has shown lactic acid to be one of the most difficult of all acids to handle from a corrosion standpoint. Iron and steel are of no use. Copper and copper alloys, including acid-resistant phosphor-bronze, have a very limited life and introduce sizable quantities of copper which subsequently must be removed. Chrome steel has very poor resistance and most of the stainless steels and Monel metal are unsatisfactory. Certain stainless steels, particularly those containing molybdenum, show reasonable resistance. Inconel and pure nickel are somewhat resistant but are still seriously attacked. Some of the relatively low iron alloys containing large percentages of nickel and chromium resist fairly well. Silver and tantalum are both very well suited for the operations except for their cost and relative unavailability.

- Peckham, 1944

Wood (Peckham: cypress or pitch pine) remained the preferred material for the construction of the fermentors as well as for the tanks used in the alkaline treatment and the acidification; the filterpresses also were equipped with wooden frames. Lead coated vessels and lead piping was used for sulfuric acid.

Inskeep et al.:

Valves on most of the process pipelines are of the diaphragm type with glass-lined bodies and rubber diaphragms. The old [1935 ?] copper process lines and bronze pumps are being replaced with Type 316 stainless steel equipment. The barrels are filled with rubber hose coupled to a silver-coated filler valve.

- Inskeep et al, 1952

The Importance of Calcium Lactate

At the end of the 1920's calcium lactate had found a number of pharmaceutical applications. It was recommended for the treatment of rachitis, scrofula (tuberculosis of the bones), of hemophiliacs prior to surgery, and for the supply of infants with additional calcium (Omega, 1929; Ullmann, 1931; Heusner, 1966).

Eisenman in a review paper on lactic acid (Eisenman, 1930) mentions the use in baking powder formulation: "It has the advantage of being a slow-rising powder."

Calcium lactate formed gas only indirectly through the double conversion with sodium bicarbonate to sodium lactate and calcium bicarbonate and the subsequent decomposition of bicarbonate into carbonic acid and calcium carbonate. Pound for pound calcium lactate was much less economical than the acid phosphates which dominated the market. Pricewise it also could not compete with cream of tartar. It seems that calcium lactate nevertheless gained a small foothold in this field (Inskeep, et al., 1952).

During the 1930's a new application was found in the firming of canned fruits and the gelling of sugar solutions (Filachione and Costello, 1952).

In 1935 the U.S. Census reports three producers with a total output of 237 t. (Total European production then was between 200 and 300 t/a, most of which was used for pharmaceutical purposes and calcium enrichment. The largest European producer was Merck with an average yearly output of 125 t.)

Price information is also scarce:

Byk-Gulden	(1935)	0.46 $/kg (delivered in lb bags, India)
Gunderson	(1937)	0.39 $/kg (for food producers, USA)
SMF	(1938)	0.28 $/kg (cwt ex works Schiedam)

Gunderson, biochemist of Quaker Oats Company, in a survey of the nutritional economics of dietary calcium (Gunderson, 1937) stresses the merits of the enrichments of foods with calcium. As examples he names pasta products, cereals and bread. Cheapest, of course, he says, would be calcium carbonate, for pharmaceutical preparations calcium lactate is the most economical.

The Position of Lactic Acid as Food Acid in America

Detailed information on the use of lactic acid is scarce (see also Table 9,I,300). Total American consumption of edible lactic grew from an estimated 300 t/a in 1929 to about 600 t/a in 1939. During World War II the volume increased further to about 2,000 t in 1945 (Inskeep et al., l.c.). Sheehan, member of the Chemical Unit of the U.S. Department of Agriculture, gave a percentage breakdown of the 1941 consumption (Sheehan, 1944).

He calculates the total use of the edible (and so-called plastic) quality at about 700 t (production minus exports), of which 400 t was sold through jobbers. The 300 t accounted for were applied in the manufacture of jams and jellies (170 t), olives (40 t), pickles (25 t), brewing (25 t), soft drinks (20 t) and a number of even less important items [Rounded-off figures].

Other applications mentioned are spreadable cheese, candy, yeast manufacture and ice-cream (Burton, 1937; Newton, private communication, 1986).

Evidently the beer and brewing industry did not develop into big consumers. This was due to the introduction of improved water treatment processes as well as to application of mineral acids in the preparation of process water and the acidification of the mash.

Typical food applications for which lactic acid was just one of the possible food acidulants did not develop because of the low price and superior quality of Pfizer's citric acid. In today's marketing slang: lactic acid was confined to its niche applications (fermentation products and products in which it was a "natural" ingredient).

Nota bene: The most important technical use was still in tanning, about 850 t in 1941; for mordanting a volume of only 64 t of technical lactic acid is calculated from Sheehan's figures.

Developments in Europe

In the 1930's a large number of new lactic acid plants were established throughout the world[10]. Consequently the German dominated "Konvention" of lactic acid producers in Europe experienced considerable difficulties in the defense of their position. This is very clearly pointed out in a November 1936 letter from Byk-Gulden to the Dutch producer Schiedamsche Melkzuur Fabriek, SMF:

> ... the position of lactic acid on the world market has deteriorated considerably since you signed the agreement with the Konvention in August 1933. You have without doubt recognized the worsening of the prices from the decrease in the tabulated Konvention export prices[11].
> For instance the prices of technical lactic acid were about 50% higher [in 1933] than they have been in the last years. Since November 3rd of this year, the Konvention had to substantially lower prices again to combat progressing outsiders from: Hungary, Czechoslovakia, U.S.A., Poland, Romania and other European countries.
>
> For chemically pure lactic acid [The German manufacturers had concluded a separate Konvention for pharmaceutical lactic acid; the agreement with SMF covered all qualities.] the August 1933 quotations were about two times the present prices. Also for this grade a further price decrease is expected as long as no friendly understanding with the firm of Howards exists.
>
> The export markets of chemically pure lactic acid experience additional disturbances from U.S. manufacturers [Du Pont] and from the Czechoslovakian outsider.
>
> On top of the lowering of the value of the sales the Konvention is confronted with a serious decline in export volumes. This decrease in volume is due to the intrusion of the following lactic acid producers into the territory of the Konvention [North America and Great Britain did not 'belong' to this territory]: Bowmans, Hungary, Czechoslovakia, U.S.A., Poland, Romania, several other European lactic acid plants. [At the end of this Chapter more information on competitive producers can be found].
>
> *- Byk-Guldenwerke, 1936*

A number of reasons can be held responsible for the decline of the dominance of German producers in the lactic acid market. For one thing the high cartel prices at a time when raw material prices were at abysmally low levels were inviting competition. The changing consumption habits, although not yet of so much importance as in America certainly also contributed to the increase of food acidulants use. This is also attested by the installation of additional citric acid producing capacity in Europe:

In 1936 Pfizer built its first overseas citric acid plant at the Kemball, Bishop & Co. location in London (Mines, 1978). The capacity of the new facility was well over 1,000 t/a; total European production including that based on lemon juice was then about 8,000 t/a (Dehio, 1966).

In 1937 C.H. Boehringer Sohn and Joh. A. Benckiser, the two leading tartaric acid producers of Germany negotiated a contract for the joint development of a citric acid fermentation process (Chem. Weekblad, 1937; Dehio, 1966). German consumption of citric acid then amounted to about 1,000 t/a, produced by Benckiser and Boehringer from imported calcium citrate.

A major factor contributing to the decline of the German lactic acid dominance were the autarkic measures imposed on German society by the Hitler regime. German lactic acid manufacturers had to pay raw material prices two to three times as high as their competitors paid (See Intermezzo 6). On top of this came trading difficulties related to the persecution of the Jews in Nazi Germany. This was particularly felt in America where most of the trading houses were Jewish (Dehio, 1966).

As a consequence of all these factors German lactic exports declined from an annual average of 2,100 t in the late 1920's to 1,000 t in 1937, and to probably still lower figures in 1938 and 1939.

Growth of Lactic Acid in Germany

When the difficult period of the French occupation of the Rhine region drew to an end (1925) the Boehringer company was ready to renew its efforts to gain a foothold for lactic acid in the food and beverage market. The sales department "Acids", while still situated in Hamburg was actively engaged in promoting the use of lactic acid in the German **baking industry** (Dehio, 1966).

In Germany rye bread and bread from mixed (wheat-rye) flour have for a long time been the staple bread of the population. The "classical", but modern yeast-leavening recipe for wheat bread applied to flours containing rye flour produces a very stiff bread, not much liked by "modern" tastes. In Germany and other central European countries rye and mixed bread was therefore still prepared by the ancient sourdough method. In this method a portion of the dough of the previous preparation is set aside to develop yeast, and lactic acid forming bacteria. These microorganisms as well as the lactic acid they produce improve the texture and the taste of bread baked the next day from flour to which up to 5% of sourdough has been added.

The sourdough method is, like other fermentation processes, susceptible to infection. An commonly occurring infection is that by *Bacillus mesentericus*, which causes “ropiness” of the bread. Moreover, the spoiled bread will have an off-taste. Formerly bakers combatted such infections by acidification with vinegar (Encyclopaedia Britannica, 15th ed., vol. 19, 1987). The sourdough process also shows large variations in bread quality depending on the storage conditions of the sourdough, causing appreciable variations in fermentation.

When bread baking gradually became an industrial activity (On the German countryside bread baking was quite a normal housewife activity up to World War I.), bakers sought more economical and hygienic methods of production than one which involved the storage of a large volume of difficult to handle sourdough.

Häussler[12] gave a short account of the history of the use of lactic acid in bread (Häussler, 1940). In World War I lactic acid had been introduced in the bakery to suppress “ropiness”. In the last years of the war the flour mixture contained ground potatoes, which made lactic acid addition almost imperative for hygienic reasons (Sander, 1931). Since those years lactic acid had remained a recognized product for the prevention of ropiness, but it was not used on a large scale because the baker could not readily mix liquid lactic acid homogeneously with flour. In 1930 this problem was solved by a number of bakery supply companies who offered bakers dry sours containing from 10 to 20% of lactic acid. [One successful brand of dry sour was “Boerol”, a Boehringer product.]

The dry sours were usually prepared by adding lactic acid to flour and other ingredients, or by mixing flour with a mash which had undergone lactic fermentation. In general the baker was advised to add sufficient dry sour to attain percentages of lactic acid (100%) on flour between 0.1 and 0.4 (Institute for Cereals, Flour and Bread TNO, 1958).

Comparatively large amounts of lactic acid must have been adsorbed by this application. When all rye and mixed bread was produced with the addition of dry sours some 2,000 t of lactic acid (100%) would have been required. The actual amount used certainly was much less, maybe a few hundred tons. (In the 1970’s when the sourdough method had been pushed back much further, total German use of organic acids supplied to the bakery industry in dry sours was estimated at more than 3,000 t/a. Private information, CCA.)

Fig. 9,7. Alfred Häussler. Courtesy: C.H. Boehringer Sohn, Ingelheim.

Beverages

Boehringer and Byk-Gulden both issued a number of brochures on the use of lactic acid in fruit wines. The advice was to adjust the acidity of the fruit juice with edible lactic to a concentration of 0.9% to ensure good fermentation, and a better taste and improved keeping qualities of the final product (Weiss, 1933). German food legislation approving the addition of at most 0.3% of lactic acid (100%) in the manufacture of fruit wines became valid on September 1, 1932.

The use of lactic acid in soft drinks and lemonades in these years also made further headway in Germany. (See Ch. 6, 182 on the success Boehringer had with its Chabeso drinks.)

Fig. 9,8. Walter Dehio. Courtesy: C.H. Boehringer Sohn, Ingelheim.

Boehringer Tackles the Food Industry

The progress Boehringer made in the food field has been well documented by Walter Dehio in his book, "The Industry Of Tartaric, Citric and Lactic Acid" (Dehio, 1966)[13]:

It should be emphasized that Ingelheim succeeded in enlarging the sales of all qualities [of lactic acid], such notwithstanding the conditions [imposed by the Konvention] of exactly equal pricing and interdiction of any trade advantages. In particular the development in edible lactic acid was more than surprising. A very large share of this was the result of the painstaking and patient labors of Dr. Alfred Häussler who advised the customers on the use of lactic acid. In a steady stream of communications to the professional press Dr. Häussler laid down the know-how gained from laboratory experiments at Ingelheim as well as from individual clients. Publication of practical recipes on the use of lactic acid supplemented these activities.

As a result Germany became in the 1930's the largest European market for edible lactic acid, a position [the acid] did not achieve in any of the other producing countries. The German eating habits which favored sour vegetables, fish marinades, etc., without doubt encouraged these developments. The systematic and science-based promotion of edible lactic by Ingelheim was the decisive factor in the increase of sales which greatly surpassed those of other firms. Our largest customer was [a] pectin manufacturer... who alone accounted for up to 300 t/a of edible lactic. [The pectin was used in the production of jams and marmalades[14].]

- Dehio, 1966

Figures 9,9 to 9,11 reproduced from the 1935 jubilee book of the Boehringer company show some details of the finishing and packing of lactic acid at Boehringer's works.

Fig. 9,9. Polishing filtration of lactic acid. Photo Dr. Paul Wolff, 1935. Courtesy: C.H. Boehringer Sohn, Ingelheim.

Other European Lactic Acid Manufacturers

GERMANY

Figures on total German production in these years are not known; for 1929 a volume of 3,000 t is given (Schmidt, 1943). Dehio reports that Boehringer's production reached the all-time high of 2,400 t/a (100%) in 1941 and 1942.

Byk-Guldenwerke. Although at the end of the 1920's this company presumably produced about as much lactic acid as Boehringer did, by 1940

Fig. 9,10. The cooper. Photo Dr. Paul Wolff, 1935. Courtesy: C.H. Boehringer Sohn, Ingelheim.

Boehringer seems to have outpaced Byk-Gulden by far. As it is known (Dehio, 1966) that the plant at Niederstriegis did not suffer any war damage, and that its postwar capacity was not (much) more than about 1,000 t/a, it can be assumed that the plant during the 1930's also had a capacity of that size (see also Ost Rassow, 1965, who reports a total German manufacturing output in the early 1960's [for the two remaining plants of Boehringer and VEB Lactacida] of over 2,500 t/a)[15].

E. Merck, Darmstadt, one of the members of the German "Konvention", gradually increased production from 232 t (as 100%) in 1929 to about 500 t in the middle 1930's. In a typical year (1935) the product mix was:

Fig. 9,11. Filling up a carboy. Photo Dr. Paul Wolff, 1935. Courtesy C.H. Boehringer Sohn, Ingelheim.

59 t pharmaceutical lactic acid (90%)
119 t calcium lactate (5 aq.)
241 t edible lactic acid (80%)
316 t technical lactic acid (50%)

In the beginning of 1940 Merck stopped production, partly because of difficulties in the procurement of raw materials and the shortage of personnel (Manufacturing Report 1940 of the company), partly also because the facilities of the plant were needed for the production of ascorbic acid – vitamin C (Dehio, 1966).

Schering-Kahlbaum, of Berlin. This pharmaceutical concern never became an important producer of lactic acid. About 1935 the firm stopped production and from then on, until 1941, bought lactic acid on a coproducers basis from Boehringer (Dehio, 1966).

Josef Pleser Söhne, Darmstadt-Eberstadt. Joseph Pleser, son of the yeast manufacturer Josef Pleser, installed around 1930 a small lactic acid plant in the family yeast works. He had studied the fermentation process at the Institute for the Fermentation Industries in Berlin, and obtained in 1926 his doctorate on a study of lactic acid fermentation with *Bacillus delbrücki* [sic] (private information, Dr. Winfried Pleser, 1987). It seems rather likely that Joseph Pleser was also familiar with the manufacturing process developed by Kownatzki, a former employee of the Fermentation Institute (Wagner, 1926). The company produced about 100 t edible and 250 t technical lactic acid per year. Pleser stopped production in one of the last years of World War II (Dehio, 1966).

H. Prinz, Melsungen near Kassel, Hesse. This company made even smaller amounts of lactic acid than Pleser. Their raw material was whey from blue cheese production. The factory was heavily damaged in World War II due to a flood wave following the destruction in May 1943 of the Eder barrage by the Royal Air Force with ricochetting torpedoes. Production of lactic acid was not resumed by the company (Dehio, 1966).

FRANCE

The **Société Normande de Produits Chimiques** which in the 1920's had become a member of the Rhône-Poulenc group of companies, was responsible for most of the French production. The 1930 production is estimated at 800 t of technical lactic acid (concentration 50% ?) (Chemical Trade Journal, Jan. 17, 1930). The "Normande" sold the output of two manufacturing plants, the larger of which was located at the Saint Fons production site of the Rhône Poulenc company, the other at the original Petit Quevilly location of the Normande itself (Dehio, private communication, 1985). During World War II the production of the Petit Quevilly plant, being located in the German occupied part of France, was moved to the Saint Fons plant in the Vichy controlled semi-autonomous part of the country (Dehio, 1966)[16].

H. Gouthière & Co., Reims. This small company stopped production during World War II (Dehio, 1966).

GREAT BRITAIN

Bowmans Chemicals. The company produced in the 1930's about 500 t (100%) lactic acid per year. The most important single outlet was the leather

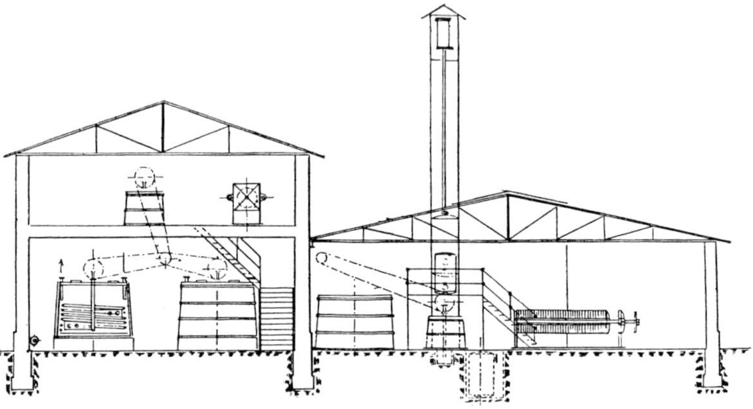

Abb. 26 b. Durchschnitt einer Milchsäureanlage für 600 t technische Säure pro Jahr.

Fig. 9,12. Cross-section of lactic acid plant. Reproduced from Wagner, 1926.

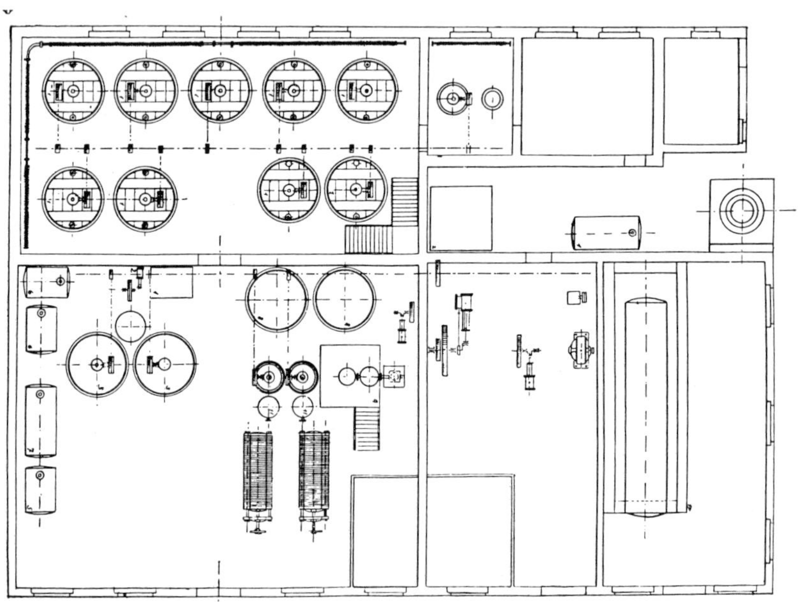

Abb. 26c. Komplette Milchsäureanlage für 600 t technische Säure pro Jahr, Bauart J. L. Carl Eckelt, G. m. b. H., Berlin W 62, System Dr. Kownatzky.

Fig. 9,13. Lay-out of lactic acid plant. Reproduced from Wagner, 1926.

industry – boots and heavy leather for conveyer belts. Production of edible lactic acid was also taken up in this period.

An important user of Bowmans lactic acid was **Howards & Son** of Ilford, near London. Howards made ethyl lactate (200–300 t/a capacity), pharmaceutical lactic acid (made by hydrolysis of captive ethyl lactate; estimated output about 100 t/a), and calcium lactate.

NETHERLANDS

The first Dutch company which tried to manufacture lactic acid on a commercial scale was the dairy company "De Meijerij" at Veghel. In the late 1920's this company produced a few hundred tons of lactic acid per year from whey. Lactic acid manufacturing was performed on an intermittent basis until 1962 when the company stopped production permanently.

Schiedamsche Melkzuurfabriek (SMF) began production in 1931 (or 1932) at a scale of less than 150 t (100%) per year[17]. By 1940 production had increased to about 250 t/a, of which almost 100 t was calcium lactate. Production of edible lactic acid was begun in 1935 or 1936. Sugar was then applied as raw material, purification was based on calcium lactate crystallization. Thanks to the use of sugar SMF made an edible lactic of good quality which was readily accepted by the Dutch food industry, particularly during the war when citric and tartaric acid were not available. The largest application was found in the bakery in the baking of rye bread and "war" bread from inferior flour compositions (SMF archives).

DENMARK

Kemisk Vaerk Køge (KVK), Køge, near Copenhagen. Around 1930 KVK was established by Sadolin and Holmblad A.S., important producers of lacquers and varnishes, as a small chemical plant. Fermentation research was carried out by the company to make butanol from cheap carbohydrate feed stock. When it became clear that the economic prospects for butanol by fermentation were rather dim, the research efforts were aimed at another fermentation process – that of making lactic acid.

A pilot plant with a capacity of not more than 100 t/a (100%) of lactic acid was installed in 1935/36 and expanded to about 600 t/a during the war to meet increasing demands from the food industry (replacement of citric acid) and the bakeries (for bread sours). Raw material: beet molasses. The company made an almost complete range of lactic acid qualities and metal lactates (Glimsholt, private communication, 1985; Arnold, private communication, 1988).

ITALY

Two very small Italian factories seem to have existed between the two World Wars and for some years afterwards (Report ILRA meeting, June 1964).

HUNGARY

A Hungarian lactic acid factory was reportedly in operation in 1935 at Budafook, near Budapest. Capacity about 400 t/a technical lactic acid; raw material used: potato starch (Die chemische Industrie, Nachrichtenausgabe, **58**, 1935).

ROMANIA

The erection of a lactic acid factory was announced in 1933 (Chemical Age, 1933).

RUSSIA

A Russian fermentation plant with a capacity of 500 t/a 50% was announced in 1932; raw material: waste liquors from the starch industry (Chemical Age, 1932).

POLAND

Poland started production of special qualities of lactic acid in 1935 (Die chemische Industrie, Nachrichtenausgabe, l.c.).

CZECHOSLOVAKIA

Czechoslovakian production of technical lactic acid was reportedly about 150 t/a (Chemical Age, 1937).

JAPAN

The first and most important producer was **Takeda Chemical Industries Ltd.**, which company took up production of lactic acid in a plant at Osaka in 1928. The company succeeded in replacing high quality German lactic acid in the sake brewery (Dr. K. Yasumatsu, private communication, 1987). Four other companies were listed in 1937 as lactic acid producers: Dai Ichi Seiyaku, Sankyo, Tanabe and Dai Nippon Industries (Chemical Age, 1937). After the war total Japanese fermentation production capacity was estimated at 500 t/a (probably 100%) (ILRA meeting November, 1965).

Notes

[1] Total U.S. citric production then was estimated at over 8,000 t; during World War II total production increased to 12,000 t (Jackson, 1958).
[2] The import prices of lactic acid are not fully comparable; the statistics include unknown quantities of much higher priced pharmaceutical lactic acid (between 10 and 50 t/a on total imports of 160 to 240 t/a).
[3] The 1928 U.S. Census Report also lists Apex Chemical Co., Elizabethport, N.J. as a producer.
[4] In 1930 the prices of many agricultural products were about 10 to 20% lower than in 1929; in 1931 some dropped to half the 1929 price; and the disaster became complete in the years 1932 and 1933 when for instance the price of corn fell to a third of the 1929 average (See Table 8,II, 242).
[5] One of the factors limiting the efficiency of the batch fermentation is the so-called "lag period, in which certain physiological changes take place in the cell, but little or no multiplication occurs." In normal manufacturing practice this period is difficult to distinguish from the subsequent time the bacterial culture requires to multiply to the cell density desired in the production period. Depending on the relative amount of inoculate, the viability of the culture and the fermentation conditions the total lag period before production really starts can be anywhere between 6 and 24 hours.
[6] Mycoderm literally means "sliming film". Presumably a combination of slime forming microorganisms belonging to the genus *Acetobacter* (acetic acid forming bacteria), in the older literature referred to as *Mycoderma*.
[7] In 1930 the parent company National Dairy acquired Kraft-Phenix, the largest

cheese maker of the United States. In 1969 the corporate name of National Dairy was changed to Kraftco after the best known brand of the company, and again in 1976 to Kraft, Inc. (Moskowitz et al., 1980).

[8] American Maize Products Company received its name in 1908 when the Royal Baking Powder Co. bought the preceding company – corn wet-miller Western Glucose Co. In 1929 Royal Baking Powder Co. merged with the large yeast producer Fleischmann Co. to Standard Brands, Inc. American Maize Products, however, was not included in this merger but became an independent company. (Haynes, Vol. 6, 1954).

[9] The Clinton Corn Syrup Refining Company of Clinton, IA, was established in 1906 by the National Candy Company of Saint-Louis, MO. The first products were glucose syrups, other products from the wet corn-milling plant, dry starch products, and dry conversion products. Production of crystalline dextrose monohydrate followed in 1928, only a few years after Corn Products Refining Company had built the first plant for this new sugar. In the course of the years the company changed its name a number of times, to Clinton Company, to Clinton Corn Products Co., when it was taken over by Standard Brands (1956), and later to Clinton Corn Processing Company. Today Clinton is part of the ADM organization under the name ADM-Clinton Co.

[10] The firm of J.L. Carl Eckelt in Berlin offered complete plants for the production of technical grade lactic acid, design Dr. Kownatzki (Wagner, 1926; see Figures 9,3 and 9,4 on p. 334). It is not known ,however, whether any lactic acid plant was built by Carl Eckelt. [11] Only a few tabulated Konvention prices have been found:

In 1926 for 50% technical 0.67 RM/kg and
for 80% edible 1.26 RM/kg (Merck, 1988)
In 1937 for 50% technical 0.48 RM/kg (SMF, 1937)
(prices of 5,000 kg lots, ex works)

[12] Alfred Häussler (1892–1958), studied chemistry at the Technical University (Hochschule) of Munich, and obtained the Ph.D. degree in 1919. He began his career at Boehringer as organic chemist in 1921. From about 1925 he acquired the knowledge and skills of a food chemist and as such he was responsible for quite a number of applications of Boehringer's edible lactic acid.

[13] Walter Dehio (1898–1988), born in Reval (Tallinn, Estland, now part of the USSR) and fluent in many languages, went to a business school in Germany. He entered Boehringer's service in 1929 as sales manager tartaric acid. In 1937 he was made responsible for the sales of all organic acids of the company, then comprising tartaric, lactic and citric acid. From this position he retired in 1964.

[14] BIOS Final Report No. 388, 1946, gives more information about this application. Lactic acid, 80% edible, was used for the extraction of pectin from dried pomace, apple pulp from apple juice and cider production. The concentration of lactic acid used in the extraction was from about 2 to as much as 7% (as 80%) on the dried pomace. (See also Maass, 1951, pp 35 and 52).

[15] After World War II Byk-Guldenwerke's lactic acid plant, being situated in the Russian occupied zone, was confiscated to become a "Volks Eigener Betrieb", literally: a plant owned by the people. As VEB Lactacida it has remained a producer of lactic acid.

[16] The "autonomy" of Vichy France lasted until November 1942 when it was also occupied by the Germans following the landing of Allied Forces in North Africa.

[17] The company was registered with the Schiedam Chamber of Commerce on October 1, 1931. Owners were Hendricus Christianus Jansen and three members of his family. The lactic acid plant was built in an old pressed yeast factory which they had closed down in 1928 under an agreement with the Nederlandsche Gist en Spiritus Fabriek which wanted to secure its monopolistic position (Oosterholt, 1952).

References

Encyclopaedia Britannica, **1987**. *The New Encyclopaedia Britannica*; 15th ed.; Encyclopaedia Britannica, Chicago.

Ullmann, **1931**. *Enzyklopädie der technischen Chemie*; F. Ullmann, Ed.; 2nd ed.; Urban und Schwarzenberg: Berlin, Vol. 7, 593.

Burton, L.V., **1937**. *Food Ind.*, **9**, 634.

Byk-Guldenwerke, **1935**. *Price Lists for India.*

Chem. Age (London), **1932**, no page.

Chem. Age (London), **1933**, no page.

Chem. Age (London), **1937**, 102, 256, 294 and 434.

Chem. Trade J., Oct. 9, **1936.**, 299.

Chem. Trade J. Chem. Engineer, Jan. 17, **1930**, 57.

Chem. Weekbl., **1938**. no page.

Defren, G., **1920**. *Chem. Age (N.Y.)*, **2**, 478.

Dehio, W., **1966**. *Die Industrie der Weinsäure, Zitronensäure und Milchsäure*, internal publication C.H. Boehringer Sohn.

Eisenman, W., **1930**. *Food Ind.*, **2**, 262.

Filachione, E.M., and Costello, E.J., **1952**. *Ind.Eng.Chem.*, **44**, 2189.

Forrestal, D.J., **1982**. *The Kernel and the Bean*; Simon and Schuster: New York.

Garrett, J.F., **1930**. *Ind. Eng. Chem.*, **22**, 1153.

Gunderson, F.L., **1937**. *Am. J. Public Health,*, **27**, 570.

Hale, W.J., **1934**. *The Farm Chemurgic*; Stratford: Boston.

Haynes, W., **1954**. *American Chemical Industry. A History*; D. van Nostrand: New York, Vols. 3, 4, 5, 6. Reprinted by Garland: New York, 1983.

Heusner, A., **197?**. *History of Various Processes of C.H. Boehringer Sohn at Ingelheim.*

Inskeep G.C., Taylor, G.G., and Breitzke, W.C., **1952**. *Ind. Eng. Chem.*, **44**, 1955.

Instituut voor Graan, Meel en Brood, T.N.O., **1958**. Communication 13-11-1958: Wageningen.

Jackson, R.W., **1958**. *Econ. Bot.*, **12**, 42.

Jenemann, J.A., **1933**. *U.S. Patent 1,906,068.*

Kirk and Othmer, **1953**. *Encyclopedia of Chemical Technology*; 3rd ed.; Interscience: New York, Vol. 10, 178.

Koudelka, **1929**. *Wochenschr. Brau.*, [17] and [18].

Lüers, H., **1929**. *Wochenschr. Brau.*, [17] and [18], [50], 16 Nov., 1.

Maass, H., **1951**. *Die Pektine*; Dr. D. Serger & Hempel: Braunschweig, 35, 52.

McMillan, W., **1939**. *Ind. Eng. Chem.*, **31**, 540.

Mines, S., **1978**. *Pfizer ... An Informal History*; Pfizer: New York.
Moore, S., Dimler, R.J., and Link, K.P., **1941**. *Ind. Eng Chem.*, **13**, 160.
Moskowitz, M., Katz, M., and Levering, R., **1980**. *Everybody's Business*; Harper & Row: New York.
National Soft Drinks Association, **1978**. *Sales Survey.*
North, D., **1939**. *Ind. Eng. Chem.*; **31**, 574.
Olive, T.R., **1936**. *Chem. Metall. Eng.*, **43**, 480.
Omega, **1929**. *Chem.Trade J. Chem. Eng.*, 581.
Oosterholt, G., **1952**. *Geschiedenis van de familie Jansen.*
Ost-Rassow, **1965**. *Lehrbuch der Chemischen Technik*; 27th ed.; Runge und Schwarze: Leipzig, Vol. 2, 1416.
Peckham, B.W., **1979**. *Economics and Invention: A Technological History of the Corn Refining Industry of the United States*; University of Wisconsin-Madison, Thesis.
Peckham, Jr., G.T., **1944**. *Chem. Eng. News*, **22**, 440.
Porter, G., Ed., **1976**. *Encyclopedia of American Economic History*; Adams Jr., D.R.; Vol. 1, *Prices and Wages*; Charles & Scribner's Sons: New York.
Rogers, L.A., and Whittier, E.D., **1928**. *J. Bacteriol.*, **16**, 211.
Sander, A., **1931** in Ullmann's *Enzykopädie der technischen Chemie*; Urban & Schwarzenbach: Berlin, Vol. 7, Milchsäure.
Schmidt, A., **1943**. *Die industrielle Chemie in ihrer Bedeutung im Weltbild*; Walter de Gruyter & Co.: Berlin.
Sheehan, M.J., **1944**. *Lactic Acid* (Synopsis of Information); Inquiry Reference Service; U.S. Dept. of Commerce.
SMF, **1938**. *Archives Schiedamsche Melkzuurfabriek.*
Spencer, G.L., & Meade, G.P., **1945**. *Cane Sugar Handbook*; Chapman and Hall: London.
SRI, **1969**. *Chemical Economics Handbook; Citric Acid*; SRI International; Menlo Park, CA.
SRI, **1981**. *Chemical Economics Handbook; Citric Acid*; SRI International; Menlo Park, CA.
Wagner, A., **1926**. *Die Herstellung von Essigsäure, Gärungsessig, Buttersäure, Zitronensäure und Milchsäure*, A. Hartleben: Vienna.
Wallach, B., **1966**. *Milchsäure und milchsaure Salze*, Internal publication of C.H. Boehringer Sohn: Ingelheim.
Weiss, H., **1933**. *Mitteilungen des Verbandes Deutscher Obst- und Beerwein Keltereien*, **1933**, [7].
Whittier, E.D., and Rogers, L.A., **1931**. *Ind. Eng. Chem.*, **23**, 532.
Windisch, Kolbach, und Banholzer, **1928**. *Wochenschr. Brau.*; [47], [48], [49].

INTERMEZZO 7

BOEHRINGER PROCESS TECHNOLOGY

Boehringer Tries to Use Cheap Beet Molasses

As pointed out in Chapter 9 (327), the autarkic measures of the Nazi Regime put a heavy burden on German lactic acid producers: Tax rebates on the sugar price were curtailed at a time world market sugar was at the lowest level ever, potato starch was becoming an expensive commodity, and other cheap (second grade) starches could only rarely be bought.

The difference between the internal German sugar price and the much lower international price was a major reason for Boehringer to investigate the use of (beet) molasses in the fermentation of lactic acid[1].

Vacuum Steam Distillation

The idea to use a vacuum steam distillation method goes back to the last century, see Intermezzo 4, 205). Boehringer began experiments to purify highly contaminated lactic acid from molasses fermentation by vacuum steam distillation in 1931. A year later the first practical installation was built (Wallach, 1966)[2].

In the beginning dilute lactic acid was passed under high vacuum over steam coils heated to about 200°C. [At this temperature caramelization and Maillard reactions take place which influence odor and taste of the product negatively!] Evaporation occurred rapidly; the condensate had a lactic strength of about 30–35%.

The first Boehringer vacuum steam distillation plant converted molasses into lactic acid for only about six months (in 1932), severe corrosion problems of the all copper installation due to the high chloride content of the molasses then forced the production management to switch back to a purer raw material (second grade starch). An additional reason for the change in raw material was the burnt taste and smell of the finished product, making it unsuitable for edible purposes.

Fig. I7,1. Bruno Wallach. Courtesy: C.H. Boehringer Sohn, Ingelheim.

The process engineers of the company fairly soon succeeded in improving the process. An important step was the application of dephlegmators (by means of which partial condensation of vapors coming from the distillation can be effected) to attain in one step the commercially desired concentration of 88–90%. Residual amounts of impurities necessitated the introduction of a second purification step – crystallization from solvents at low temperatures (see next page).

It is not unlikely that the improved method was inspired by a Du Pont patent on the same method (Macallum, U.S. Patent No. 2,013,104, application date January 30, 1933). In this patent steam of a maximum temperature of 120°C was blown through the lactic acid solution (in the one example of a strength of 44%). The vacuum was kept at between 13 and 33 mbar, and the first condenser at a temperature of about 45°. The total yield (first plus second condensate) was about 86%.

A similar process [also based on Macallum's patent?] was worked out for SMF by Shell Research in Amsterdam in 1941[3]. Details of this process have been preserved in the archives of the Dutch lactic acid producer:

The intake was 75% lactic acid [probably of edible, or near edible quality] which was preheated to 120°C and then sprayed over the top of a column filled with Raschig rings. Into the bottom of this column superheated steam

(140°C, 1 bar) was blown countercurrently. In the column a vacuum of 80 mbar was maintained. The condensate was separated by means of fractional condensation into a concentrated fraction containing about 90% lactic acid (condensation temperature 60°C) and a very dilute fraction of about 2% concentration. The amount of steam used was five times the amount of intake of lactic acid. Yield of the process 90%, after recuperation of lactic acid from the second distillate 95%.

A design of an installation for the continuous production of pharmaceutical lactic acid was presented by the laboratory of N.V. De Bataafsche Petroleum Maatschappij [Shell] in the summer of 1941.

The installation was never built. The war period was not the right time for the investment, large scale continuous production of lactic acid also was not an interesting option for a small lactic acid producer. Instead a plant for the manufacture of ethyl lactate, also designed by the Shell laboratory in Amsterdam, was constructed. Shell technical personel assisted in installing the plant; a Shell chemical engineer, J.A. Meyer, put it into operation (Ch. 10, 379).

Purification of Lactic Acid by Crystallization

History of the crystallization process: Krafft and Dijes of the University of Heidelberg were the first to study the crystallization behavior of racemic lactic acid in detail (Krafft and Dijes, 1895). They made pure lactic acid from a commercial fermentation product by high vacuum distillation (about 1 mbar)[4]. The distillate appeared to crystallize when received in a freezing mixture. The crystal mass obtained was very hygroscopic, it rapidly deliquesced when exposed to humid air. Krafft and Dijes then further purified the acid by repeated vacuum distillation. The product finally obtained did not show strong hygroscopic properties at low temperatures; its melting point was 18°C. [This is the melting point of the racemic product; see, however, Borsook et al., 1933, who report a melting point of 16.8°C for the racemic mixture.]

The experiments of Krafft and Dijes further showed that internal condensation products of lactic acid [lactoyllactic acid] had a negative influence on the crystallization process.

In subsequent years when efforts to prepare chemically pure lactic acid were intensified, crystallization phenomena were observed rather frequently. A few examples:

Noerdlinger also obtained crystalline lactic acid (melting point 18°C) when lactic acid was purified by distillation. (Intermezzo 4, 206).

Merck (Internal Report Lactic Acid Plant, May 21, 1930) reports crystallization of a distillate fraction at room temperature, (and a lactoyllactic acid content of the fraction of 2.5%).

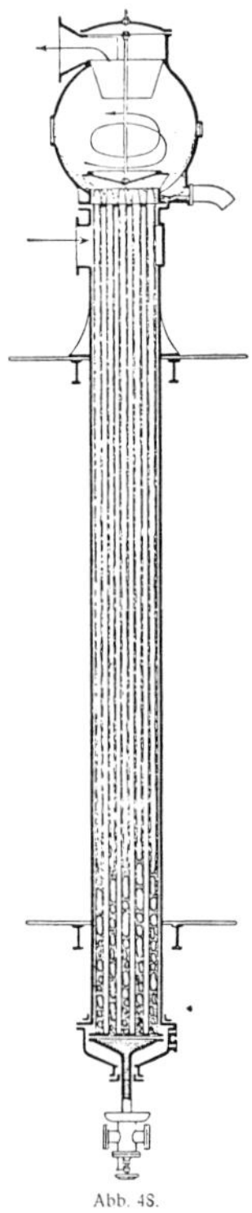

Fig. I7,2. Principle of the Kestner climbing film evaporator.
Explanation: The heat exchange pipes through which the dilute lactic acid flows have a small diameter and a relatively large heating surface, making an energetic development of steam bubbles in the liquid possible. As a consequence the liquid is forced upwards and spreads in a thin layer along the walls of the tubes. Concentrating takes place very rapidly; the concentrated liquid is carried off at the bottom.
Source: Ullmann's Enzyklopädie der technischen Chemie, 1914, Vol. 1, 21
Reproduced with permission from Verlag Chemie, Weinheim.

Boehringer began a study how to use the crystallization behavior of lactic acid for its purification in the fall of 1932. The first technological problem the company had to solve was the preparation of a concentrated lactic acid with a low content of lactoyllactic acid. This was achieved with a high capacity three stage (Kestner) evaporator, installed in March 1933 (Figure I7,2). This evaporator concentrated dilute lactic acid rapidly to a concentration of almost 100% lactic acid with a lactoyllactic content of less than 5–6%. [Which is much below the thermodynamic equilibrium content of monomeric lactic acid, which is 32.3% at 100% total concentration; Holten, 1971, 201.]

Crystallization of lactic acid was carried out from a solution of the fresh concentrate in a mixture of butyl acetate and toluene at -15 to -20°C. The crystals were centrifuged and washed free from adhering mother liquor at -10 to -15°C. Pharmaceutical lactic acid (90%) was prepared by dissolution in water.

A pilot installation was put into operation in April 1933, commercial production began in September of the same year (capacity of 20 t/month). The heat stability and organoleptic quality of the acid were excellent, even after thirty years samples passed all quality tests (Wallach, 1966).

Crystallization of Enantiomeric Lactic Acid.

In 1931, Otto Dietlen and Heinrich Felser of the Niederstriegis plant of Byk-Guldenwerke applied for a patent on a process for the manufacture of solid lactic acid. The first claim of this German Patent 593,657, 1934 reads:

> Process for the manufacture of a solid lactic acid, in the pure form melting at about 53 °C, which is characterized by the rapid evaporation at the lowest possible temperature of water or organic solvent from a solution of lactic acid with a high content of [L(+)-lactic] acid and a low to zero content of anhydride. The residual lactic acid is cooled as quickly as possible. [Purification of the crystals is performed by recrystallization from ethyl acetate or chloroform.]

Two years later American investigators (Borsook et al., 1933) published a method to prepare the enantiomeric forms of lactic acid. The method consisted in fractional crystallization of distilled lactic acid (at 105°C and about 0.13 mbar) from a mixture of equal volumes of diethyl and diisopropyl ether. The method only worked well when one of the isomers was in excess. (All commercial samples collected by the authors contained an excess of the L(+) isomer. The D(-) isomer was obtained from the corresponding zinc ammonium lactate.) The melting points of the enantiomers found were: L(+)-lactic acid, 52.8°; D(-)-lactic acid, 52.8° [5]. These were almost 30 degrees higher than the values reported for the enantiomers about thirty years earlier by Jungfleisch and Godchot (Jungfleisch and Godchot, 1906). Wallach, in his survey of technological progress at Boehringer, also observed a much higher melting point for the enantiomeric form of lactic acid:

> In the beginning of 1934 highly concentrated lactic acid crystallized spontaneously in the plant at 30 to 40°C. The lactic acid at issue had a high content of an optically active component with a melting point of 59°C.
>
> The fermentation must have taken another route, allegedly with preponderant production of one of the optically active forms [enantiomers]. Immediately experiments were made to use this property to elaborate a method for the purification of lactic acid through crystallization from water. These experiments failed to give good results due to problems with purification and further processing [compare Table I4,III, 202]
>
> \- *Wallach, 1966*

Allied Intelligence Reports on Lactic Acid Manufacture in Germany

In the final weeks of World War II Ingelheim was occupied by the American Army (March 20, 1945). In the first three months of the occupation chaotic conditions prevailed, for the population as well as for most of the enterprises. The works of Boehringer were confiscated and used for all kinds of military "purposes".

On the first day of July 1945 the American soldiers marched off and Ingelheim became part of the French zone of occupation. The Boehringer works were placed under the authority of the French captain E. Villain.

Already during the War the Allied Governments had decided to carry out a general inspection of the German industry. The background and reasoning for this inspection were manifold, and conflicting. For one thing it was intended to confiscate and demolish all establishments which had played a major role in the German war efforts. But what was major? Some of the Allied Nations, in particular those nations which had been robbed and plundered by the Germans wanted machinery and equipment, even complete factories as part of the compensation payments.

Other Nations (the British and American) were content with the procurement of secret German process know-how, and the undoing of the German war industry and the giant chemical combine I.G.

Dehio (Dehio, 1966) pictures the events as follows:

It was a distinguishing feature of the years 1946 and 1947 that a never ending stream of visits was paid by French and American – less frequently by British – interested parties from the pharmaceutical and chemical industry to learn about productions and processes of the firm of Boehringer. On some occasions they appeared at official missions in uniform, at others with permits from the French military command in plain cloths. Anyhow their inquisitive questioning did put much stress on the leading personnel of the company.

The company had to relinquish all process specifications, including flow sheets, to the "Field Industrial Agency Technical" [FIAT], which published the collected information of all German enterprises [visited], and sold this through the booksellers. When these reports later became available in Germany as well many German firms also became acquainted with the painstakingly guarded processes of their German competitors.

- Dehio, 1966

The lactic acid and citric acid plants were most thoroughly inspected on March 6, 1946, by a French mission. The head of this mission was Mr. Graber, managing director of the Société Normande de Produits Chimiques, Mr. Juin, sales director, and Mr. Pichon, chemical engineer of the Saint Fons plant (Compte rendu de la visite des Usines Boehringer à Nieder-Ingelheim par la mission de la Société Normande de Produits Chimique, Mars, 1946).

The receiving party consisted of capitain Villain and Messrs. Albert Boehringer, Dr. Ernst Boehringer, Julius Liebrecht, and Walter Dehio[6].

Boehringer Know-how in French Eyes

The French mission made a rather complete and reliable report about Boehringer's lactic acid technology:

We have visited the part of the plant in which we are interested, that is the manufacture of technical, edible and pharmaceutical lactic acid and the derivatives: sodium lactate and aluminum lactate.

General impression:

The installations for technical lactic acid are very old. The edible production has been renovated some dozen years ago using much old equipment. The pharmaceutical production is the most modern being about ten years old.

The fermentation plant is equipped with 24 cubical concrete fermenters, internally lined with [acid resistant] tiles: 18 vats of 40 m^3 each and 6 vats of 24 m^3. The vats are provided with stirrers with wooden paddles.

For inoculation 1,000 l of a running fermentation are taken; total fermentation time averages 10 days (at the time of the visit the factory was not in operation, "due to lack of raw material").

Acidification, filtration of crude acid, removal of heavy metals (with sodium ferrocyanide) and decolorization (with 1% activated carbon on lactic acid content) do not present any particulars.

Edible lactic acid

a. Principle: steam distillation of technical lactic acid at very reduced pressure (13 mbar).

b. Steam distillation: the process is carried out in eight copper apparatuses each consisting of a heating kettle in which bundles of serpentines are placed. The lactic acid to be evaporated is [finely] divided over the serpentines which are heated with steam. At the same time a small amount of steam is passed through to permit desodorization.

The inflow [of technical acid] is continued until 4,000 l of distillate of 25% concentration has been collected ... The residue is sold at a low price. The most volatile products are condensed and collected separately.

Removal of heavy metals, decolorization and concentrating is required to finish the products.

Yield on technical acid: 85 to 90%.

Pharmaceutical lactic acid

a. Principle: Crystallization of edible lactic acid at low temperatures and separation of the crystals by centrifugation.

b. The process: The crystallization is performed in six stainless steel (V 2 A) vessels of 200 l, equipped with very slow turning stirrers. Cooling is effectuated at -35° with liquid ammonia.

Crystallization is carried out at -25°C; a centrifuge is used to separate the crystals from the mother liquor, and for washing. The crystals are taken up in water, the solution treated with potassium ferrocyanide and active carbon, and after concentrating to 90% passed over a filter.

Yield of production calculated on edible: 90 to 95%. Production capacity: 12 to 15 t per month.

Conclusions of the French Party

Generally speaking the works are equipped with very ancient installations but the scale is such that a large production of lactic acids [of different qualities] is possible. Moreover, as a consequence of the ancientness the equipment has been fully amortized, which indisputably will have a favorable influence on the sales price.

The manufacture itself does not yield any secret. [In the view of the author a remarkable, maybe politically inspired statement considering the advanced technology in steam distillation and low temperature crystallization the Boehringer company had mastered.]

Sundry Observations of Other Visiting Expert Parties

Other visits which had the lactic or citric acid plant as the main subject of investigation were reported in the BIOS Final Report No. 236 (visit of 24 September, 1945), BIOS Final Report No. 766 and the BIOS Final Report No. 1.652 [BIOS = British Intelligence Objectives Subcommittee].

BIOS Final Report No. 766:

The output of the company was reported at 240 t (100%) lactic acid per month [Dehio, 1966: during the war a maximum yearly production of 2,400 t was attained.] of which between 50 and 80 t were of a technical grade. The yield of the primary production was 85 to 90%; in the analysis lower fatty acids were included, however. Edible acid only contained minor proportions of volatile acids. Raw material costs: potato flour 25 Marks per kg (starch content 60%). [This must have meant 25 Marks per 100 kg.] Manufacturing costs of edible lactic acid 80% (exclusive of raw material cost): 86 Marks per 100 kg.

BIOS Final Report No. 236:

No definite advances in technology were noted, but a point of interest was the installation of Jena glass pipelines, similar in construction to those of Pyrex glass now manufactured in the U.S.A. The installation of these was stated to have been necessitated by the shortage of suitable metals rather than by a preference for glass. A plastic ("Vinidur", manufactured by I.G. Farbenindustrie) was also being used for pipelines and fume ducts for a similar reason [Vinidur is a type of hard polyvinyl chloride].

Notes

[1] Byk-Guldenwerke also used molasses. This is attested by a letter from Werner Ulrich, former employee of the company (Ch. 8,263), to SMF (February 2, 1953) in which he offered know-how on the use of molasses as feedstock.

[2] Bruno Wallach (1900–1976) studied chemistry at the Technical University of Munich with Professor Hans Fisher. In 1926 he entered Boehringer's service as chemical engineer. Until his retirement in 1966 he held positions in various research and production departments of the company.

[3] During World War II when the Shell company was cut off from its overseas

operations the laboratory in Amsterdam solicited the Netherlands chemical industry for research and development work to keep the lab busy and to prevent the occupying German military administration from recruiting research personnel for forced labor in the German war industry.

[4] The product was probably 80% grade from the then new producer, C.H. Boehringer Sohn of Ingelheim. In the literature around 1900 a distinction usually was drawn between pharmaceutical lactic acid (density 1.22 kg/l) and the fermentation grade (density of the 80% w/w solution 1.20 kg/l). Krafft and Dijes experimented with commercial lactic acid of densities 1.21 (80%?) and 1.16 (probably 60%).

[5] A still higher melting point of 54°C was reported in 1965 (Lockwood, 1965).

[6] Albert Boehringer (1891–1960), oldest son of the founder of the Boehringer company, received a professional training in business and administration. In the company he was charged with the alkaloids department, later with the sales of pharmaceuticals generally, and with purchasing.

Ernst Boehringer (1897–1965), second son of the company founder Albert Boehringer, studied chemistry at the universities of Würzburg and Munich, obtained his doctoral degree in 1926 and entered the family enterprise in 1927. Professionally he did much to extend the activities of the Boehringer company in the pharmaceutical field.

Julius Liebrecht (1891–1974), son-in-law of the founder of the company. Organic acids were his prime responsibility. By the end of his career he was generally referred to as "Senior of the European organic acid manufacturers".

(Private information from Boehringer Ingelheim, 1988).

References

Borsook, H., Huffman, H.M., and Yun-Pu Liu, **1933**. *J. Biol. Chem.*, **102**, 449.

Dehio, W., **1966**. *Die Industrie der Weinsäure, Zitronensäure und Milchsäure*; Internal publication of C.H. Boehringer Sohn: Ingelheim.

Holten, C.E., **1971**. *Lactic Acid*; Verlag Chemie: Weinheim (F.R.G.).

Jungfleisch, E., and Godchot, M., **1906**. *C.R. Hebd. Seances Acad. Sci.*, **142**, 515.

Krafft, F., and Dijes, W.A., **1895**. *Ber. Dtsch. Chem. Ges.*, **28**, 2589.

Lockwood, L.B.,, Yoder, D.E., and Zienty, M., **1965**. *Ann. N.Y. Acad. Sci.*, **119**, 854.

Wallach, B., **1966**. *Milchsäure und milchsaure Salze*, Internal Report of C.H. Boehringer Sohn: Ingelheim.

CHAPTER 10

PETROCHEMICALS PREVAIL OVER FERMENTATION CHEMICALS

LACTIC ACID FROM 1940–1965

During World War II the chemical industry of the United States was fully mobilized to provide the chemicals, fuels, rubbers and many other products needed for the all-out effort the American Nation was to make to defeat Nazi-Germany and Imperial Japan. What the process industry achieved in a very short time span is one of the most impressive feats of the war: production of aviation gasoline was upped by a factor of 15, synthetic rubber, a pilot plant product in 1941, was made at an amount of 935,000 long tons in 1946, to give two of the most illuminating examples (from Spitz, 1988).

Only one feedstock could in such a short period of time supply the quantities of raw material required for these immense productions: petroleum. Fortunately the American oil and chemical industry were ready, or almost ready, to expand from small scale units to the large operations now demanded.

The story how this happened has recently been presented by Peter H. Spitz in an interesting, albeit highly technical account of the history of the petrochemical industry (Spitz, 1988). In his book Spitz also pays attention to the relative decline of the fermentation industry: from 1930 to 1950 the volume of oxygenated solvents (alcohols, esters, ethers, and ketones) produced in the U.S. increased more than fourfold; the percentage based on fermentation, however, dropped from 85 to 20 (compare Ch. 9, 309).

During World War II huge petrochemical plants were built with daily capacities over 200 t, whereas in the 1930's capacities of 50 t/day were still considered very large. The emphasis American institutes of higher learning had put on chemical engineering and the study of unit processes paid off highly during the war! As a result costs of production came down dramatically. The war taught the chemical industry the same lesson the automotive industry had learned thirty years earlier: large scale production meant low cost price and rapid expansion of use. Then just as Ford motor cars became the normal means of transport for every American family in the years following World War I, now, after World War II, products based on the new synthetic polymers found their way into every American household. (One of the first products and certainly one of the most popular was nylon stockings, first put on the market by Du Pont in 1940.)

Why Did the Fermentation Industry Stay So Much Behind?

Several factors determine whether the economics of a certain production favor carbohydrate or petrochemical feedstock. The cost of the feedstock, although of much importance is not the prime factor. The number of reaction steps required to convert the substrate into the final product, and the overall efficiency with which this conversion is carried out is often more crucial. This efficiency can be enhanced further when unavoidable by-products are upgraded to saleable chemicals, or used to improve fuel efficiency of the plant. Such efficiencies were realized on a grand scale by the petrochemical industry in the years of World War II and in the decades that followed.

The basis for this success was, of course, laid in the 1920's when ever increasing amounts of gasoline were needed for the ever growing number of cars. In this period the industry learned how to economize production, how to use by-products. Then the concept of a refinery as a plant which only produces marketable end products and no waste became accepted as the ultimate goal (except for smokestack emissions, to be fair).

The concept of full utilization of all products (and by-products) from a plant was also proposed for chemical plants based on agricultural raw materials – the Chemurgy concept of William J. Hale (Hale, 1934. See also Ch. 9, 309). On the basis of Hale's ideas a National Farm Chemurgy Council was founded which sought "to advance the industrial use of farm products through applied science" (McMillen, 1939).

Chemurgic Concepts Applied to the Dairy Industry

In 1939 the U.S. Department of Agriculture began to implement the plans for the establishment of four regional research centers devoted to the development of industrial uses of farm crops. One of these centers, the Eastern Regional Laboratory (ERL) had as one of its subjects the valorization of milk products. This was a continuation of the work done earlier under the direction of Dr. Rogers, director of Research Laboratories of the Bureau of Dairy Industry (Ch. 9, 312), a research institute of the Department of Agriculture which was absorbed by the new ERL.

The project “lactic acid from whey” had been continued by studies of the purification economics of the crude product of the fermentation, and by investigating the preparation of chemical derivatives. In a press release of March 22, 1939, the United States Department of Agriculture proudly announced the results of this research under the heading:

> Dairy Scientists Perfect Process for Converting Lactic Acid of Whey into a Chemical Product that Has Many Promising Uses.

The two chemists, Lee T. Smith and H.V. Claborn, presented the full results of their investigations 12 days later at a symposium on the Industrial Utilization of Dairy Products (ACS Meeting, Baltimore, MD, April 1939) (Smith and Claborn, 1939a and 1939b). In the beginning of their talk they stressed the fact that large amounts of lactose are available:

> Cow’s milk contains about 5 per cent of lactose. In the United States over [45,000,000 t] of milk are produced annually, 46 per cent of which is separated to obtain cream. The skim milk from this operation contains [1,000,000 t] of lactose. In addition, the quantity of lactose in buttermilk and in whey from the manufacture of cheese amounts to [180,000 t] of lactose, a large percentage of which could be made available for fermentation to lactic acid.
>
> - *Smith and Claborn, 1939a*

The Bureau worked out methods for the preparation of 18 different esters of lactic acid, from water miscible esters of the lower alcohols to water insoluble esters of higher (fatty) alcohols. Esters of glycol and glycerol were also prepared. The reaction of lactic acid with alcohol was still performed in the rather primitive way patented by Chemische Fabrik Güstrow in 1906 (Smith and Claborn, 1940. See also Ch. 8, 246).

Many uses for the esters were suggested by Smith and Claborn at the ACS symposium:

> All the esters have utility as solvents, plasticizers, and modifiers in lacquers, plastics, varnishes, cements, inks, etc.

With acid anhydrides or ketene the hydroxy group of the ester can be further esterified. This again leads to a large number of interesting compounds with similar applications. [The way Smith and Claborn tried to promote the use of lactic acid in the production of derivatives much resembled the activities of Union Carbide chemists about one decade earlier when they – more successfully from a commercial point of view – made a whole variety of ethers and esters of ethylene glycol (Compare Ch. 8, 251)].

The vision of Smith and Claborn even went further. Citing an article of three ICI researchers (Burns et al., 1935) they remark:

> By pyrolysis these lactic acid di-esters [produced by the reaction of acetic anhydride with lactate ester] can be converted into the corresponding acrylic acid derivative. For example, the methyl alpha acetoxy propionate
>
> will yield monomeric methyl acrylate $H_2C{=}CH{-}COOCH_3$ and acetic acid CH_3COOH. This monomeric material polymerizes readily with heat or peroxide catalysts into poly(methyl acrylate). Laboratory results indicate that poly(methyl acrylate) can be produced as cheaply by this procedure as by the cyanhydrin process now commercially applied.
>
> \- *Smith and Claborn, 1939a*

Is Whey the Right Raw Material for Lactoprene?

In the ensuing years the Eastern Research Laboratory, later called the Eastern Regional Research Laboratory (ERRL), continued research on lactic acid properties, process technology and economics, and derivatization. An appraisal of the ERRL research and some other studies on lactic acid and lactates was published in 1949 in "Sugar" (Needle and Aries, 1949)[1].

The paper of Needle and Aries pays much attention to a synthetic rubber called *Lactoprene*, developed by the ERRL:

> This product is highly superior to natural rubber in some respects and inferior in others. Lactoprene has special properties that make it particularly valuable for uses that require flexibility and resistance to oxidation, oils and aging at normal or elevated temperatures. Lactoprene is polymerized from an acrylic ester such as ethyl acrylate as the principle monomer, it is not a butadiene or diene copolymer like many of the synthetic rubbers .... Not having olefinic linkages lactoprene is resistant to oxidation and weathering.
>
> ...
>
> Lactoprene rubber has become more than a laboratory curio. One of the large rubber companies has been manufacturing it on a substantial basis and subjecting the resulting product to all sorts of tests to establish its reliability and effectiveness as a synthetic rubber of the better variety. It is believed that in time, when lactic acid can be made available at a low enough price, this rubber will compete very favorably with others on the market.
>
> \- *Needle and Aries, 1949*

The authors go to appreciable lengths to show that blackstrap molasses are the cheapest and best suited raw material for large scale production of lactic acid and chemical derivatives. They also conclude that "the synthesis of methyl acrylate from methyl lactate is competitive with the synthesis from ethylene chlorhydrin or ethylene oxide", provided "blackstrap molasses can be obtained for 1 cent a pound."

Milk whey, Needle and Aries say, is hampered by the fact that it is not available by any one company in amounts to ensure the economy of scale desired (7,250 t/a of lactic acid 100%). "Furthermore whey is perishable and would have to be concentrated if stored or shipped, or treated with chemicals to prevent spoilage." Whey is only suitable for small volume production of lactic acid, the authors conclude. [A fact which helps to explain failures by earlier users of this material.]

Lactoprene Was Made from Starch

A fermentation plant with a capacity exceeding 7,000 t/a did not exist, nor did any company have it in the planning stage when Needle and Aries wrote their paper. Most existing plants had capacities of only about 1,000 t/a and were based on starch hydrolysates. [When the starch companies in the course of years learned to avoid "production" of second grade dextrose hydrate, they had switched their lactic acid feedstock to cheaper starch hydrolysate. This adversely affected the quality of the edible material, which during the war years and the first decade thereafter was not too serious as no competition was felt from imported lactic acid.]

Quite ironically the methyl lactate bought by "one of the large rubber companies" (Needle and Aries) was neither based on whey nor molasses. It was the product of a starch company, the Clinton Corn Processing Company of Clinton, Iowa. In 1943 this company had built a pilot plant for the production of chemically pure methyl lactate using a process developed by Filachione and Fisher of the ERRL (Filachione and Fisher, 1946). The installation of the plant was executed in close cooperation with the two USDA scientists. For about three years regular shipments of 55 gallon drums of the ester were dispatched to a chemical processor, who converted it into methyl acrylate. This monomer was used in the manufacture of a new type of adhesive tape. This went on until about 1947 when cheaper methyl acrylate became available from purely synthetic sources (Newton, interview, 1986).

In the process of the ERRL methyl acrylate was produced by passing methyl 2-acetoxy propionate in a stream of nitrogen gas through a packed combustion tube at a temperature of about 550ł C. At contact times below 10 seconds up to 80% conversion into methyl acrylate and acetic acid could be obtained, about 10% of the starting material was further decomposed into more volatile substances. Reaction equation:

$$CH_3\underset{|}{\overset{OCOCH_3}{C}}HCOOCH_3 \rightarrow CH_2{=}CHCOOCH_3 + CH_3COOH$$

[in weight a maximum of 47 g of methyl acrylate was obtained per 100 g of the acetoxy compound together with a maximum of 33 g of acetic acid.]

Clearly the process developed by Filachione and Fisher was a complicated process, it not only involved extreme conditions of temperature (high) and reaction time (very short), but also the preparation of the acetoxy derivative as intermediate and the recovery of large amounts of acetic acid. The latter could only be used again after reconversion into acetic anhydride.

A chemist working with the Minnesota Mining and Manufacturing Company succeeded in 1952 in producing acrylic acid and acrylates directly from the corresponding lactic compounds (Holmen, 1958). Holmen passed an aqueous solution of lactic acid over a dehydration catalyst composed of alkaline earth salts of sulfates or phosphates at temperatures of about 500° C. In most of the examples presented in the patent, yields are low, at best about 50% (and in one case 68% of the theoretical). This seems rather low compared with an overall maximum yield of 70% obtained by Filachione and Fisher for the transformation of lactic acid into methyl acrylate. Maybe lack of interest can be held responsible for the absence of further progress in reaction efficiency.

([The lack of interest, of course, was due to lack of economics: methyl acrylate and acrylic acid were generally produced from ethylene cyanohydrin, which in turn was obtained by reacting ethylene oxide with hydrogen cyanide. Ethylene oxide itself was made from ethylene via the addition of hypochloric acid to ethylene chlorohydrin and subsequent removal of hydrochloric acid with calcium hydroxide. In the 1950's new processes in which ethylene was directly oxidized to ethylene oxide replaced the earlier process. These new processes were cheaper due to the elimination of the use of rather costly chlorine. It is a remarkable repetition of what happened 25 years before when ethylene glycol derivatives pushed the solvent ethyl lactate back into oblivion. The same now happened with methyl lactate as an intermediate in the production of acrylics. This time it was not ethylene glycol which was the malefactor, but an improved process of manufacture of the closely related ethylene oxide. As a matter of fact the new process was so cheap that ethylene glycol now became a derivative of ethylene oxide (Waddams, 1978; Spitz, 1988).]

The conclusions of Needle and Aries regarding the use of starch as raw material for lactic acid help to explain the situation:

> The present market price for lactic acid, plastic grade, 80% strength, is around [0.53 to 0.66 $/kg] on 100% basis. This is too high for consideration to any large extent as a component for alkyds [one of the bulk applications of lactic acid advocated by Aries and Needle[2]].

A lactic acid factory based on molasses at 1 cent per pound [0.022 $/kg] and a capacity of 7,250 t/a would bring lactic acid prices in line with the other ingredients used in alkyd production. Needle and Aries point in particular to the manufacture of glycerol lactates as interesting intermediates. [The rationale of this proposal is found in the poor solubility of glycerol in standard alkyd reaction formulations.]

Cereal grains are also out of the picture because of price:

> Unless an artificial price is maintained, they cannot compete with blackstrap molasses. Furthermore, the grains require more elaborate and involved treatment before the sugars derived from them can be made available for fermentation.
>
> *- Needle and Aries, 1949*

Albeit on paper only, Needle and Aries certainly proposed an attractive production process for large scale production of plastic grade lactic acid. The main features of the process were based on the work of Filachione and Fisher of the ERRL:

Fermentation in 38,000 l fermenters, or larger ones; acidification with sulfuric acid; filtration over a rotary filter to remove calcium sulfate; concentrating to about 50%; conversion into methyl lactate according to the process published by Filachione and Fisher in 1946 (see below); continuous hydrolyzation of methyl lactate, followed by distilling off methanol and excess water. The result is an aqueous solution of chemically pure lactic acid.

> No commercial application of the above esterification process has been, as yet, accomplished, so that a pilot plant set up is indicated before large scale commercial involvement can be recommended.
>
> *- Needle and Aries, 1949*

A New Purification Method for Lactic Acid

Although the research of Filachione and Fisher was published in 1946, the actual work had been done much earlier, in or before 1943. This is evident from the fact that Clinton had built the pilot plant in 1943 with the aid of the two researchers of the ERRL. It can also be concluded from a remark in the 1946 paper of Filachione and Fisher:

> Wenker [U.S. Patent 2,334,524, 1943], and Schopmeyer and Arnold [U.S. Patent 2,350,370, 1944] also purified lactic acid by passing alcoholic vapor through crude acid; their patents appeared after the completion of this [ERRL] investigation[3].

The drawing in the patent of the two American Maize scientists Schopmeyer and Arnold showing the main features of the process is almost self-explanatory, compare Figure 10,1.

Crude lactic acid of a concentration between 40 and 60% is fed continuously into the heated reactor 10, where it reacts with methanol under the influence of small amounts of sulfuric acid. The molar ratio of lactic acid to methanol is kept at about 1:1.5. The vapors distilling from the reactor consist of a mixture of methanol, methyl lactate and water, with traces of lactic acid. This mixture is conducted to the middle of a fractionating column, and separated into a top fraction of methanol which is returned to the reactor and a bottom fraction containing methyl lactate, water and lactic acid. Hydrolysis of the methyl ester already takes place in the fractionating column and is completed in kettle 18.

The reaction is semi-continuous. To quote the patent:

> It may be and is carried on almost indefinitely. Only after long operation does the residue of unfermented material, salts, etc., from the crude acid feed accumulate in the kettle to such an extent that the operation must be stopped.
>
> - *Schopmeyer and Arnold, 1944*

Here the inventors appear a little too optimistic as shown by a closer look at the methods and results given by Filachione and Fisher. Although the paper of these authors does not report how these reactions were carried out, the method itself appears to be as suitable for continuous execution as that of Schopmeyer and Arnold.

The most important difference between the ERRL method and that of Schopmeyer and Arnold is the use of a counter currently operated packed

June 6, 1944. H. H. SCHOPMEYER ET AL 2,350,370

LACTIC ACID PURIFICATION

Filed Feb. 16, 1943

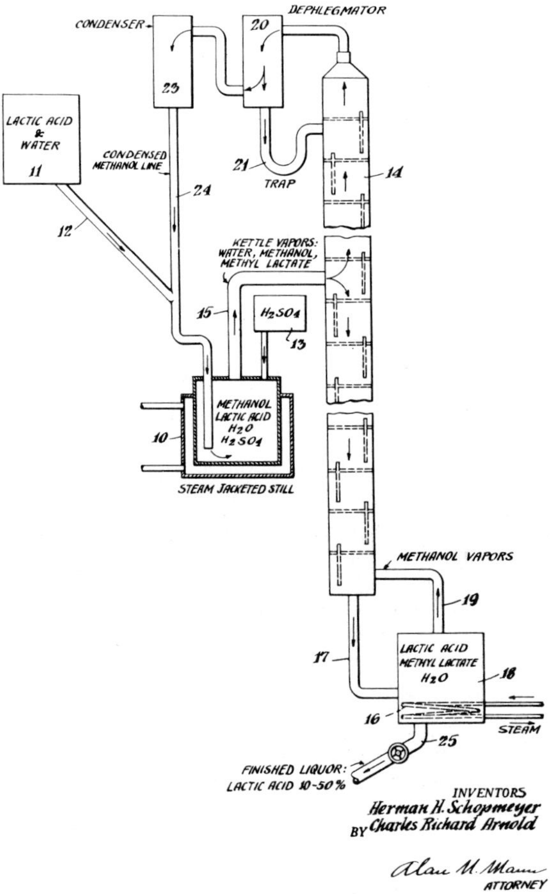

Fig. 10,1. Lactic acid purification via distillation of the methyl ester (U.S. Patent 2,350,370).

tower by Filachione and Fisher, in which crude lactic acid was fed over the top and methanol vapor via the bottom. The vapors coming from the top of the column consisting essentially of methanol, water, and lactate ester were fed to a continuous stripping still in which the major portion of the alcohol was separated and automatically returned to the alcohol vaporizer. The methyl ester was then hydrolyzed to lactic acid. Optimum results were obtained with a feed of 80 to 90% crude acid, a ratio of 9 moles of methanol to 1 mole of lactic acid, and a residence time of about 50 minutes. [When the dimensions of the laboratory equipment of Filachione and Fisher are extrapolated to the 7,000 t/a lactic acid production Needle and Aries consider the economic minimum to compete with functionally "related" petrochemicals a reaction tower with the dimensions of 1.5 to 2 m diameter and 20 to 25 m height is required. This would seem to be an acceptable proposition in the 1940's.]

A drawback to which Filachione and Fisher pay little attention is the decomposition of residual amounts of sugars and proteins present in crude lactic acid. When they say, however, that the purified acid sometimes became pale yellow or amber, entrainement of minor quantities of foreign material originating from the crude acid must have taken place. When this already occurs in one passage of the crude acid over the column – in their experiments the residue is collected at the bottom and disposed off – what will happen when the column is in operation for some time? [Within days (or weeks?) the packing material in the column can be expected to become fouled with impurities from the crude lactic acid and entrainement of volatile degradation products will become a serious matter.]

The thoroughness with which Filachione and Fisher carried out their research is borne out by the large number of experiments performed on crude lactic acid from all American lactic acid producers: Du Pont; Sealtest, Inc.; Sheffield Farms Company, Inc.; American Maize-Products company; and Clinton Company. No appreciable differences apparently were found. Some experiments were even performed with crude unfiltered fermentation liquor from the nearby Du Pont fermentation plant at Grays Ferry, Philadelphia. These were carried out in a discontinuous manner, the product of the esterification, methyl lactate, was recovered by vacuum distillation.

Corrosion Problems Are Finally Surmounted

Nearly all process studies until about 1950 carry remarks about the highly corrosive nature of lactic acid. A summary of these problems was given by Peckham (Peckham, 1944) and also by Inskeep et al (Inskeep et al., 1952); see Ch. 9, 320.

The solution to most of the corrosion problems with lactic acid came when better stainless steels were developed in the years of World War II and the decade thereafter. In particular to meet the needs of the chemical and nuclear industries a great diversity of stainless steels – alloys – was investigated, and as a result many new steels were put at the disposal of chemical and mechanical engineers.

The history of stainless steels, however, had begun much earlier. As a matter of fact a discoverer of stainless steel cannot be singled out. Many investigators developed at about the same time (1900–1920) alloys with rust resisting properties. A historical survey given in "The Epic of Steel" (Fisher, 1913) tells that

> commercially stainless steel was born and reared in the cutlery trade of Sheffield [U.K.]. The ability of the steel to withstand oxidation at high temperature was recognized at an early date, resulting in its application in steam turbine blades in 1914. The first large scale commercial order for stainless steel in the United States was placed in 1924 by E.I. du Pont de Nemours for tanking and piping nitric acid.
>
> *- Fisher, 1963*

In Germany stainless steels were also introduced in the early 1920's. For instance Merck's yearly Manufacturing Report 1923 says about the lactic acid plant:

> I [Dr. Gruschwitz[4]] have many complaints about the enameled ether distillation equipment. The enamel of the covers and at the sharp edges is damaged easily, causing considerable pollution of the lactic acid with iron .... The experiments with V2A steel of the firm of Krupp have been extended further. It would be much welcomed if this company would succeed in offering [corrosion] resistant equipment for lactic acid. Such a metal would also find use in the concentrating of the pure acid.
>
> *- Gruschwitz, 1936*

In 1928 a distillation kettle of stainless steel V2A was installed for trial purposes. Because of corrosion this kettle had to be replaced within two years by a copper kettle. A new evaporator, bought in 1931, was again completely made from copper.

The 1936 report writes:

> To prevent the high costs involved in the frequent leading of the condensers of the lactic acid vacuum [installation], a V4A Supra condenser has been installed to replace the most affected condenser. Until the end of the year no corrosion was observed.
>
> *- Gruschwitz, 1936*

Peckham wrote in his survey (Peckham, 1944) "most of the stainless steels ...

are unsatisfactory. Certain stainless steels, particularly those containing molybdenum, show reasonable resistance".

Inskeep, Taylor and Breitzke (Inskeep et al., 1952) still call corrosion a major problem in the production of lactic acid. At the same time they advocate the use of stainless steel 316 as the solution to most of these problems:

> Both the acid evaporators are constructed throughout of Type 316 stainless steel. The evaporator bodies were annealed after fabrication to reduce the possibility of intergranular corrosion. The evaporator steam jet ejectors are constructed of Type 316 stainless steel.
>
> \- *Inskeep et al., 1952*

Table 10,I constructed from figures given by Rabald and Bretschneider shows the relative corrosion resistance of the metals and alloys mentioned above:

Table 10.I. Corrosion resistance of metals to lactic acid.

metal or alloy	composition	corrosion when exposed to 30 – 60% lactic acid at 50 – 80° C in mm/a
copper		0.025 – 0.3
lead		dissolves slowly
steel V2A	18 – 19% Cr, 9.0 – 10.5% Ni, 0.07% C	0.3
steel V4A	16.5 – 18.5% Cr, 11.0 –12.5% Ni, 2.0 –2.5% Mo, 0.07% C	< 0.1
AISI 304	19% Cr, 9% Ni	0.08
AISI 316	17% Cr, 13% Ni, 2,4% Mo 0.046% C	0.0025
AISI 316	18.5% Cr, 12.8% Ni, 3,0% Mo, 0.018% C	< 0.0025

Source: Rabald and Bretschneider, 1961.

For further evaluation of the corrosion phenomena with lactic acid and the danger of taking figures like those of Table 10,I at face value, the reader is referred to Holten (Holten, 1971).

Large Scale Production of Lactic Acid from Molasses?

As Needle and Aries pointed out in 1949:

> [I]f lactic acid could be obtained for less than [0.22 $/kg] and preferentially at [0.18 $/kg] on a 100% basis, a potential market of approximately [90,000 t/a] would be tapped. Arrival at the estimate given can be very readily understood when one notes that the alkyd resin field alone enjoys a volume sale of over [60,000 t/a] and is expected to continue to rise.
>
> Another field that would utilize large volumes of acid is that of cast phenolic resins[5].
>
> In addition, its greatest potential use is perhaps in the manufacture of acrylic resins.
>
> - *Needle and Aries, 1949*

Despite the highly qualified work of the Eastern Regional Research Laboratory of the U.S. Department of Agriculture not even a tiny fraction of these impressive estimates was realized in the succeeding decade.

The largest U.S. manufacturer of lactic acid (1,000 t/a), the Clinton Company, installed around 1950 a pilot plant to gain experience in the production of heatstable lactic acid by hydrolysis of methyl lactate. The growing activities of the company in other fields asked so much of the company resources, however, that an investment to produce heatstable lactic commercially was never approved (Newton, interview, 1986)[6].

The Sugar Research Foundation which had sponsored the study of Needle and Aries in 1949 also published a review paper on the chemical production of lactic acid by the alkaline degradation of sugar. The paper (Montgomery, 1949) begins with a short historical paragraph which refers to the pioneering research of Hoppe-Seyler in 1871 and Kiliani in 1882, in the years in which industrial production of lactic acid first came to the fore (Ch. 5, 129).

The conclusion of the literature study of Montgomery is ambiguous: The yield of the alkaline degradation process is at best 60–67%, whereas by fermentation yields of 80–95% can be attained.

> However, [in the words of the author] from a manufacturing point of view fermentation demands a large amount of space and operational time as well as close control of conditions. In view of the versatility of and therefore potential demand fotic acid in chemical industry it would be advantageous to have a process in which the above disadvantages are overcome or at least reduced.
>
> - *Montgomery, 1949*

Although Rex Montgomery, in his review paper, does not mention one study involving alkaline degradation of molasses, the scientific director of the Sugar Research Foundation, Robert C. Hockett, considers in his introduction that blackstrap molasses is "one of the most promising raw materials" for such a process.

The Effect of USDA Research on Lactic Acid Sales

Until the start-up of Monsanto's synthetic lactic acid factory in 1964 the U.S. Tariff Commission reported yearly figures of lactic acid production. [After the start-up of this plant only two lactic acid producers remained active; under the rules of the Commission production figures are confidential when the number of producers is two or less.]

Table 10,II shows yearly productions and imports over the period in which the Eastern Regional Research Laboratory devoted much of its activities to the development of lactic acid.

Keeping in mind that the ERRL research was primarily aimed at opening

Table 10.II. U.S. production and consumption of lactic acid (production in tons 100%, imports recalculated to 100% under the assumption that lower than 55% is mainly 44%, and higher than 55% is 80% concentration).

year	technical	edible and medicinal	total production	imports	total consumption
1940	757	676	1433	1	1434
1942	1328	1415	2743	0	2743
1944	2019	1885	3904	–	3904
1946	1005	1732	2737	–	2737
1948	1049	1212	2261	–	2261
1950	832	1639	2471	–	2471
1952	538	1362	1900	–	1900
1954	517	1780	2297	–	2297
1956	507	1745	2252	–	2252
1958	547	1812	2359	–	2359
1960	459	2021	2480	202	2682
1962	669	1842	2511	651	3162

Source: Official U.S. Import Statistics and Census Reports.

up markets for lactic acid as a chemical intermediate, it is clear that the results were negative. Only in food applications, and, maybe for pharmaceutical uses can modest increases be noticed from about 1950 onwards. Only by 1960 was the maximum war use of food grade acid of about 1900 t/y reached again. The growth in lactic acid use from 1960 onwards is mainly due to the C.J. Patterson Company in Kansas City, which developed a completely new application for the acid (Ch. 11).

Ethyl Lactate from Petrochemical Feedstock

Around 1950 synthetic ethyl lactate was offered by the Industrial Chemicals Division of American Cyanamide as "an excellent solvent for cellulose esters" (company brochure, no date, received by SMF in February 1951). To quote further:

> Aero Ethyl Lactate is a truly synthetic product made from lactonitrile and ethanol and is superior to the material obtained from the older process of esterifying fermentation lactic acid[7].

The course of the synthesis is given by Holten (Holten, 1971): Lactonitrile is reacted with alcohol in the presence of anhydrous hydrochloric acid; the intermediately formed hydrochloride of the iminoester was subsequently hydrolyzed with water to form ethyl lactate:

$$CH_3\text{-}CHOH\text{-}CN + C_2H_5\text{-}OH \xrightarrow{HCl} CH_3\text{-}CHOH\text{-}\overset{\overset{NH_2^+,Cl^-}{\|}}{C}\text{-}OC_2H_5$$

$$\xrightarrow{H_2O} CH_3\text{-}CHOH\text{-}COOC_2H_5 + NH_4Cl$$

The McGraw Hill Directory of Chemicals and Producers of 1951 only lists American Cyanamide as producer of ethyl lactate. The other entry (Stoney-Mueller) is a trading company.

Synthetic Lactic Acid Production Begins in Japan

The first laboratory preparation of synthetic lactic acid was performed by Strecker (Strecker, 1850), who obtained lactic acid (2-hydroxypropionic acid) by the reaction of alanine, 2-aminopropionic acid, with nitrous acid. A more suitable synthesis was found by Wislicenus in 1863. Wislicenus had discovered lactonitrile as the reaction product of acetaldehyde with hydrogen cyanide

$$CH_3\text{-}CHO + HCN \rightarrow CH_3\text{-}CHOH\text{-}CN$$

and found that lactonitrile under the influence of hydrochloric acid would readily hydrolyze to lactic acid (Holten, 1971):

$$CH_3\text{-}CHOH\text{-}CN + 2\ H_2O \rightarrow CH_3\text{-}CHOH\text{-}COONH_4$$

The result of this synthesis is of course a racemic mixture of the two enantiomers of lactic acid.

Although both reactions, the cyanohydrin formation of the lactonitrile, as well as the hydrolysis of the nitrile can be carried out with very high yields, efficient purification processes had to be developed before the method would be suitable for commercial manufacture of lactic acid. The availability of the raw materials acetaldehyde and hydrogen cyanide in sufficient volume and at reasonable prices was, of course, another prerequisite.

The first patent for such a process was requested by I.G. Farbenindustrie A.G. in 1930 (German Patent Application J.82.30 Kl. 12 o). In the IG process lactonitrile was saponified with sulfuric acid, the resulting solution treated with lime, and the calcium salts of sulfuric and lactic acid separated by means of their different solubilities in water. The poor crystallizing properties of calcium lactate could be one of the reasons why this process never attained commercial interest.

The Du Pont company also was interested in making hydroxy acids from cyanohydrins. In 1933 the company was granted a patent (Macallum, U.S. Patent 1,984,415) on the preparation of cyanohydrins in general. The only example given relates to lactonitrile, which indicates that the hydroxy acid of prime interest was lactic acid. The invention claimed is the use of the reaction product itself – lactonitrile – as solvent for the reacting products – aldehyde and hydrocyanic acid. In this way the large amount of heat generated by the reaction can be more easily kept under control.

The first company to actually manufacture synthetic lactic acid by the reaction of acetaldehyde with hydrogen cyanide was Musashino Chemical Laboratory Ltd. of Tokyo, Japan. This company was established in 1949 as a research and development company for the purpose of manufacturing lactic acid for the brewing industry (Company brochure, about 1970; Yamazaki, 1978). In the same year (or in 1950) small scale manufacture of lactic acid was begun. Some historical information on the Japanese venture was revealed in 1964 as a result of publicity around Monsanto's announcement of a synthetic lactic acid plant in America (Chemical Engineering, 1964, 84):

In 1962 (O.P.D. Reporter, September 10, 1962) Monsanto Chemical Company announced its intention to build at Texas City, TX, "the first plant in the world to manufacture synthetic lactic acid". This publication came to the attention of Musashino, who reacted in a letter to Monsanto claiming their priority of being the first producer of synthetic lactic in the world (Private information, Marnett, 1987). The news editorial of Chemical Engineering, referred to above, acknowledging the fact of Musashino's priority, wrote then:

> Actually, Musashino's crude lactic acid is produced some [240 km] north of Tokyo, then brought to the city for the purification steps. The crude production is by Kyowa Gas Chemical Industry Co., ..., which produces acetaldehyde, methanol and other chemicals from natural gas[8]. Like Monsanto, Kyowa Gas Chemical makes the lactic acid from lactonitrile, which in turn is derived from acetaldehyde and hydrogen cyanide. The crude acid is shipped to the 330 t/year purification unit at Tokyo. The first step during purification is to hold the acid at about [146° C] while letting it absorb a vapor mixture of water and methanol. The water prevents the polymerization of the acid, while the alcohol reacts with the acid to form methyl lactate [methyl lactate boils at 145° C and consequently evaporates immediately].
>
> Methyl lactate is then purified by fractional distillation to remove "undesired color bodies and odor. The final step produces pure lactic acid by reacting ... methyl lactate ... with water".
>
> \- *Chemical Engineering, 1964*

Musashino Chemical Laboratory's first successful application was in the very demanding sake market, where they succeeded in replacing the fermentation based product from the Takeda company in 1955.

Until 1968 the Tokyo [pilot] plant supplied sake customers and, maybe, a few other small users with lactic acid. In that year an industrial installation with a capacity of 3,000 t/a (88%) was built in Isohara cho, Kitaibaragi, on the east coast of Honshu and about 150 km north-east of Tokyo. In the same year Takeda closed its fermentation plant; the company from then on supplied its customers (including most of the sake brewers) with synthetic lactic acid from the new Musashino plant. [The production of fermentation grade lactic acid of the company had been rather modest; ILRA estimated this in 1964 at about 500 t/a.]

The development and commercial success of the Musashino company has been much influenced by the introduction of the petrochemical industry in Japan. Spitz describes the history of the advent of the Japanese petrochemical industry in considerable detail (Spitz, 1988):

> MITI, the Ministry of International Trade and Industry, initiated in 1954 a fundamental study how to establish a petrochemical industry. Within three months the committee consisting of 20 executives of leading Japanese chemical firms brought out its recommendations, the most important of which was that all petrochemical technology should be imported rather than developed locally.

Four petrochemical complexes based on imported technology were built in a short period of time, the first one of these became operational in 1958. In (steam) crackers naphta was converted into ethylene and propylene, basic products for petrochemical conversions. Spitz:

> Over the 1960's, Japan's chemical industry continued to import a great deal of foreign petrochemical technology. Some examples are ..., acetaldehyde from Wacker, ... Over this period a number of other petrochemical complexes were constructed, including ... that of Mitsubishi Chemical at Mizushima [about 160 km east of Osaka in the Okayama Prefecture].
>
> - *Spitz, 1988, 381*

As a matter of fact it was at the Mizushima complex where know-how supplied by the Wacker company of West Germany was applied in the production of acetaldehyde from ethylene[9]. In a letter from Mitsubishi Chemical to Schiedamsche Melkzuur Fabriek of October 1964 the company offered lactonitrile as a raw material for the production of lactic acid. As the letter says

> we are the largest manufacturer of acrylonitrile in Japan with a daily capacity of 140 t (60 t by the common Monsanto process, and 80 t by the adjustable Sohio process). In this way we are able to obtain lactonitrile with [a] low boiling point by the Monsanto process, and to prepare lactonitrile of high purity synthetically by coupling by-product HCN from the Sohio process with CH_3-CHO manufactured in the same installation in accordance with the Wacker procedure.

As far as is known Musashino (later?) also used lactonitrile from the Mizushima petrochemical complex. In the 1980's total Japanese lactonitrile production capacity was estimated to be between 7,000 and 11,000 t/a (Shigematsu, personal communication, 1984). Total Japanese production of acetaldehyde in 1979 was about 500,000 t (Japan Chemical Annual, 1980).

Spitz summarizes the events in Japan in the words:

> Petrochemical technology obsolesced Japan's fermentation units when it was decided to start producing ethanol, acetone, and butanol from petroleum feedstocks rather than from agricultural wastes. The switch from acetylene to ethylene also proceeded fairly rapidly in Japan, due to the high cost of coal and electric power and government incentives to increase steam-cracking capacity.
>
> - *Spitz, 1988, 380*

European Developments From 1945 to 1960

Right after World War II economic conditions in Europe were abysmally poor compared with those in the United States: whole cities had virtually been wiped out, major industrial complexes, particularly in Germany, damaged beyond repair. The American industry on the other hand had received a tremendous impetus from the war, entirely new industries had been created, almost overnight.

In his famous address of June 5, 1947, the U.S. Secretary of State, General George C. Marshall announced the intention of the United States Government to launch a European Recovery Program (soon to be called the Marshall Plan). The plan was put into operation in 1948, and carried out over the next four years. The plan was highly successful, in particular for the European chemical, engineering and steel industries.

In the early 1950's, when economic conditions and trade relations had returned to normal, the European chemical industry began to expand rapidly. Looking back it seems incredible that almost every new venture met with success, the market seemed to be insatiable. (The European Recovery Program lasted from 1948 until the end of 1951. In these three years overall industrial production of Western Europe increased by 64%, most sectors of industry achieved yearly production increases between 10 and 20%.)

Lactic acid producers, many of whom also produced a range of other products, did not experience a similar growth in their markets, however. Exact figures to substantiate this view are only available for the Netherlands

Table 10.III. Per capita consumption of lactic acid (in g/a/capita of population).

year	U.S.A.	Germany	Netherlands
1940	5	35	14
1945	15	–	–
1950	12	?	11
1955	12	11	10
1960	11	?	13

(SMF Archives). These data, converted into use per capita, are compared in Table 10,III with U.S. per capita use of food grade lactic over the same period, and two German figures for 1940 and 1955 (from Dehio, 1966).

Production figures over the period are scarce and unreliable, the author guesses 1940 European production at 4,500 to 5,000 t (100%), and 1960 production at 6,000 to 6,500 t. Important consumption increases over the prewar level were only noticeable (in the confectionery industry) in the United Kingdom, (in olive fermentation) in Spain, and (in baby food) in the Netherlands. It was also observed that lactic acid lost most of its war gains in the lemonade and soft drinks industries to citric acid; also in jams and sourbread applications lactic acid was largely replaced by citric (in particular in Germany).

The stream of publications from USDA's Eastern Regional Research Laboratory was well noted by the research people, one may suppose, but commercial interest only awoke when the paper of Needle and Aries pointed at the large commercial potential of lactic acid (Needle and Aries, 1949). Leading British, Dutch and German sugar companies, members of the Sugar Research Foundation, established research departments in those years to imitate and evaluate some of the "sucrochemical" investigations sponsored by the Sugar Research Foundation.

In Holland a research group was set up by Centrale Suiker Maatschappij N.V. (CSM), one of the two sugar producing companies in that country. In the early 1960's this group developed on the basis of the study of Montgomery (Montgomery, 1949) a method to prepare technical lactic acid by alkaline degradation of molasses (Breen, interview, 1988). Apart from the low yield of about 40% on sugar present in the (cane) molasses, the method had the drawback that low percentages of glycolic acid and of an isomeric butyric acid formed in the degradation process were almost impossible to remove from the technical grade lactic acid.

Europe Counts Too Many Producers of Lactic Acid

In the early 1950's Western Europe counted nine lactic acid producers (Glimsholt, interview, 1985; Klein, 1963): C.H. Boehringer Sohn, Bowmans Chemicals, Société Normande, Schiedamsche Melkzuur Fabriek, Luis Ayuso, Kemisk Vaerk Køge and three small Italian producers.

C.H. Boehringer Sohn of Ingelheim, Federal Republic of Germany. Although the company had survived World War II without major damage, production of lactic acid was not resumed until 1951. As Dehio (Dehio, 1966) reports:

The Ingelheim lactic acid works had been closed down in the fall of 1944 .... In the first postwar years restarting the plant was not an immediate priority, other matters deserved more attention. [The citric acid plant was restarted in March 1946, the French occupying administration had agreed to this under the condition that most of the output was exported to France under special conditions, that is at prewar prices]. The raw material situation [for lactic acid production] was extremely difficult. Potato starch either was not available or only in insufficient quantities, and at high costs. The sugar price had risen to 100 DM per 100 kg, which made it much too expensive for lactic acid (Intermezzo 6, 283).

The only suitable raw material under these conditions was molasses. A thorough investigation [of a manufacturing process] using molasses had been worked out by Dr. Wallach of the Process Engineering Department .... [See Intermezzo 7].

At the close of 1951 lactic acid production on the basis of molasses was put into operation, after a costly revision of the seven years idle plant.

... beginning in 1948 we had sold imported Dutch edible lactic acid from Schiedamsche Melkzuur Fabriek to keep in touch with lactic acid users in Germany.[10] In this way we were able to prevent German trading houses ... from establishing themselves in the lactic acid market, which would have hampered our return to the market.

As our own edible lactic acid in the beginning did not always comply with the quality requirements of the clients, we continued the import from Holland for a certain period. The very good Dutch material was delivered to [more] demanding customers.

... in comparison with prewar and war years lactic acid consumption was drastically reduced. The leather industry had new synthetic agents for deliming, and the large production of jams for the army, the [Nazi] labor service and other mass consumers, the main body of our sales, had disappeared. Pectin production had been reduced in line.

Unfortunately it soon became clear that molasses in the long run could not meet our expectations.

It took ... about three years before we achieved a yearly production of about 1,000 t/y, all qualities taken together. The major part of this was edible lactic acid, about 600 t (100%), for which we again established an appropriate market.

- Dehio, 1966

Bowmans Chemicals of Widness, United Kingdom. Since about 1930 this company produced technical as well as edible grade lactic acid. The edible was made by the old and not very efficient process of crystallizing calcium lactate and purification by chemical removal of heavy metals.

A major problem for the company arose from the about 50/50 output of technical and edible lactic acid (Ch. 8, 268). The problem how to sell the 50% output of technical lactic acid became very serious after the war when heavy leather, in particular that used for conveyer belts, was replaced by artificial materials. The stocks of technical lactic acid at times amounted to a half year's production or more (Arnold, interview, 1988).

In 1948 the Bowmans company became public under the name Bowmans Chemicals Ltd., with a share capital of £ 120,000. The number of employees then was 130. Production was still not more than about 500 t/a, calculated as 100%. The figure includes a small amount of calcium lactate.

Bowmans Introduces Buffered Lactic Acid

(Sources of information: T.F. Reid, interview, 1987;
M.H.M. Arnold, interview, 1988)

In the 1950's Bowmans was responsible for the successful introduction of buffered lactic acid, a combination of pure lactic acid and sodium lactate (solution), in the manufacture of hard candies, high boilings or what name may be given to the translucent hard sweets so many people are fond of.

Traditionally hard candies were made from a combination of sugars, flavoring extracts, coloring ingredients, and citric or tartaric acid. The addition of about 1% of acid is required to impart the fresh and crisp taste. The acid also supports and enhances the flavor of the candy.

> The old technique was to cook the sugar, pour onto a slab, sprinkle the acid over – and knead in by hand, finally pulling [the mass] out into a rope and cutting [the rope] into pieces. Air bubbles trapped during kneading, etc. were pulled out in the rope into 'fishmouths', which spoiled the clarity. But if [citric or tartaric] was added in the pan during cooking to evade the kneading process, the acidity caused too much inversion, leading to a sticky product[11]. The use of buffered lactic acid meant that addition could be made in the cooker without excessive inversion and the poured mass could be pulled out and cut without introducing fishmouths. The technique ultimately became that of 'depositing' whereby the acidulated [buffered] cook is pumped into separate moulds to give a 'star-bright' drop. The chemistry involved may be summed up as:

pH controls inversion, but acid taste is controlled by total titratable acidity. Buffering to a large extent enables the two factors [pH and acidity] to be separated.

- Arnold, private communication, 1988

The history of the introduction of buffered lactic acid in the candy industry has been told in the Autumn 1961 issue of AIM, a commercial and very readable quarterly magazine of the Bowmans company:

At Pascall's [the Mitcham, Surrey, works of James Pascall, Limited] Mr. J.H. Mallows, formerly Chief Chemist and now Production Director, began experimenting with the use of lactic acid in high boilings as long ago as 1954, and two years later saw his findings adopted as standard production practice. The smoother taste and greater clarity of the sweet resulting from this new formulation have boosted sales and created new standards in the minds of many customers.

- Aim, 1961

In 1964 Mallows himself gave another not fully consistent account of what happened (Mallows, 1964):

Enough has been said ... to substantiate the view that production of this type of hard confectionery [fruit drops] is very largely a chemical engineering problem and it is to the credit of the Scientific Department at Needler's Ltd., that this firm was the first to produce sweets containing lactic acid on a commercial basis. Undoubtedly the kind of 'Fruit Drop' which was produced has been an outstanding success and for a number of years was the only one of its kind marketed in this country, but other firms engaged in this type of sugar confectionery production have produced similar sweets containing lactic acid, and these sweets are also deservedly popular.

- Mallows, 1964

Reid, formerly managing director of Croda Bowmans Chemicals Ltd., the successor company of Bowmans Chemicals, remembers his involvement in buffered lactic acid very well:

I was seconded from plant research [in the 1950's] to what we then called the sales development section, and one of the projects I was given was the extension of the use of lactic acid in sugar confectionery.

The first major company to use lactic acid [in confectionery products] in England was Needler's of Hull. They originally used to purchase the pure, straight lactic acid from us but following all our developmental work we were eventually able to supply them with the ready made buffered mixture just like the ones we were supplying to other major manufacturers in the U.K. Mentioned earlier is the involvement of Pascals, Mitcham, and they certainly, from the fieldwork point of view, helped a great deal to move from batch to continuous production of sugar confectionery via the use of lactic acid. There is no way continuous production with the old methods of adding powder acids could ever have been devised. So the advent of the clear, glace fruit drop eventually became a reality and the smooth, bland flavour of lactic acid became an accepted one in the sugar confectionery industry.

Within a decade this new use of lactic acid amounted to several hundred tonnes annually in the United Kingdom alone. This development had been helped by the fact that Baker Perkins Ltd. of Peterborough, a major manufacturer of machinery for sugar confectionery purposes, introduced their 'Microfilm' continuous cooker. This was still meant to be continuous in its cooking operation but batchwise in the actual production, and a company called Confectionery Developments of Hemel Hempstead saw the opportunity, working in conjunction with Baker Perkins, to use the Microfilm cooker as the feed to their special piece of machinery which was known as the continuous depositor. The two combined undoubtedly helped put lactic acid in amongst the major acidulants being used in the sugar confectionery industry in the Sixties.

- Reid, private communication, 1989

In the former British Dominions buffered lactic acid also soon became the standard acidulant in the manufacture of high boiled sweets. The introduction of fully automated machinery did much to promote the cause of buffered lactic, as did the world wide sales activities of Bowmans. Other countries accepted the new method more slowly depending on the type of automated depositing machines introduced.

Bowmans Becomes the Leading Lactic Acid Manufacturer

During the 1950's the success of buffered lactic acid compelled Bowmans to increase production of edible lactic acid (and of primary fermenting capacity as well). Head of the research activities then was Clifford George Childs (M.Sc. University of London), a diligent and ingenious chemist.

An interesting and rather complete overview of lactic acid production technology installed at Bowmans under the responsibility of Childs was presented by Grenville Machell in 1959 (Machell, 1959). Bowmans, unlike its European colleagues, had continued to use starch as raw material, which was converted with mineral acid into glucose. The fermentation procedure was rather similar to the one used by Clinton and American Maize.

Arnold in a 1988 interview told the author about the starch used: The starch came from a very special source, from a producer of battery separators. These separators were made from pvc into which maize starch was mixed. The separator sheets were made by extrusion and thereupon the starch was dissolved and removed from the plastic by the action of a mineral acid,

Fig. 10,2. Microscopical control at Bowmans. Reprinted from AIM with permission from Croda Bowmans, Widnes.

leaving empty holes in the place occupied by the starch granules. The dilute starch hydrolyzate together with cheap sago starch were Bowmans' main raw materials. They were hydrolyzed together to near completeness, and the solution purified to some extent before it was moved to the fermenters.

The manufacture of a good quality edible lactic acid was still a major problem, Machell writes. In the past Bowmans had purified lactic acid through crystallization of its calcium salt, a method with many drawbacks. Bowmans, Machell speaks of the United Kingdom producer, "now has decided to install a continuous extraction plant using [di]isopropyl ether". This was the process developed and patented by Jenemann of the Grasselli division of Du Pont in 1933 on which the patent rights had expired (Ch. 9,

306). Childs was responsible for the development of a reliable extraction installation at Bowmans.

A few months after the appearance of Machell's paper more details about the all-glass continuous counter-current extraction and purification plant were published in Chemistry and Industry (October 3, 1959). The system worked well and Bowmans had fully converted to the new purification method, the paper said. Total costs of the installation had been low, less than £ 25,000.

According to market rumors of the time the investment in the solvent extraction plant was pushed by Howards of Ilford, one of the share holders of the Bowmans company. The company probably had the intention of closing down its own lactic acid purification plant and to transfer its main customer Beechams (Lucozade) to the improved edible quality to be made by Bowmans (Glimsholt, interview, 1985).

Machell also described the use of ion exchange resin for the removal of cations from crude lactic acid. The importance of this innovation in lactic acid purification technology can hardly be overestimated, because from then on it was possible to remove heavy metals (cations) from the crude product without the use of unpleasant and polluting chemicals like sodium sulfide and potassium ferrocyanide (more about ion exchange technology on p. 379).

Developments at Bowmans would gain further momentum when M.H.M. Arnold, a very competent chemist with a lot of experience in many branches of industry, was appointed to the position of technical director in 1958. A year later he succeeded S.H.W. Pert as managing director[12].

When Arnold joined Bowmans total lactic acid production was about 500 t/a but steps had already been taken to enlarge production. The company had just ordered two new stainless steel tanks of 70 m^3 each, to replace old wooden fermenters. Under the management of Arnold the renovation plan was completed with much cheaper rubber coated steel tanks of the same size. Within a few years total fermentation capacity was upped to a production level of 2,000 t/a of lactic acid, making Bowmans the largest producer of lactic acid in the world.

In the same period of time Bowmans began to ship lactic acid in bulk. Wooden barrels, traditionally used for smaller deliveries were replaced by plastic-lined 40-gallon drums (182 l) (AIM, Autumn, 1961). Arnold was also one of the first U.K. company executives to convert his company to the metric system!

Fig. 10,3. Fermenter at Bowmans. Reprinted from AIM with permission from Croda Bowmans, Widnes.

Société Normande de Produits Chimiques (Rhône-Poulenc)

In France edible lactic acid was not accepted as a food additive and hence no incentive existed for the sole French producer to make an edible grade of lactic acid. Production, during the war moved from Petit Quevilly near Rouen to Saint Fons near Lyons, had increased nevertheless in capacity from a prewar level of about 800 t/a to 1,300 t/a in 1960 (Klein, 1963). Raw material reportedly was still molasses, the method used is unknown, but: "the Boehringer plant [which had used molasses during the war] did not yield any secret" when it was visited by an expert team from Saint Fons in 1946 (Intermezzo 7, 349).

Schiedamsche Melkzuur Fabriek

During the war when lactic acid sales in the Netherlands rose to about 250 t (100%) per year it was decided to enlarge and rebuild the fermentation department. Because the right type of wood was not available it was decided to build 18 m^3 concrete tanks to replace the old 8 m^3 tanks which had been inherited in 1931 from the pressed yeast production of the distillery firm "Herman Jansen".

Between 1942 and 1944 16 glass tiled concrete tanks with contents from 10 to 18 m^3 were installed, and the capacity of the plant increased to about 1,000 t/a (100%). In August 1944 lack of raw materials and coal forced the owner to close the plant until the end of the war.

In August 1945 the fermentation was started up again, with a *Lactobacillus delbrueckii* culture from the nearby Heineken brewery at Rotterdam (Meyer, personal communication, 1976). The raw material used was, as before and during most of the war, beet sugar, which could be obtained free of sugar tax.

The adhesion of the glass tiles to the concrete of the fermentation tanks soon appeared to be of poor quality; within a few years tiles began to fall off, and the concrete became exposed to the acid conditions of the fermentation process. Cracks and pits appeared in the concrete, making frequent repairs mandatory. The cracks provided favorable (anaerobic) growth conditions for butyric fermentation, which often infected the lactic fermentation. Thorough cleaning was required to keep such infections at bay. The period is remembered by the personnel as one of much cleaning and the irritating, nauseating smell of butyric acid.

When sales in the late 1950's neared capacity new wooden 20 m^3 tanks were installed to replace the old concrete tanks. By 1961 production capacity had been enlarged to about 1,300 t/a (100%). [The extra capacity was almost completely used to supply Boehringer with about 600 t/a lactic acid 100%.]

The use of sugar as raw material was continued, an important aspect of "Schiedams" efforts to upgrade lactic acid to a high quality product for food uses. Economic considerations certainly played a major role in the decision not to work with starch – small amounts of which were used during the war when sugar was on allocation. Technological advantages inherent in the use of sugar enabled SMF to produce a quality of edible lactic acid equal or superior to the products of most of its European competitors.

The ever increasing demands of SMF's customers for high quality edible lactic acid kept plant superintendent Meyer and his small staff quite busy[13]. From their haphazard trials resulted an improved purification process based on the use of ion exchange resins which would give "Schiedam" a decisive edge in quality and in production efficiency. The growth of the company to become a major producer of lactic acid is as much related to the use of the purest carbohydrate raw material as feed stock as to the very practical

Fig. 10,4. Johannes August Meyer. Courtesy: Mrs. E. Meyer-Segaar.

technology introduced by the indefatigable production manager Meyer (When he was in the midst of an important series of measurements, Meyer often stayed awake one or two nights to supervise the experiments. One reason for this conduct lay in the fact that Meyer had hardly any laboratory equipment at his disposal, most experiments were conducted in the production plant!)

Introduction of Ion Exchange Resins in Lactic Acid Purification

(Sources of information: SMF's Archives; J.D.F. Companus, interview, 1988)

Synthetic ion exchange resins had been known since their introduction in Germany in 1938 by IG-Farbenindustrie under the tradename Wofatit. The first use of these resins was in the softening of boiler feed water. In this application a synthetic resin to which a large number of acidic groupings are attached is used. This polymeric resin, called cation exchange resin, is supplied in the neutralized condition, usually as the sodium salt. When tap water is passed through a column filled with small beads of the resin, calcium ions (responsible for the hardness) dissolved in the water will completely replace the sodium ions of the resin. The water so treated will be devoid of calcium, with the result that scaling of the boiler pipes does not occur anymore.

Before the war the IG company also started to sell anion exchange resins, which had basic groupings attached to the polymeric molecule, and consequently were able to remove anions from aqueous solutions. When columns of cation and anion exchange resins were placed in sequence even total demineralization of process water became a possibility, which earlier could only be done by complicated chemical precipitation techniques or by distillation.

A special application of anion exchange resins was the **total adsorption** of organic acids from fruit juices (Bleuler, 1944). In this case the resin is used in the basic form, the side groups attached to the polymeric structure usually are tertiary amino or quaternary ammonium groups.

As Bleuler's patent says:

> The solutions from which the acid has to be extracted are best passed over a layer of beads of the anion exchange resin. [When saturated] the [adsorbed] acids are extracted with alkali, or expelled by stronger acids. [According to the rule: a stronger acid pushes a weaker one out of its salts.] The resulting pure ... solutions can be concentrated or dried if so desired.
>
> *- Bleuler, 1944*

The first time SMF applied ion exchange resins was in 1959, when the need was felt to use demineralized process water in the production of chemically pure lactic acid. This product, obtained by hydrolysis of ethyl lactate with tap water, was until then treated with potassium ferrocyanide to remove copper and iron, and with silver lactate to precipitate small amounts of chloride originating from the tap water.

A small installation consisting of two 1 m^3 ebonited steel columns was put into operation in 1959. The first column was filled with a strongly acidic cation exchange resin (Dowex 50), the second one with a strongly basic anion exchange resin (Dowex 2). The apparatus worked well for the supply of demineralized water to the hydrolysis unit.

The next trial performed was the extraction of cations from concentrated edible lactic acid. This also posed no major problems – apart from the much reduced flow of the concentrated acid, due to its high viscosity (at a concentration of 85% the viscosity is about 3 times that of water). Then it was tried to remove chloride and sulfate anions in a similar way by passing the acid over the anion exchange column. Now a seemingly strange phenomenon was noted: the volume of the resin increased by something like 20%, and flow was even more restricted than with the cation exchange column. [The cause of the expansion was the adsorption of voluminous lactate ions by the basic groups of the anion exchange resin. The volume expansion appeared to be a cyclic phenomenon – when the resin was regenerated with strong alkali, the beads contracted again.]

When passing more crude lactic acid over the resin, lactate groups were gradually replaced by sulfate and chloride ions which were present in low concentrations in the crude solution. In this way the resin finally became "loaded" with the contaminating anions which in this way were extracted from the acid.

To solve the problem of the restricted flow it was decided to pass dilute, crude lactic over the ion exchange columns, the viscosity of crude acid as it came from the filterpress being hardly more than that of water. This worked, and in 1960 an installation to demineralize all crude lactic acid was ordered. The problems were not over, however. Within a year the flow through the

Fig. 10,5. Demineralization of lactic acid (Campos, Brazil, 1969). The Campos installation (Chapter 12, 422), was built with SMF know-how. Courtesy: Indústria Química de Sínteses e Fermentaçôes.

anion columns had been reduced to such an extent that the equipment no longer had sufficient capacity to treat all the crude acid. When the columns were opened to look for the obstruction it was found that the beads had been turned into powdery material. After some difficult exchanges of complaints with the supplier of the installation – who was not informed about the nature of the process carried out – the conclusion was reached that a combination of fairly high operating temperatures together with the cyclic expansion of the resin was responsible for its mechanical breakdown.

It took about five years before Meyer and his staff knew the most suitable type of commercial resin for the extraction of strong anions from crude lactic acid. Much time also was devoted to finding the best sequence of the ion exchange columns, and the most suitable regeneration methods.

Luis Ayuso S.A.

Luis Ayuso Utiel (1912–1980), a chemist with much industrial experience, founded the company that bears his name in 1947. The purpose of the company was to produce technical lactic acid for the leather industry.

Fig. 10,6. Luis Ayuso Utiel. Courtesy: Luis Ayuso S.A.

The first production site of the company was Cornellá de Llobregat, a suburb of Barcelona. As raw material the company used locust bean husks. (See about the composition of locust beans for instance Whistler and Smart, 1953.) The husks were hydrolyzed; the extract of the hydrolysis, mainly consisting of C6 sugars, filtered off and pumped to the (small wooden) fermenters. The filter cake served as fuel for the boiler.

The initial capacity of the plant was about 100 t/a (100%). In the 1950's some edible lactic acid was produced by means of the calcium lactate crystallization process (Private communication, Arnold, 1988).

In the 1960's the company stopped using locust bean husks, and switched to starch hydrolyzate as raw material. In the 1970's sugar came into use.

Fig. 10,7. Cooperation between Bowmans and Ayuso.
From right to left: Cliff Childs (Bowmans), Michael Arnold (Bowmans) and Eusebio Diaz (Ayuso), in front of Bowmans solvent extraction plant. Reprinted from AIM with permission from Croda Bowmans, Widnes.

In 1962 Ayuso concluded a ten year agreement with Bowmans to supply technical knowhow in return for a royalty based on production volume. Under this contract Eusebio Diaz, technical director of the company and José Luis Ayuso, son of the founder, received a technical training of about three months in Widnes to become acquainted with the latest purification technology.

One result of the know-how agreement was the installation in 1965 of an extraction plant to make a better quality of edible lactic acid. Instead of diisopropyl ether as extracting solvent Ayuso introduced (some years later) diethyl ether for this purpose. The first big use of edible lactic acid was in the production of olives (see below).

Lactic Acid in Olive Production

The main reason for Ayuso's interest in edible lactic acid was the introduction of the acid in olive fermentation and canning.

The use of lactic acid in olive fermentation is the result of studies begun at the University of California at Berkely in 1937 (Vaughn, 1973). In olive fermentation lactic acid had already earlier been recommended as an adjunct to stop the growth of unwanted microorganisms during fermentation (Eisenman, 1930).

Vaughn and collaborators found the addition of acid of particular value in the production of Spanish-type olives which are first treated with lye, washed and then brined. Formerly the removal of excess lye took 24–48 hours of repeated washings. The industry had tried to shorten this period to prevent graying of the color of the olives. A natural consequence of less washing were fairly high residual amounts of alkali which impeded the subsequent fermentation. To neutralize the excess of alkali the effect of addition of lactic acid to the brine on subsequent fermentation was studied.

3% Lactic acid added to the salt solution proved to be sufficient to take care of the neutralization and a drop in pH to about 6 or lower. The addition of lactic acid also proved to extend the shelf life of the packed product, enhance the flavor (compared with packing in dilute acetic acid) and prevent clouding of the brine.

The American olive industry began using lactic acid during the war: for 1943 Sheehan (Sheehan, 1944) estimates the total use at about 20 t (80%).

The application of lactic acid in olive brining was introduced in Spain around 1960. The first lactic acid Ayuso sold to this industry was their normal grade, a rather dark product. At that time fermented olives were shipped to the U.S.A. in wooden vats, and repacked in that country in small cans. The packer replaced the brine in which the olives were delivered by a fresh solution of 5% common salt, adjusted with lactic acid to a pH of about 4 (Casas, personal communication, 1989).

Then, in the 1960's, Libby, one of the big American canning industries, installed a packing plant in Spain. The lactic used in the packing operation was synthetic material imported from U.S.A. Imports began with 21 t (50%) in 1967, and reached 311 t (50%) in 1970 before it was gradually replaced by edible lactic acid from Ayuso's plant.

By 1970 Ayuso was estimated to produce between 800 and 900 t/a lactic acid (80%). Exports were insignificant; by far the largest local consumer of lactic then was the olive industry.

Kemisk Vaerk Køge A.S.

Like SMF Køge made lactic acid from sucrose. The company applied a type of bacterium, based on *Bacillus coagulans*, different from the *Lactobacillus* types used by most other lactic acid producers (See Intermezzo 3, 168). *B. coagulans* grows on a poor substrate, organic nutrients not being required. Only minor amounts of vitamins were added. The bacterium, however, is of the heterofermentative type, it coproduces with lactic acid up to 7% mannitol and 3% succinic acid (Private information based on a works visit, van Velthuijsen, 1972).

Edible lactic acid could not be made directly from the fermentation broth, the Jenemann extraction process with diisopropyl ether was used to remove impurities, in particular mannitol.

Since the installation of the small lactic acid plant just before the war, Kemisk Vaerk Køge had developed into a medium sized chemical plant with quite a variety of products. Most of the products were more profitable and showed better growth potential than lactic acid and derivatives. As a matter of fact from about 1950 on the lactic acid operation almost continuously operated in the red.

In 1953 Knud Sadolin, owner of the Køge works, decided to engage Edward Glimsholt, an old acquaintance of his, to make a study of the lactic acid operations of the company[14].

Glimsholt, more or less dropped without the consent of the management into Køge's lactic acid business, soon realized that lactic acid would remain a product of minor importance unless the quality could be improved and standardized, the manufacturing costs lowered, and the volume increased by searching for and developing new applications. The most important area of new uses, he thought, would be in the food field (Glimsholt, interview, 1985).

European Lactic Acid Producers Establish a Research Association to Promote Lactic Acid

Glimsholt's ideas about promoting lactic acid were not welcomed by Køge's management. The company did not intend to provide the funds and manpower for his program. Indeed it seemed the only thing the management was waiting for was how to get rid of the lactic acid operation as well as of Mr. Glimsholt.

Glimsholt then took a different and unexpected initiative: he decided to approach the other European lactic acid producers about their interest in doing together something about the position and status of lactic acid. Maybe he had in the back of his mind the idea that Køge would follow the other producers in case a breakthrough could be accomplished?

The first company Glimsholt approached was the outsider Howards of Ilford. In March 1954, Glimsholt visited Howards' managers in Ilford, who gave him the advice to go to the primary producer Bowmans to discuss the matter – they certainly were in favor of the idea, but their position was too unimportant to be of much help.

A month later, with the positive reaction of Geoffrey Howards and E.G. Turner of Bowmans in his pocket, Glimsholt went to Schiedam, where he had a lengthy discussion with Dirk Stapel and a brief encounter with Hans (H.C.) Jansen, the owner of the lactic acid plant. Stapel also could be convinced of the advantages of some form of cooperation to promote lactic acid.

The toughest job came next: to convince the Boehringer company, which everybody still believed to be the leading producer of Europe, maybe of the world. Boehringer, also visited in April 1954, did not like the idea very much. As Glimsholt said (interview, 1985):

> Boehringer forwarded the view that they were the leading company [in lactic acid], the first producer [sic] and the biggest. They were a very demanding company, an important producer of pharmaceuticals with high regard to quality.
>
> \- *Glimsholt, 1985*

Dehio (1966), about the same meeting:

> [Glimsholt] informed us about the negotiations with other foreign producers about the founding of a research association for new areas of application of lactic acid and standardization of the quality of edible lactic acid. Increases in sales and stabilization of the market could ensue; ... price agreements would not fall within the scope of the association. The research association would carry the name "International Lactic [Acid] Research Association [ILRA], the two English producers Bowmans and Howards, the Dutch Schiedam, the Danish Køge, and Boehringer, Ingelheim, were envisaged to participate.
>
> .... The explanations of Køge [Glimsholt] left us with the feeling that [the other manufacturers] were intent on using ILRA as a frame to profit from Ingelheim's experience and particular know-how. We reacted accordingly and protracted the negotiations for a number of months.
>
> - *Dehio, 1966*

Glimsholt himself believed that the main reason for this meager response were conflicting interests caused by the important stakes Boehringer had set in the promotion of citric acid. [The production level of the Ingelheim citric acid plant had by now surpassed 2,500 t/a. Citric had replaced lactic acid in many of its prewar fields, the most important of which was in sour bread acidulant mixtures. Captive use by Boehringer's bakery supply company alone amounted in 1952 to about 800 t of citric acid!]

Another reason for Boehringer's reluctance were the difficulties encountered in the purification of crude lactic acid from the molasses fermentation operation. The company could not meet the requirements of its most demanding customers without product assistance from SMF.

In September 1954 Boehringer had a separate meeting with Howards and Bowmans on potential cooperation. Again with poor results:

> The only point on which we reached some agreement was in the exchange of samples of edible lactic acid, to elaborate a basis for a common [future] standard of quality.
>
> One difficulty already then became evident [in the form of] the many yellowish edible qualities of the firm of Bowmans which in itself were chemically safe but did not conform to the almost waterwhite edible lactic acid produced by Schiedam and Boehringer. [The extraction plant which placed Bowmans' edible quality on equal footing did not become operative until 1959.] Bowmans sold large volumes of this second grade edible acid to Canadian brewers, who did not object to the color and were not prepared to pay the full price of the waterwhite acid.
>
> - *Dehio, 1966*

To show their good will Boehringer invited the other potential participants in ILRA to a meeting in Ingelheim in May 1955. Dehio:

Apart from improved personal relations between the [participating] companies the meeting did not produce noticeable results. Unity of understanding between the five companies about the purpose and intention of ILRA was not achieved, mainly as a consequence of the diverging point of view of Ingelheim. The result was that the English, the Dutch and the Danish associated themselves in ILRA and concluded an agreement while Boehringer abstained.

Fig. 10,8. Participants of the May, 1955, meeting.

1: K. Thomsen, Køge
2: D. Stapel, SMF
3: W.H. Bellamy, Bowmans
4: W. Goehtz, Boehringer
5. S.H.W. Pert, Bowmans
6: G.E. Howard, Howards of Ilford
7: interpreter
8: J. Liebrecht, Boehringer
9: W. Dehio, Boehringer
10: E.G. Turner, Bowmans
11: H. von Hahn, Boehringer
12: E. Glimsholt, Køge
Courtesy: W. Dehio.

In the ensuing years we stayed in touch with the ILRA members. The secretary, E. Glimsholt (K ge), took great pains to put life into the research efforts of the Association, but with little avail. The wind started to blow from another corner when Mr. Pert, the longtime managing director of Bowmans, was replaced by Mr. Arnold, a chemist with a wide background who had ambitious world encompassing plans for lactic acid. Quite soon [Arnold] took over the position of chairman of the board of ILRA and initiated a number of research plans with scientific institutes which, however, only matured at snail's pace.

- Dehio, 1966

The First Five Years of ILRA

Until the election of Arnold to the presidency (1960?) ILRA was only moderately active. Glimsholt gave a survey of this period under the title "A Review of the International Reconstruction of a Chemical Industry and its Markets", with the subtitle "Lactic Acid 1954–59" (1959). The list of references to this paper reveals some of the activities. Howards of Ilford had then already left the lactic acid business and ILRA after the start up of Bowmans' extraction plant in 1959. Boehringer had stayed in contact, but had not yet joined.

Technical meetings took place twice a year, the first one was at Howards in 1956. Until 1959 Ilford remained the meeting place, from then on meetings took place at different locations. For technical investigations the founders of ILRA had allocated a yearly sum of £3,000. The areas on which most of the money was spent, were:

Basic data [of physical and chemical properties], applications in jams and fruit preserves, and in soft drinks.

The Basic Data Book

Although only a small secondary producer of (mainly) pharmaceutical lactic acid the Howards company had made considerable efforts to promote lactic acid. One of the means the company used for this purpose was an extensive

39 page brochure, first issued around 1948, and giving a detailed description annex list of literature references of all known applications in medicine, pharmacy, food and beverage industries, cosmetics, and technical industries.

When ILRA was organized it was Howards who in the beginning supplied most of the initiative. Howards advocated to update the data on applications and to extend the knowledge laid down in the 1948(?) brochure.

In addition to the applications survey Howards proposed to collect all existing data of the chemical and physical properties of lactic acid and its more important derivatives. In ILRA this venture bore the name "Basic Data Book". The activities to collect the required data would become the most important task of the technical committee of the Association.

At some time, and only after much discussion, the Board of ILRA decided that the Basic Data Book should not include essential information about production methods, nor about applications. The first subject officially never appeared on the agenda of the meetings of the Technical Committee; applications, research carried out on behalf of ILRA was to be published in scientific and trade journals, in brochures and leaflets. From 1955 until 1959 more than 40 publications of such a nature appeared, either from one of the participating companies or originating from institute research funded by ILRA. The Basic Data Book appeared after much delay in 1971 under the title: "Lactic Acid, Properties and Chemistry of Lactic Acid and Derivatives", by C.H. Holten.

Arnold Chairman of ILRA

The new chairman appeared to be much more open minded regarding the exchange of application and general information than the ILRA members had experienced so far. This probably not only relates to the character of the chairman, but also to the fact that in the beginning of the 1960's Bowmans undoubtedly had become the leading lactic acid producer of the world.

Arnold, who had a thorough knowledge of the coatings industry, added to the applications list studies aimed at the use of lactic acid in the manufacture of alkyd resins. Another addition to the research program of the technical committee was a general study of the "Synthetic Chemistry of Lactic Acid". [The sample of lactic acid Bowmans supplied in 1964 for this study was rather ironically a sample of synthetic lactic acid from Monsanto's new Texas City plant which had just started production. See Ch. 11, 408.]

Although Arnold for nearly 10 years (1959–1968) did much to blow more life into ILRA, the results were disappointing. No new applications arose from ILRA's activities. Total sales of lactic acid were not noticeably affected by the works of the Association. As Arnold himself said much later (private communication):

It was difficult to get the members down to constructive discussions. New applications were not taken seriously. The secretary (Glimsholt) was pushing things very hard, but he obviously preferred university type research and had little feeling for commercial aspects.

One important result was the standardization of specifications. The most important achievement was the publication of the Basic Data Book [Holten, 1971].

- Arnold, 1988

Boehringer Temporarily Halts Lactic Acid Production

In the same years Bowmans grew to the position of leading manufacturer and Schiedam followed as second largest European producer, Boehringer was experiencing large difficulties in maintaining its position on the German market. Dehio (1966):

The decision to base lactic acid production entirely on molasses increasingly proved to be far from the best. The edible specifications were becoming tighter. The purification required new investments, year after year, augmenting manufacturing costs. Proceeds, however, went down, as we were facing cheap foreign competition in every export market.

- Dehio, 1966

The company, of course, was well aware of the use of sugar as raw material by SMF and Køge. In Germany such use was impossible because of the high inland sugar price and the refusal by the government to reimburse the special "equilibrium" import duty on world market sugar when sugar based lactic acid was exported (the so-called export refund). Dehio gives some figures:

German sugar price 1.00 DM/kg, including a tax of 0.21 DM/kg; world market sugar price 0.30 DM/kg.

Dehio about the requests made from 1953 onward by the company to secure sugar at world market prices:

Our requests were turned down with every possible pretext, although the relatively small amount of sugar [about 1,000–1,200 t/a] did not constitute any endangerment for the German sugar industry. Most of the bureaucracy of the [prewar] Berlin administration had been reassembled in Bonn, and in particular the Finance Department proved to be completely adamant .... The Department of Agriculture, responsible for export refunds, initially also was deaf to our arguments.

- Dehio, 1966

For economic reasons a return to starch as raw material was not possible either. Boehringer was forced to halt production in the fall of 1958. This step was only taken after delivery contracts had been negotiated with Schiedamsche Melkzuur Fabriek [for about 750 t (80%) edible lactic acid per year] and with Société Normande (Rhône-Poulenc) for about 200 t/a technical grade. Schiedam also supplied Boehringer with calcium lactate (30–40 t/a) and Société Normande helped out with ethyl lactate. Dehio:

The agreement with Schiedam enabled us to lay down our own production and still supply the German market; in general the clients did not notice that they received Dutch product instead of Boehringer material.

The agreement ... was continued for five years. The volume grew steadily and brought us net returns between 5 and 7%, a profit we could not have achieved with own production. The ability of Schiedam to ship [edible lactic acid] at low prices had a number of reasons: good process technology, sugar at international prices, very low overheads due to low personnel and administrative costs.

In particular the last point should be stressed as our commodity type chemicals also are charged for the huge overhead costs inherent to high priced specialties [pharmaceuticals].

- Dehio, 1966

The lobbying in Bonn to obtain sugar at world market price had some partial success in 1959 when the Ministry of Agriculture approved the restitution of the import levy, for exports as well as for local sales. Inland sugar tax, however, still had to be paid, and, although the tax had been reduced to 0.10 DM/kg, it still made own production financially less attractive than continued imports of Dutch and French lactic acid. Protracted lobbying finally helped in removing this last obstacle to the use of low cost sugar in the fall of 1961.

Production and research departments of Boehringer had in the meantime made extensive studies and preparations to reintroduce production based on sugar, a raw material which had not been used since the years of World War II, and which, in the past, had not given optimum results (Dehio, 1966). The fermentation chambers [from World War I !] had to be refurbished, a number of other renovation investments were required as well. In view of Boehringer's historical position in lactic acid production, supported by a market analysis of the sales department, the decision was taken to restart production, however.

Dehio:

The second 'return' of Ingelheim to the lactic acid market in 1963 was not an easy matter. Foreign manufacturers had established themselves since our production stop of 1958 more firmly in [various] markets, and prices were depressed. From America came reports about synthetic lactic acid production by Monsanto, which in a short time would take away our remaining exports to that country.

- Dehio, 1966

Boehringer Joins ILRA

Under these conditions, [Dehio continued], we were of course interested in every possible extension of the lactic acid market, and, for that reason, viewed the efforts of ILRA in another perspective than in 1954. Schiedam and Køge also were keen to admit Boehringer to ILRA to provide some counterpoise to Bowmans and its 'strong man', ILRA chairman Arnold. The wish to have Ingelheim participate also had a financial background as the research expenditures of ILRA had increased substantially [to about £ 9,000 per annum]. A fourth member would diminish the costs for existing members.

... after having accepted the statutes of ILRA, we were welcomed [as the new Member] and from then on participated in the Board and technical meetings, respectively twice and three times per year.

Efforts by ILRA with Boehringer as intermediary to induce the French producer Société Normande/Rhône-Poulenc to participate did not succeed. The company said not to be interested in the goals of ILRA, ... Quite probably Rhône Poulenc's plans made public in early 1966 to build a combined synthetic lactic/oxalic acid plant influenced their decision against participation in an association of companies employing old [sic] fermentation processes.

- Dehio, 1966

Boehringer proved to be a valuable member. Scientists of the company made important contributions to the Basic Data Book. Together with Bowmans they were the only company in ILRA carrying out a substantial amount of research. In this way Boehringer was more a welcome addition to ILRA than a counterpoise to Bowmans.

Notes

[1] Robert S. Aries operated a consultancy agency under the name R.S. Aries & Associates, Polytechnic Institute of Brooklyn, N.Y. Haskell C. Needle was a former employee of American Maize, where he had been involved in lactic acid research. The study reported in the paper mentioned had been commissioned by the "Sugar Research Foundation", Inc., New York, an organization of sugar producers set up to promote the use of sugar as a chemical.

[2] Alkyd resins are reaction products of polyhydric alcohols (hence the alk in the

name) and polybasic acids (responsible for the yd = id). These polymeric resins had been introduced around 1930 as vehicles for the production of fast drying oil based paints. Oil paints used before were based on drying oils like linseed oil, which dried slowly by oxidation when exposed to air. Alkyd resins in which as polyhydric alcohol monoglycerides from drying oils were incorporated, dried much faster because of a much higher initial molecular weight. Through variation in fatty acid composition, degree of condensation (molecular weight) and types of acids and alcohols used in the manufacture a very large range of properties and possible applications could be offered to the trade. Within a decade alkyd resins for instance took a large share of the car paint market. In refrigerator enamels nitrocellulose lacquers were completely displaced by alkyd based enamels. Alkyd resin production rose from 4,500 t in 1933 to 32,000 t in 1939! (Haynes, vol. 5, 1954).

[3] Wenker was first in submitting a patent application on the purification by means of methyl lactate (April 14, 1942) in the name of Apex Chemical Co., Inc., Elizabeth, NJ). The essential features of his method are not very different from those of Schopmeyer and Arnold, nor from Filachione and Fisher. His method, however, has the disadvantage of being more a batch-type process as crude lactic acid is not added continuously to the esterification reactor.

[4] Dr. W. Gruschwitz (1876–1944) headed the lactic acid factory from 1920 until 1924, when he was charged with the supervision of a number of plants of the Merck, Darmstadt works. He was killed in an American air-raid on Darmstadt on December 12, 1944, in which 55% of Merck's production facilities were wiped out (Löw, 1952).

[5] This was the only use of lactic acid in the field of synthetic resins existing at the time. The so-called plastic grade lactic acid was designed for this application. The advantage of lactic acid over other weak acids or acidogens seems to be that it acted as a catalyst as well as a plasticizer (Barron, 1946, p. 222). Total U.S. usage was given as 30 t/a in 1941 (Sheehan, 1944), and as 180 t/a in 1948 (Watson, 1948).

[6] Clinton had a strong position in the production of amylases, enzymes which were about to replace hydrochloric acid as hydrolyzing agent in the production of starch syrups and dextrose hydrate from corn starch. The first enzymatically prepared starch syrup, called Sweetose, was produced by the Staley company in 1939 (Forrestal, 1982). When the Staley patent expired in the late 1950's the starch industry switched to enzymatic conversion on a large scale.

[7] The oldest paper found about an industrial synthesis of ethyl lactate from acetaldehyde is from 1926 (Armstrong, 1926). The author remarks that the synthetic production of ethyl lactate is under study, the fermentation process for making lactic acid being too laborious.

[8] A Musashino company brochure of about 1970 said that Kyowa Gas Chemical Industry Co., Ltd., was established as a joint investment of Musashino with Nippon Mining Co., Ltd., and Kyowa Hakko Kogyo Co., Ltd. The plant was probably located near the natural gas field of Niigata Prefecture, on the west coast of Honshu, 240 km north of Tokyo.

[9] The direct liquid phase oxidation of ethylene was developed in 1957–1959 by Wacker-Chemie and Farbwerke Hoechst. The catalyst is an aqueous solution of $PaCl_2$ and $CuCl_2$ (Kirk-Othmer, 1976).

[10] From 1948 until 1950 SMF exported about 200 t/a lactic acid (80%) to Germany, derivatives included. This amount diminished to less than 100 t in 1951 and subse-

quent years.

11 Inversion is the process whereby sucrose is hydrolized to glucose and fructose, The name is related to the fact that the optical rotation changes from plus (dextrorotatory) to minus (levorotatory). The inversion should only be partial, sufficient to prevent crystallization – "graining" – of the sucrose. When too much sucrose is converted the fructose concentration becomes too high, and the candy will easily attract moisture from the air. Stickiness is the result, a property not liked by the customer either.

12 Michael Henry Miller Arnold, BA MSc (Oxford) LLB (London), chartered chemist, chemical engineer and biologist, was born in 1915 in Liverpool. From 1937 until 1947 Arnold was employed at the ICI Billingham research center, where he worked on catalytic gas reactions and biological problems (fish farming). In 1947 he joined BX Plastics at Manningtree, Essex, as research manager (nitrocellulose and cellulose esters). Arnold's next job was with the paint firm of Postans at Birmingham. Here, as technical manager, he was responsible for the introduction of manufacturing processes for oil modified alkyd resins, urea-formaldehyde and melamine-formaldehyde resins, and for the development of antifouling coatings for shipbottoms. He then also became honorary editor of JOCCA, the scientific British journal for the paint and varnish industry.

13 Johannes August Meyer (1909–1982) studied chemical engineering in Bucharest, Romania, in which country his father had a position with a Shell affiliate. In the 1930's he was employed at Shell's first petrochemical plant in Europe (Pernis, Rotterdam). In 1942 SMF "borrowed" Meyer from Shell to build an ethyl lactate plant; in 1945 he became superintendent of the SMF plant.

14 Edward Glimsholt (1908) studied medicine for a number of years at Copenhagen University, then was employed by the paint company Hempel. Later he travelled much in South America to introduce a vaccin for foot and mouth disease.

References

Armstrong, E.F., **1926**. *J. Soc. Chem. Ind., London, Trans. Commun.*, **45**, 425.

Barron, H., **1946**. *Modern Plastics*; Chapman & Hall: London.

Bleuler, H., **1944**. *Swiss Patent 233,394.*

Burns, R., Jones, D.T., and Ritchie, P.D., **1935**. *J. Soc. Chem. Ind., London, Proceedings*, Prt I, 400.

Chem. Eng., **1964**. *Ind. Eng. Chem., Int. Ed.*, **56**, [2], 55.

Dehio, W., **1966**. *Die Industrie der Weinsäure, Zitronensäure und Milchsäure*; internal report C.H. Boehringer Sohn.

Eisenman, W., **1930.** *Food Ind.*, **2**, 262.

Filachione, E.M., and Fisher, C.H., **1946**. *Ind. Eng. Chem.*, **38**, 228.

Fisher, D.A., **1963**. *The Epic of Steel*; Harper & Row: New York.

Forrestal, D.J., **1982**. *The Kernel and the Bean*; Simon and Schuster: New York.

Hale, W.J., **1934**. *The Farm Chemurgic*; Stratford: Boston.

Haynes, W., **1954**. *American Chemical Industry*; D. van Nostrand: New York. Reprinted in 1983; Garland: New York.

Holmen, R.E., **1958**. *U.S. Patent 2,859,240.*

Holten, C.E., **1971**. *Lactic Acid*; Verlag Chemie: Weinheim, F.R.G.
Inskeep, G.C., Taylor, G.G., and Breitzke, W.R., **1952**. *Ind. Eng. Chem.*, **44**, 1955.
Jpn. Chem. Ann., **1980**, no page.
Kirk-Othmer, **1976**; *Encyclopedia of Chemical Technology*; 3rd. ed.; Interscience: New York, Vol. 1, 103.
Klein, M.D., **1963**. *Marktoverzicht Melkzuur*; internal HVA report.
Löw, C., **1952**. *Die chemische Fabrik E. Merck, Darmstadt*; E. Merck: Darmstadt.
Machell, G., **1959**. *Ind. Chem.*, June 1959, 283.
Mallows, J.H., **1964**. *Confect. Prod.*, August 1964.
McMillen, W., **1939**. *Ind. Eng. Chem.*, **31**, 540.
Montgomery, R., **1949**. *The Chemical Production of Lactic Acid from Sugars*; Sugar Research Foundation: New York.
Needle, J.C., and Aries, R.S., **1949**. *Sugar*, December 1949, 32.
Peckham Jr., G.T., **1944**. *Chem. Eng. News*, **22**, 440.
Rabald, E., and Bretschneider, H., **1961**. *Chemische Beständigkeit der Werkstoffe; Milchsäure*. Dechema-Werkstoff-Tabelle.
Sheehan, M.J., **1944**. *Lactic Acid* (Synopsis of Information); United States Department of Commerce, Bureau of Foreign and Domestic Commerce; Report 2/11096.
Smith, L.T., and Claborn, H.K., **1939a**, **1939b**. *Ind. Eng. Chem. News Ed.*, **17**, 370 and **17**, 641.
Smith, L.T., and Claborn, H.K., **1940**. *Ind. Eng. Chem.*, **32**, 692.
Spitz, P.H., **1988**. *Petrochemicals. The Rise of an Industry*; John Wiley & Sons: New York.
Strecker, A., **1850**. *Ann. Chem.*, **75**, 27.
Vaughn, R.E., **1973**. *Lactic Acid Fermentation of Olives*; in Carr, J.G., Cutting, C.V., and Whiting, G.C.; Eds.; *Lactic Acid Bacteria in Beverages and Food*; Academic Press: London.
Waddams, A.L., **1978**. *Chemicals from Petroleum;* 4th ed.; John Murray: London.
Watson, P.D., **1948**. *Ind. Eng. Chem.*, **40**, 1393.
Whistler, R.L., and Smart, C.L., **1953**. *Polysaccharide Chemistry*; Academic Press: New York, 300.
Yamazaki, N., **1978**. *Jpn. Chem. Week*, April 27, 1978.

CHAPTER 11

C.J. PATTERSON FINDS A NEW USE FOR LACTIC ACID

THE STORY OF STEAROYL LACTYLATES, CONDITIONERS OF BREAD

Sources of information:
Lawrence F. Marnett, interview, 1986.
Brochures, leaflets, etc., C.J. Patterson Co.

The use of additives in bread baking is a development closely related to the rise of mass production and mass marketing in the United States. It is appropriate, therefore, to first pay some attention to the history of the American baking industry. In this history Kansas City, situated in the center of a vast wheat growing area, occupies a special position.

Around the turn of the century bakers in Kansas City started to replace manual labor by machinery. Dough mixing machines were installed, chain conveyors for carrying the dough through temperature controlled areas during the raising periods followed soon. Then came automatic baking ovens, and, with the advent of the supermarket, slicing and wrapping machines.

Bread in the Supermarket

Since the introduction of the Ford Model T in the early twenties the American Way of Life had changed profoundly. The next big change came in the early 1930's with the introduction and quick acceptance of the Supermarket. In the 1920's old fashioned grocery stores (the "shops around the corner") had yielded to self-service shops, cash-and-carry shops belonging to grocery chains (MacFayden, 1985). Although the cash-and-carry stores already offered much of their merchandise in packed standard amounts, meats, vegetables, and some other grocery products had still to be cut, weighed and packaged by store personnel. Shoppers used baskets to carry their purchases.

According to MacFayden the first real supermarket in the present sense of the word was opened in August 1930 by Michael Cullen in Queens (New York). From its very beginning the idea of self-service, ample choice, low prices (on most goods) made possible by large turnovers, and large, convenient parking facilities met with tremendous success. Competitors soon entered the market: by 1936 a total of 1,200 supermarket stores were in operation in 85 cities.

With the advent of the new marketing system prepackaging of grocery products became a must.

> Where previously many items had been distributed to stores in bulk containers and broken down, or packaged individually for each patron, most groceries now had to arrive at the store already wrapped and boxed in what producers judged to be convenient sizes.
>
> *- MacFayden, 1985*

Bread posed a difficult product in the supermarket system. The bakery industry did not have large problems with the desired slicing and prepackaging, but the staling of the bread within a day made the product not very suitable for the type of shopping that was emerging: only once or twice a week.

The first answer to this problem was an increase in the amount of fat (baking margarine) to much higher levels than before, additions up to 6% and more became common. In the 1940's cereal chemists began to investigate other methods of preventing staleness.

Flour being an inhomogenous composition of starch particles, gluten, fat and other materials like bran, cannot be well defined along the lines of traditional chemistry. When in the preparation of the dough water, salt, yeast and margarine are added phenomena occur which belong to the science of colloid chemistry, the science which studies the formation of and reaction in interphases.

Cereal chemists began to look at what happened – or might happen – during the various steps of the bread making process, how the different materials and phases influenced each other, how they reacted, and how the final product was affected. In the light of the demands of the Supermarket System, the keeping qualities of bread, of course, were a subject of lively interest.

C.J. Patterson, Cereal Chemist and Industrialist

Curtis J. Patterson (1889–1960), born in the State of Kansas, was one of the early cereal chemists of the United States. In 1913 he had entered the bread industry as a chemist, and a year later he was involved in establishing the

American Association of Cereal Chemists. In those years the few professional chemists employed by the bakery industry were mostly engaged in quality control, and so was C.J. Patterson. When C.J. Paterson, or for short "Pat", as he became known in the bakery trade, in 1919 became head of Campbell Baking Co.'s laboratory in Kansas City, he also began to study the bread making process as such, including engineering and organizational aspects. Nutritional aspects did not escape his interests either.

While working for Campbell Baking C.J. Patterson invented a number of products for use in the bakery. One of these products, baptized Paniplus, was a flour mixture containing a small percentage of calcium peroxide. The product was quite successful in preventing stickiness in machine-made doughs. The trade name Paniplus and the application of the product were sold to Continental Baking Co., which organized a separate division to handle the sales of the new product. The Paniplus Division of Continental Baking was one of the first bakery ingredient manufacturers in the United States.

A number of years later we find Patterson with Campbell Taggart Bakeries, Inc., as vice-president of research, production and engineering. From this position he retired in 1944 when the company headquarters were moved from Kansas City to Dallas, TX. Patterson did not want to leave the Kansas City area[1].

Patterson Forms His Own Company

So at the age of 55 Patterson found himself without a job. On the basis of his wide knowledge of the baking industry and his acknowledged name as one of the foremost cereal chemists of the country he decided to start his own company. In January 1945 the consulting firm of C.J. Patterson Company was formed with the aim of providing service of a technical nature (quality control, baking formulas, advice on equipment) to smaller bakeries. Advice on commercial and economic matters, including advertising also was offered by the new firm.

In the course of the next six years Patterson built up a complex 12-million dollar a year business, the backbone of which was formed by a number of bakeries he had acquired since 1945 and incorporated in 1950 into a separate bakery division called "Holsum Bakeries".

C.J. Patterson Establishes a Bakery Research Laboratory

Right from the onset C.J. Patterson wished to apply a scientific approach to the engineering and chemical problems of the baking industry. He established a "Technical Group" with expertise in both fields (Baker, 1969). This group included a research laboratory headed by Dr. R.J. Summer. When Larry F. Marnett joined the company in January 1947 six people were working in this lab[2]. The engineering staff was even larger than that. In the course of time the C.J. Patterson Company became a breeding place for cereal chemists and bakery experts. Twenty years later experts trained or one time employed by Patterson were found in almost any American institute doing bakery research and in many associations of cereal and baking chemists. Quite a few companies producing bread improvers also engaged former Patterson employees.

> Concurrently with its own research activities, the company gained an enviable reputation as service consultants for independent bakers throughout the country in the fields of plant construction, equipment, engineering, chemical research, production control, advertising and management counselling. Extensive pilot baking facilities were made available, regular service bulletins were published and, for a number of years, periodic seminar-type conferences were held for the company's clients. Gradually, however, more and more independent bakers merged into the larger wholesale chains which provided their own management services, and this aspect of the company's business has virtually been eliminated.
>
> \- *Baker, 1969*

The chemical research staff was involved in investigating the action of dough emulsifiers and conditioners, and in developing bread softeners and yeast food[3]. These research activities showed so much promise that the company decided to staff a sales organization, later called Patco, to sell its own products and services nation-wide.

An important product of the sales organization was a bread softener "Sta-Soft", a polyoxyethylene derivative of stearic acid, first produced in the 1930's by the Atlas Powder Company for use as a technical emulsifier. During the war years when bakers' shortening was in short supply this product became accepted as an extender and partial replacer of bakers fat. Cereal chemists had observed that polyoxyethylene stearate had additionally an excellent influence on the quality of machine produced bread. From the study of the influence of this emulsifier on the machineability of the dough to the investigation of the influence on the properties of the finished bread was only the next logical step. So it was discovered that polyoxyethylene stearate had a remarkable and positive effect on the freshness and the keeping quality of bread (See for instance Edelmann and Cathcart, 1949). The product was put on the market by Atlas Powder and by C.J. Patterson. The latter company had the product manufactured on a toll basis by Glyco Chemicals of Greenwich, CT. Mono- and diglycerides also came into use, primarily for their softening properties. A special ester was put on the market by Hachmeister, Inc. of Pittsburgh, PA, in 1948. It was a diacetyl tartaric acid ester of monoglycerides, called "TEM" (Birnbaum, 1955). This product apparently met with only limited success.

C.J. Patterson, of course, was in the midst of these developments. Through his many contacts with cereal chemists and technologists in industry and government he was also aware of the possibility that one day ethoxylated products like Sta-Soft might be banned from the use in bread baking. In 1948, he, therefore, initiated a research program for a replacement of the product. A laboratory group was formed to screen systematically most existing emulsifiers as well as newly synthesized ones. The first major breakthrough of this research group headed by Marnett came in 1950 "with the discovery of new and patentable applications for a family of acyl lactylate compounds" (quoted from Baker, 1969).

> The initial product was an additive for use in baking which was named "VERV". Even before the first batch of bread dough was made the technical staff knew it would work. That first loaf baked with VERV was black in color – but it was two inches higher than normal. The color problem was soon solved and [in the same year] the new product was submitted to the Federal Food and Drug Administration [for approval]. Discouragingly, it was a time of turmoil in the FDA. Congressional Committees were intensively studying the testing standards to be applied to food additives before they could be used in food products. The company could not market its new product until Federal Approval was obtained but even the FDA did not know what constituted adequate evidence of safety.
>
> \- *Baker, 1969*

The patent applied for on December 12, 1951, by the inventors Thompson and Buddemeyer was titled "Acyl lactylic acid products, and the use thereof in bread baking to improve the softness and keeping qualities of baked products" (U.S. Patent 2,789,992, granted April 23, 1957). Reading the patent the invention does not look very practicable: lactic acid first was polymerized to a very viscous glasslike melt, which in the next step was reacted with stearoyl chloride. After the removal of the by-product hydrochloric acid at an elevated temperature the stearoyl lactylic acid was converted into its calcium salt. The color of the product was deep black, due to impurities in the lactic acid and to the extreme conditions of the reaction. But, as already said, the product was highly effective, even at the low concentration of 0.5% on the flour, and the manufacturing process was soon improved[4].

Fig. 11,1. The pioneers of stearoyl lactylates (about 1951). Left to right: C.J. Patterson, L.F. Marnett, J.B. Thompson. Courtesy: C.J. Patterson Company, Kansas City.

The F.D.A. Scrutinizes Bread Emulsifiers

The "Standards of Identity for Baking Products" were issued by the Food and Drug Administration in 1953. As anticipated by Patterson and others, these new standards forbade the use of ethylene oxide derivatives in bread baking, including that of polyoxyethylene esters of fatty acids like Sta-Soft. The reasons given were: there was lack of proof of non-toxicity of these products, and "the use in bread was deceptive to the customer, because the bread appeared fresher than it really was".

At the time, however, the FDA did not really know how adequate standards of safety for the consumer could be precisely defined, nor how the required information could be obtained from animal experiments. It was, of course, evident to the responsible officials that additives to be used in a staple product like bread should be scrutinized profoundly before permission for their use could be granted.

Patterson had initiated the first biochemical and toxicological studies of "Verv" in 1950 with the Kansas City-based Midwest Research Institute of which he had been a cofounder in 1944. The study at M.R.I. included a lethality study (The determination of the LD 50 of rats, the dosage of the product at which 50% of the rats die.) and a 90-day subchronic study, also with rats. None of these studies showed any toxic properties of the new emulsifier at levels substantially exceeding those contemplated for use in bread. The application for the approval of the use of "Verv" in bread and bakery products was made soon afterwards. FDA, however, wanted more information, to be collected through more studies and after completion of these, additional investigations were demanded. The studies and their evaluation by physiologists, toxicologists and statisticians became a very protracted and very costly affair, requiring in total some 11 years to complete.

All-in-all, until FDA clearance was finally obtained in April 1961[5], the total number of investigations had risen to 18, and the costs had surpassed two million dollars. Stacked on top of each other the toxicological reports had reached a height of about 60 cm! All of this to prove the safety of a product that once ingested in the stomach and duodenum would totally break down into its constituent components, completely harmless stearic and lactic acid, as one of the early toxicological investigations had already shown.

The high cost of the toxicological studies had forced C.J. Patterson in 1955 to take a most dramatic decision: he summoned the complete resarch staff in a meeting and told them about the predicament the company was in, it either had to give up further funding of the toxicological studies (and "Verv"), or it had to stop spending any money on research by the company itself. He, Patterson, had already made up his mind: a product which showed so much promise, a product in which everybody of the research department had such a strong belief, and which everybody thought to be completely safe, could not be abandoned. What would otherwise be the purpose of research at all? So he had to say farewell to almost all the people who had worked so hard for so many years to make a success of the new product. The inventors of the product were among those who had to find new jobs.

C.J. Patterson himself parted in an even more tragic way from his company, shortly before the struggle for the FDA clearance of "Verv" finally was won: in September 1960 while taking photographs on a country road in

Canada, he was struck and killed by an automobile.

Introductory marketing efforts, supported by papers in the Baker' Digest and the Journal of Cereal Chemistry started around 1954. Real sales, of course, could not begin until clearance from FDA had been obtained. When this happened, in 1961, the ground had been well prepared indeed, and "Verv" almost overnight became the big success C.J. Patterson had foreseen but very sadly did not live to see.

The Supply of Lactic Acid

Market studies based on the broad experience of the company in the bread market promised a large market for the new product, a market so large that the procurement of sufficient amounts of one of the basic ingredients, lactic acid, was expected to become a major problem. The quality of the lactic acid required posed another problem. Because of the high reaction temperature of about 200° C needed for the formation of the esterbond between the carboxyl group of the stearic acid and the hydroxyl group of the lactic or lactylic acid a number of unwanted side reactions between contaminating compounds present in lactic and stearic acid could take place. These side reactions caused coloring of the reaction mass, and, hence, of the final product unless a rather complicated and costly purification procedure was applied. A more logical attack of this problem was to eliminate all or most of the color forming impurities from the raw materials. For stearic acid this was a matter of the right specification, this acid had already been in use for other chemical conversions for some time, and as a consequence the production of a chemical grade had already been solved by the manufacturers. The main culprits of the dark color were, however, the impurities in the lactic acid of the trade: proteinaceous materials and residual sugars, both originating from the fermentation process. Under "Verv" reaction conditions these two materials react easily with each other by the so-called Maillard reaction[6].

Response of American Lactic Acid Manufacturers

Marnett and associates tried, of course, all commercially available sources of lactic acid:

The edible grades of Clinton and American Maize were not acceptable; Du Pont had a good quality of limited quantity for which it, however, charged double the price. (This was probably extraction quality, produced in a small plant since 1933; see Ch. 9, 308.) Discussions with these companies neither led to improvement in the quality of their existing products nor to a sizable investment in the production of heat stable lactic acid in sufficient quantity. All three of the companies knew very well how to make heat stable lactic acid, either by means of the extraction process of Jenemann, or through purification by distillation of the methyl ester of lactic acid, the Steffens process. In view of the large potentialities of lactic acid for the chemical industry propagated by scientists of the Eastern Regional Research Laboratory of USDA (Ch. 10, 353 ff), this seems very surprising. Apparently neither the breakthrough suggested by the ERRL was believed to be very imminent, nor did these companies see very much future in Patterson's new product.

Clinton, the largest U.S. manufacturer of lactic at the time, certainly had some interest to invest in what amounted to a number of times their existing production (Newton, interview, 1986). Since about 1948 the company had a pilot plant in operation to learn how to optimize the production of heat stable lactic acid by hydrolysis of methyl lactate (an ERRL process).

When Patterson approached Clinton about large scale production of heat stable lactic acid, the focus of attention had already shifted away, however, from lactic acid to enzymatic conversion of starch to starch syrup and dextrose[7]. Research on this subject had led to commercially viable methods for the production of the two (amylolytic) enzymes required to manufacture all kinds of starch syrups and hydrolysates, up to those of the high dextrose content needed for producing crystalline dextrose.

Clinton's management chose to apply the resources of the company in expertise and money in the first place to enzymatic starch conversion instead of to the smaller and less promising heat stable lactic acid venture[8].

Further developments would show how right the decision of Clinton was to choose for enzymatic starch conversion. Quite ironically producers of fermentation lactic acid would, in the very long run, also profit from the negative decision of their (then former) competitor[9].

Patterson Approaches European Lactic Acid Producers

European producers were also contacted, to supply additional quantities. This helped Patterson for some time but did not prove to be an acceptable solution for their long term demands.

C.H. Boehringer Sohn had the best edible quality, and limited amounts of a good heat stable quality (respectively bought from Schiedamsche Melkzuur Fabriek and made by reprocessing Société Normande product).

Schiedamsche Melkzuur Fabriek had too much variation in quality. [Here Schiedams lack of systematic quality control revenged itself; only shipments to Ingelheim were checked on conformity with the high standards set by Boehringer.]

Bowmans Chemicals initially was not able to supply a product of acceptable quality. The company, however, was the only European producer which answered the new prospective customer actively, its managing director, Arnold, personally went to Kansas City to find out what possibilities the new application could have for Bowmans. The meeting (in 1961) is well remembered by the Patterson gentlemen, it was the first time they met Arnold, who in appearance and attitude was not at all the standard British businessman they were expecting. But, more important, he answered Patterson's questions regarding the supply possibilities of Bowmans in the affirmative: Bowmans was prepared to ship a suitable product within 30 days and would be able to deliver sizable quantities of the same quality in the future.

Table 11.1. U.S. production and imports of lactic acid.

year	local production	total imports (conc. 55% or more)	increase in consumption (%) (%)	imports from from Britain (conc. 55% or more)
1957	2.273	143		–
1960	2.493	328	6	76
1962	2.510	693	16	409
1963	2.400	abt.700	4	abt.380
1964	?*	abt.400	?	?

* From 1964 onward local production figures (U.S. Census Reports) were no longer published because only two producers were left: Clinton, and Monsanto, which had brought on stream a 10 million pound synthetic lactic acid plant at Texas City in February 1964. Import figures from U.S. Import Statistics.

In Arnold's recollection (personal communication, 1988):

We had the [Jenemann] extraction plant [with a capacity of over 1,000 t/a] in operation since about a year. Material from this plant had been rejected by Patterson, however. At the meeting it occurred to me that a double extraction might yield the desired heat stable quality.

- Arnold, 1988

From this time onward Patterson heavily relied on deliveries from Bowmans and used the English product admixed with more or less acceptable grades from other sources, depending on availability and price. The data in Table 11,I give an indication of the growth in use of lactic acid in the United States in the first years after the approval of "Verv" (1961). All figures are in metric tons, U.S. production is reported as acid 100%, the import figures are taken as such, they probably mainly refer to 85 and 88% material.

Synthetic Lactic Acid Enters the Picture

From the beginning Patterson were convinced that stearoyl lactylates and in particular the calcium salt (csl for short) would become a big product in the baking industry, and, with a lactic acid content of about 32%, production would require large amounts of lactic acid – maybe up to 2,000 t, or more. This amount was the total usage of lactic acid in the United States at that time. But, as said, the existing U.S. producers contacted in the 1960's about the possible supply of so much extra lactic acid of a much improved quality were reluctant to commit the required investments. Then Patterson was approached by several large chemical manufacturers, some of whom had already been playing with the idea of making lactic acid from petrochemical raw material. The laboratory process to make lactic acid from lactonitrile by acid hydrolysis was a well known and simple reaction. A small factory making synthetic lactic acid from lactonitrile had been operated by Musashino Laboratories near Tokyo since the early 1950's (Ch. 10, 367). This, however, was known only to a few insiders; the American petrochemical industry was in general not very interested in this kind of limited venture, they preferred projects with anticipated product markets of at least 5,000 t per year.

After sales started in 1961 it soon became evident that "Verv" indeed was a product with great prospects; when the extra consumption of lactic acid is translated into production of "Verv" over the first three years annual increases of 350 to 400 t can be calculated. By 1964 sales reached or surpassed 1,500 t/a.

By the end of 1962 Patterson could already show real progress in sales. In the spring of 1962 the Monsanto Chemical Company discussed the status of the product with Patterson, the expectations of the company and the possibility of Monsanto becoming the prime supplier of lactic acid. An important factor in Monsanto's decision to take up the production of lactic acid was the valuation of by-product lactonitrile from its large acrylonitrile plant in Texas City. On August 10, 1962, Monsanto made the first announcement of the synthetic lactic acid venture in the Oil, Paint and Drug Reporter. Their claim to become the first producer of synthetic lactic acid was immediately objected to by Musashino. Monsanto's claim to be the first was not repeated in a September, 1962, publication in the same journal . The announcement then made gives a good presentation of the situation at the time the crucial decision to make synthetic lactic acid was taken:

Lactic Acid: A New Bread Additive May Push It into Chemical Big Time

Two Missouri firms are out to change the image of lactic acid. A few trade people think they can manage it, but others long-acquainted with this low-volume, high-cost, modestly growing chemical say "show me". Thus the understandable surprise when Monsanto Chemical Company disclosed a fortnight ago that it is building a 10-million-pound-a-year synthetic lactic acid plant at Texas City, which is scheduled to be in operation by the last quarter of 1963.

Sudden enthusiasm for lactic acid, which has also captured the interest of Miles Chemical Company, is due primarily to a bread additive tradenamed "Verv" and known chemically as calcium stearyl -2 lactylate [correct chemical name: stearoyl-2 lactylate].

Manufactured by Patco Products, Inc., a division of C.J. Patterson Company, Kansas City, operator of the Holsum bakery chain in the Midwest, "Verv" is about 40% lactic acid[10]. It won its GRAS-label [meaning approval] for use in yeast-raised products (one-half of one percent of flour weight) from Food and Drug Administration about a year ago, and in the interim sales efforts have paid off handsomely.

For bread makers, "Verv" offers decided advantages. Its use offsets over and under mixing, over and under fermentation and variations in time and temperature factors. The result is a closer, more uniform textured bread, which, due to Verv's incidental softening effect, has better keeping quality and a more tender crumb.

A fine, white, dry powder, "Verv" is added to the mix separately. It combines with the gluten, producing a stronger-than-average, more extendable structure, which means, in the last analysis, a fatter loaf of bread with no holes.

Most of the major baking concerns in the country, in fact, are already using the additive. Market potential is far from realized, however, according to Patco's enthusiastic vice-president James Kirkpatrick.

Clues to Lactic Requirement

While citing no figures, Mr. Kirkpatrick offers a few clues as to what "Verv" can mean in terms of lactic acid. He says that his firm is probably the largest single buyer of lactic acid; furthermore, that it has been forced to go overseas to fill requirements, having exhausted the available domestic supply.

The problem is that a very high-grade lactic acid is needed in order that the product does not become discolored by the extremely high temperatures of modern baking processes [meant is: the high temperatures of the "Verv" manufacturing process].

To fill this supply-demand gap is obviously Monsanto's intention. The two Missouri firms, however, have made no arrangements, though presumably Patco will get first crack at Monsanto's production, some of it tailor-made for use in "Verv" and other emulsifiers.

Today, "Verv" has no competition, but there are a few products reportedly that can enhance its activity. One is "Seqol 140" (stearyl monoglyceridyl citrate), manufactured by Chas. Pfizer & Co.

...

Besides "Verv" there are several logical reasons why Monsanto is interested in lactic. Most important perhaps, is that it is an addition to a long list of food and beverage chemicals, and, in the industrial area, rounds out a line of plasticizers (for phenolaldehyde resins).

One probable result of the availability of synthetic lactic acid will be lower prices. This is what inclines some trade people to think that lactic might be able to pick up some business from such food-grade acids as adipic, phosphoric, fumaric and citric.

An obvious limiting factor is taste, and for this reason it's doubtful that it will ever be a strong competitor for citric and fumaric markets. Where only acidity is required, such as in the packing of olives and pickles and in canned sauerkraut, on the other hand, it could win increasing favor.

- OPD, 1962

The long "editorial", out of which only the most relevant paragraphs have been quoted, ends with mentioning the three existing U.S. producers of (fermentative) lactic acid, Du Pont, Clinton, American Maize, and one prospective producer: Miles.

Monsanto's Raw Material Position

Monsanto's Texas City plant was built for the U.S. Government during World War II to produce styrene (for synthetic rubber production). In 1946 Monsanto bought the plant (Forrestal, 1977), a year later it was totally destroyed when a French freighter loaded with ammonium nitrate blew up on a nearby quay, setting Monsanto's works on fire. 227 Monsanto employees died in the burning holocaust.

In April 1949 the styrene plant was started up again. A year later construction for a large acetylene based acrylonitrile plant was begun. The process was the catalytic reaction of acetylene with hydrogen cyanide in an aqueous medium (Faith et al., 1957). One of the by-products of this reaction is lactonitrile (Ullman's Encyklopädie, 1953, mentions a lactonitrile yield of only 2%.)

Because of the high price of acetylene and the stark demand for acrylic fibers a new method to produce acrylonitrile, the Sohio process, became in the early 1960's the favorite process with a number of important petrochemical industries of the U.S.A (Spitz, 1988). One of these was Monsanto, which started up a Sohio installation in 1965 (Forrestal, l.c.).

In the Sohio process a mixture of propylene and ammonia are catalytically oxidized with air to form in about 80% yield acrylonitrile. An important by-product is hydrogen cyanide (about 15%) (Kirk-Othmer, 1978). It is this by-product from which by reaction with acetaldehyde lactonitrile is now made (See further Ch. 10, 366 ff).

Fate of U.S. Producers of Fermentative Lactic Acid

In the fall of 1963 Du Pont closed the old lactic acid plant at Gray's Ferry, PA, which was at the time the only manufacture still going on in one of the oldest chemical plants of the United States, the Harrison Brothers plant established more than a century before.

American Maize waited until Monsanto had proven that its new plant worked well. They finished lactic acid production in March 1964. Clinton Corn Syrup Refining Co. also had plans to stop production, but postponed this when they were approached by former customers of Du Pont and American Maize. One of these customers was Du Pont's Cellophane department. Clinton would continue production for another 18 years, but never made a sizable investment in the old lactic acid plant anymore. Instead they concentrated on corn syrups and on high fructose corn syrup of which they were to become the first and leading producer[9].

The First Ten Years of Verv

Thanks to Monsanto's new synthetic lactic acid plant Patterson now was assured of a steady supply of heat stable lactic acid for the production of

"Verv". The quality problems being solved the future of stearoyl lactylates had become very bright indeed.

In 1968 sales of calcium stearoyl lactylate had reached a volume which made it economical to start own production. Research in the pilot plant at Grandview, MO, had provided the optimal conditions for large scale production. The plant, proudly announced in the 1968 Annual Report of the company as "a highly automated and versatile specialty chemical plant", came on stream in 1969. Initial production capacity was 9,000 t/a of lactylate esters. (At maximum output requiring about 3,500 t of lactic acid!)

It can be estimated that around 1968 Monsanto sold about 60% of its synthetic lactic acid production to the Patco Division of C.J. Patterson. In view of the continuing growth of the new application Monsanto decided in 1969 to double the Texas City lactic acid plant to a capacity of 9,000 t/a.

The tremendous success of the new application of lactic acid had world-wide impact:

In 1965 the Distillers Company Limited studied together with Monsanto the prospects for synthetic lactic acid production in Britain (ILRA minutes, December 9, 1965). A year later Distillers informed ILRA members, however, that they had dropped plans to produce synthetic lactic acid.

In the same year Rhône-Poulenc announced plans to build a large, combined oxalic/lactic acid plant (European Chemical News, March 1, 1966). In a discussion with Schiedamsche Melkzuurfabriek Rhône-Poulenc gave a few clues on their new process: propylene was subjected to "nitrous oxidization" to form a mixture of oxalic and lactic acid [precursors]. The capacity of the new plant would be 15,000 t oxalic and 5,000 t lactic acid. Initial production to start in the second half of 1987, would be much below installed capacity as the company did intend to enter the market with "reasonable" prices. (For the fate of this venture see Ch. 12, 446 ff.)

HVA, the new owner of Schiedam, also in 1966, leaked information about enlarging their plant to produce 5,000 to 6,000 t/a. [The figures were exaggerated for preemptive purposes by expressing the envisaged production in lactic acid 50%. Monsanto and Rhône-Poulenc were calculating on the basis of 88%.] In reality production increases in the following years were much more modest, by 1969 a capacity of about 2,000 t/a lactic acid 88% was reached. In the meantime the company had taken up pilot production of calcium stearoyl lactylates (csl). A petition to allow the use of csl in baking had already been filed with the Dutch food authorities one year before.

Musashino, the worlds first producer of synthetic lactic acid, in the mean time was not sitting idle. In 1968 a really commercial plant with a capacity of

3,000 t/a (88%) was constructed at Isohara. Small exports of synthetic lactic acid were made from 1967 onward. In 1969 the export volume surpassed 100 t, a few years later when Patco had become a regular buyer of Japanese synthetic lactic acid, export levels of 500 t/a were reached.

Monsanto also became a regular exporter of lactic acid. Total exports to Europe reached about 350 t in 1969, 500 t in 1970 and 750 t in 1971 (Internal estimates CCA, 1972). One of their important distributors was Boehringer-Ingelheim! In a discussion with Schiedam (April 10, 1969) Monsanto said they would consider again the prospects for a 5,000 t/a European synthetic lactic acid plant. One of the reasons given was that the plant expansion now en route in Texas City would be the last one possible on that site.

Notes

[1] Biographic data on Curtis John Patterson are from:

Dick Fowler, Kansas City Star, 1951;
The Southwest Miller, September 27, 1960;
Herman Steen, Milling & Baking News, February 27, 1973.

[2] Lawrence (Larry) F. Marnett (1920) obtained his BS in chemistry at Rockhurst College, Kansas City, MO, in 1942. After military service and a year with Nutrena Mills (Cargill, Inc.) he was engaged by C.J. Patterson Company in 1947. He held many research and technical functions in the company, acted for many years as one of its vice-presidents and was active in the American Association of Cereal Chemists and a number of other associations in the food and baking industries.

[3] Dough conditioner: an ingredient capable of complexing with or directly reacting with one or more major flour components when added to bread dough and which results in significant improvements in dough machineability and in some factors associated with final bread quality.

Bread softener: a surface active agent which when added to bread dough produces baked products which are softer than products not containing this agent. The softening aspect is generally believed to be the result of starch-complexing which reduces the swelling of starch during baking and impedes the crystallization of starch during ageing.

Yeast food: a combination of ingredients usually consisting of a readily soluble nitrogen source for yeast nutrition, minerals to compensate for water deficiencies and one or more oxidizing agents (like potassium bromate) for dough improvement.

Dough emulsifier: a product with emulsifying properties which when added to bread dough has a beneficial influence on the softness and the prevention of staleness of bread (see bread softener).

4 Manufacturing method (Holten, 1971):
Lactic acid is reacted with mixed fatty acids (stearic and palmitic acid) at 150–200° C in vacuum to remove the water of reaction, followed by neutralization to form the calcium salt.

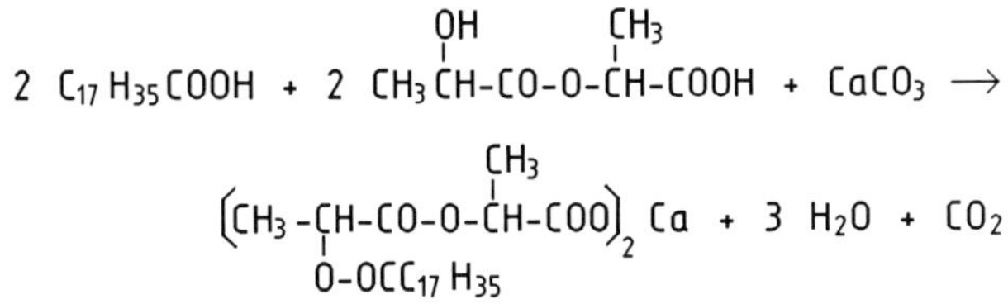

The product obtained in this way is chemically not uniform, but consists of a mixture of stearoyl and palmitoyl esters of lactic acid, lactoyllactic acids and of higher polylactic acids containing up to six lactic acid units.

5 On April 12, 1961, the Commissioner of Food and Drugs issued the following food additive regulation:
§ 121.1047 *Calcium stearyl-2-lactylate* [the correct scientific name is calcium stearoyl-2-lactylate].

The food additive calcium stearyl-2-lactylate may be safely used in or on food in accordance with the following prescribed conditions:

a. The additive is a mixture of calcium salts of stearyl lactylates and minor proportions of other salts of related acids, manufactured by reaction of stearic acid, lactic acid, and calcium carbonate.
b. The additive meets the following specifications:

Acid number	82–103%
Calcium content	4.2%–4.4%
Lactic acid content	32%–35%
Saponification number	140–188

c. It is used or intended for use as follows:

1. As a dough conditioner in yeast-leavened bakery products in an amount not to exceed 0.5 part for each 100 parts by weight of flour used.
Etc., etc.

6 A similar type of reaction takes place during the baking process in the crust of the bread, giving it an attractive brown color.

7 Until about 1960 starch syrups were made by hydrolyzing starch at temperatures above 120° C with hydrochloric acid. It had since long been known that hydrolysis could also be done with so-called amylolytic enzymes, but good preparation and purification methods for one of the enzymes required had been lacking.

The biggest advantage of enzymatic hydrolysis was in the manufacture of dextrose, with acid hydrolysis the maximum yield of glucose (= dextrose) attainable was about 75%, with double enzymatic conversion 90 to 95% became possible. Later the process was further improved to a yield of 96 to 98% by applying a third enzyme. Another factor hampering the development of enzymatic conversion constituted a 1938 patent of Staley for a process to manufacture a high maltose starch, called Sweetose (Forrestal, 1982). The Staley patent expired about 1955, which opened the way for other starch manufacturers to freely use enzymatic conversion processes.

[8] The Glyco production chemist, T.T. Lithgow, Jr., who was responsible for "Verv" toll production until C.J. Patterson Company established its own in 1968 (compare note 10), even tried to convince Clinton's parent company, Standard Brands, of the feasibility of a heat stable lactic acid plant. He was turned down: the project was too big [sic!] and too complicated. Everything had to be rebuilt, and the space for expansion had already been allocated to enzyme production (Marnett, interview, 1986).

[9] Continued interest in enzyme technology brought Clinton in 1965 in contact with Japanese laboratories which had developed an enzymatic process for converting glucose to fructose (Forrestal, 1982). Clinton bought the process and succeeded within a short time in producing the new enzyme, isomerase. The first shipment of high fructose corn syrup (hfcs) was made in February 1967. Within ten years hfcs would completely upset the U.S. sugar market, and, in turn, the international sugar market as well:

year	amount of sugar replaced (in 1,000 t)
1973	160
1978	1,000
1983	2,900
1988	4,200 (estimate)

Clarke (Clarke, 1988) estimates the total market loss of U.S. cane sugar refiners between 1977 and 1987 at 5,000,000 t. Imports decreased further in 1989, making the loss for world market sugar producers even greater (The difference with the 4,200,000 t taken away by hfcs is due to increased U.S. sugar production, which is favored by a high, government regulated domestic sugar price; see also Intermezzo 6, 290.)

[10] In the first years the C.J. Patterson Company did not produce "Verv" itself. Instead it was produced by the Glyco Company at Wiliamsport, PA, on a toll basis. The Patco division of C.J. Patterson, established to promote and sell food additives and chemical specialties, built in those intervening years a pilot plant in Grandview, Missouri, to determine the optimal conditions for the production plant.

References

Baker, J.A., **1969**. *The C.J. Patterson Company Corporate Story*; Internal Report of C.J. Patterson Company.

Birnbaum, H., **1955**. *Bakers Digest*, **29**, [10], 101.

Clarke, M.A., **1988**. *The United States Sugar Industry*; paper presented at A.C.S. meeting, September 25/30, 1988.

Edelmann, E.C., and Cathcart, W.H., **1949**. *Cereal Chem.*; **26**, 345.

Faith, W.L., Keyes, B.K., and Clark, R.L., **1927**. *Industrial Chemicals*; Chapman & Hall: London, 47.

Forrestal, D.J., **1977**. *Faith, Hope and $5,000. The Story of Monsanto*; Simon and Schuster: New York.

Forrestal D.J., **1982**. *The Kernel and the Bean*; Simon and Schuster: New York.

Holten, C.H., **1971**. *Lactic Acid*; Verlag Chemie; Weinheim/Bergstrasse (F.R.G.)

Kirk-Othmer Encyclopedia of Chemical Technology, **1978**. 3rd ed.; John Wiley & Sons: New York, Vol. 1, 415–417.

MacFayden, J.I., **1985**. *American Heritage*, October 1985.

Spitz, P.H., **1988**. *Petrochemicals. The Rise of an Industry*; John Wiley & Sons: New York.

CHAPTER 12

FERMENTATION LACTIC ACID CONSOLIDATES ITS POSITION

In the second half of the 1960's the foundations were laid for dramatic changes in the lactic acid scene. Most important, of course, was the success of Patterson with stearoyl lactylates in the United States (Ch. 11). In not much more than one decade this new application of lactic acid would almost double its use. It also would provide a strong impetus for other producers to improve on quality standards.

The envisaged opening of the European market for stearoyl lactylates, as well as the penetration of this market by producers of synthetic lactic acid strongly supported concentration efforts in Europe. The continuing efforts of one company, N.V. Chemie Combinatie Amsterdam C.C.A., later renamed to CCA biochem b.v., to establish a dominating position as supplier of lactic acid will be the major theme of this Chapter. An important problem confronting CCA in its struggle to achieve a leading position – the raw material position of the company within the EEC – will be another topic of much interest.

The success of calcium stearoyl lactylates in America also was, at least in part, a strong motive for the French chemical giant Rhône-Poulenc to try to develop a new route for the synthesis of lactic acid – by nitrous oxidation of propylene. This venture would also exert a profound influence on the developments in lactic acid production in the 1970's.

Between 1972 and 1982 three European lactic acid producers stopped the fermentative production of "normal" lactic acid. (Køge in 1972, Boehringer in 1973 and Rhône-Poulenc, which closed down the old fermentation plant in Saint Fons in 1982.) The French producer, Rhône-Poulenc, did not pull out completely, however, but started up a new plant in the same year in Melle for the production of D(-)-lactic acid.

Verenigde HVA-Maatschappijen N.V.

In a certain sense the HVA company, of which the full long name appears in the subheading above, bears much of the responsibility for initiatives which in the long run brought about the concentration of fermentative lactic acid production in a small number of companies.

The HVA company, founded in 1878 with the purpose of trading in colonial products, had in the course of half a century of activities in the Netherlands East Indies (now the independent Republic of Indonesia) become an important producer of cane sugar and a number of other "colonial" agricultural products (such as sisal, rubber and tapioca. HVA, 1975).

When World War II was over HVA made a vigorous attempt to repair the damage the war had inflicted upon their agricultural estates and factories in the East Indies. At the same time, feeling insecure about the prospects of exploiting large Dutch-owned estates in independent Indonesia, the company started to explore the possibilities for tropical agricultural enterprises in other countries. Attention was focussed on Brazil and Ethiopia, in both countries the prime interest was in cane sugar production.

In 1951 a cane sugar plant was set up in Ethiopia; twenty years later the company operated three cane sugar mills in that African country with a total yearly output of about 125,000 t (HVA, 1975).

Exploratory activities in Brazil were only taken up again in 1958 when the company was forced to leave Indonesia as a consequence of the dispute with the Dutch government about the future status of West New Guinea (Now Irian Jaya).

Cornelis van der Made (1922–1975), the top ranking official of HVA in Indonesia, was, immediately upon his return to Amsterdam in 1958, charged

Fig. 12,1. C. van der Made. Reproduced from HVA Annual Report 1975. Courtesy: Mrs. C.J. van der Made-Six.

with new exploration studies in Brazil. The company felt an urgent need to find new opportunities for the considerable number of experts repatriated from Indonesia, as well as for its financial resources.

The first business acquired in Brazil was in the sisal trade. The main interest of HVA was still in cane sugar, however, the first and most profitable product in the company's history. A group of four experts of the company made a thorough investigation of the Brazilian sugar industry in 1960 and 1961. One of the three sugar growing regions they visited was the Campos region, about 250 km east of Rio de Janeiro. Here, in the fall of 1961 the HVA experts by some happy [?] coincidence got acquainted with one of the two owners of a small lactic acid plant, Chrystovam Lysandro de Albernaz.

Indústria Química de Sínteses e Fermentaçôes

The company, called Sínteses or IQSF for short, which owned the lactic acid plant, had been established a few years earlier, in 1955, by Henrique Teixeira Sence, his son and another gentleman. Sence, an engineer by education, and his son, a chemical engineer, were in the possession of a butanol-acetone fermentation plant and intended to use their experience also in another, related field of fermentation, lactic acid. It was their purpose to produce ethyl lactate for which solvent a good market was thought to exist in Brazil. The new fermentation plant was installed in an old distillery, Distilaria Central de Campos, as much of the old equipment they thought usable remained in use, even though some of it dated back to the nineteenth century.

The plant did not work well, lactic acid fermentation appeared to be more difficult than expected and the owners of the plant did not daily supervise how it worked, their main business – butanol production – being elsewhere. In 1958 the company was sold to Messrs. Lysandro and Rubens Arêas Venâncio, both Campos residents and hence in a better position to look after the operation. The new owners hired a chemist and tried to get things to work, but again this proved difficult. Fermentation often took the wrong road, giving a butyric smell to the (technical) product and causing problems in the esterification process. Maximum production was of the order of 200 to 300 t/a of ethyl lactate and very small amounts of technical lactic acid.

One day in the fall of 1961 Lysandro and Rubens decided to stop the business and sell everything, the equipment of the plant as scrap, and buildings and terrain separately for real estate value (Personal communication, Lysandro, 1984).

Later on that very same day Reinder Meindertsma (1916), the chemical and sugar expert of the HVA survey team, happened to have his first contact with Lysandro, who informed him that the lactic acid plant was for sale. He did not mention that the owners had decided to close the plant in the near future!

Fig. 12,2. The IQSF plant at Campos (1966). Courtesy: Indústria Química de Sínteses e Fermentaçôes.

Meindertsma immediately recognized that a lactic acid plant might make a good start in the field of the chemistry of agricultural products [chemurgy]. Van der Made and he had been discussing this kind of diversification rather systematically since their return from Indonesia, and had established contacts with outside experts on the subject. Meindertsma was well aware of the studies made by the ERRL (Ch. 10, 353 ff). Moreover, as former technical technical supervisor of a distillery in Java, owned by the HVA company, he had gained some practical experience in fermentation processes.

Negotiations were easy, which in hindsight is not surprising. Van der Made was prudent enough, however, to negotiate a partial take over; the two Campos gentlemen were to keep 40% of the shares to assure HVA of their continued interest in the lactic acid enterprise. The contract was signed in June 1962, almost immediately the daily management of Sínteses was thereupon taken over by three HVA employees.

Within months it became clear that running a (very small) chemical plant was something that had to be learned from scratch. Many of the problems that had plagued lactic acid manufacturers since the years of Charles Avery in Littleton, now came down upon the HVA organization. It must be said that the problems were faced head-on, and with much skill:

- All new technical personnel were first trained at the Delft Technical University (at the renowned microbiology department founded by Beyerinck, today called the Kluyver Laboratory of Microbiology) before going to Brazil;
- They also had to take intensive Portuguese language training courses;
- A former Shell employee, Dr. Smithuysen, who had been involved in installing (ethyl lactate) process equipment in the SMF plant at Schiedam during World War II, was hired as consultant and, later, sent to Campos to advise on installations and procedures.

In the summer of 1963 the fermentation process was changed drastically: Cane sugar replaced molasses as substrate; malt sprouts which supplied nutrients as well as lactic acid bacteria was replaced by bakers yeast and a *Lactobacillus* culture.

Despite extensive cleaning and renovation operations the fermentation still did not work well, however, and production of ethyl lactate was much below capacity, and the product of poor quality. Fortunately expert help from SMF appeared on the horizon.

HVA Buys Schiedamsche Melkzuurfabriek

About 1962 Hendricus Cristianus Jansen (1897–1972), the owner (major share holder) of SMF, approached his bank to act as an intermediary in the sale of his company. He had reached the age of 64, his health was ailing, so he considered it wise to sell the company at the moment returns and balance sheet were showing the best results ever (thanks to the Boehringer contract; see Ch. 10, 392).

In the spring of 1962 the bank first contacted the sugar producer Centrale Suiker Maatschappij, CSM, which, being not interested themselves, referred the bank to HVA. [CSM knew about the interest of HVA in sucrochemistry.] Negotiations with Jansen took quite some time, the final contract was only signed in December 1963. Already before the signing SMF supplied some know-how to HVA experts to help them in renovating the Campos lactic acid plant.

After the take over the complete know-how of Schiedam became available, which was of much help in selecting the right type of corrosion resistant material for new equipment. It was then also decided to produce edible lactic acid according to processes developed by SMF. For this purpose amongst other things a battery of ion exchange columns was installed to decolorize and demineralize crude acid.

By 1965 the Campos plant was almost completely rebuilt, the fermentation department now worked quite well, but the manufacture of ethyl lactate was still a major problem due to some design faults as well as to lack of training of supervisory personnel and operators, many of whom were illiterates. The production manager of Schiedam, Meyer, was then sent to Brazil to bring the plant up to "Dutch" standards. In 1966, four years after the take over, production finally was brought to an acceptable level – about 900 t/a of edible lactic acid 100%.

Fig. 12,3. IQSF celebrates the 10,000th drum (1968). Courtesy: Indústria Química de Sínteses e Fermentações

HVA Establishes a Research Laboratory

The expensive lessons of the Brazilian venture (in four years time the amount of additional investments had risen to twice the 1962 asset value of Sínteses, with no income at all in this period due to lack of production) led van der Made to the conclusion that progress in chemical diversification required the backing of a well-staffed and well-equipped research laboratory. The participation of Meindertsma in the technical meetings of ILRA provided much support for his conclusion (Ch. 10, 389 ff).

The laboratory was opened in 1966, it was headed by Dr. John Adriaan van Velthuijsen (1934). The program of course focussed on lactic acid and derivatives, but from its beginning the HVA Research Department also had the task to follow agrochemistry in a wider sense. One of the first successes of van Velthuijsen and his team was the conversion of the fermentation process to the production of enantiomerically pure L(+)-lactic acid.

Why Was HVA Interested in the Manufacture of L(+)-Lactic Acid?

HEALTH ASPECTS OF LACTIC ACID

When World War II was over the ideals of the democratic western nations for a free and prosperous world had been translated into a number of transnational organizations which were given the task of implementing these ideals. Two of these organizations, operating under the umbrella of the United Nations, were specifically created to fight hunger and disease:

1. The Food and Agricultural Organization of the United Nations with the objective of affording "freedom from want" (F.D. Roosevelt to the U.S. Congress on January 6, 1941) on a world scale by "raising levels of nutrition and standards of living ... and securing improvements in the efficiency of the production and distribution of all food and agricultural products..."
2. The World Health Organization with the objective of "attainment by all peoples of the highest possible level of health."

The two organizations overlapped each other to some extent in the area of providing healthy food to the ever growing world population, and for this reason joint expert committees were set up to deal with problems relating to health and nutrition (1958). One of these groups is the Joint FAO/WHO Expert Committee on Food Additives. Members of this expert committee are selected on the basis of "their ability and technical experience, with due regard to adequate geographical distribution" (WHO, 1972). The Joint Expert Committee on Food Additives has since gained the status of the world's most expert and unbiased group of experts on health aspects of food additives.

In 1965 the Joint FAO/WHO Expert Committee on Food Additives discussed at its annual meeting in Rome among other subjects the use of lactic acid in baby foods. One of the members of the Expert Committee informed ILRA's Chairman, Arnold, about the results of the discussion. Arnold brought this to the attention of the Secretary on January 3, 1966:

> In confidence I [the Committee member] can tell you that the Committee took a very strong line on the hazards of the "unnatural" isomer, namely D(-)-lactic acid . It considers that this isomer should under no circumstance be given under the age of 6 months because of the stage of development of the human liver is inadequate to cope with it. There was also a strong suggestion that a certain proportion of adults, by reason of enzymatic defects or other causes, cannot tolerate the D(-) isomer and a stringent limit was set for this isomer in foods.

Arnold recommends that the Secretary of ILRA (Glimsholt) tries to find out exactly what the Committee said, and then

> arranges for the preparation of an authoritative case to dispose of this error. ... it may be worth pointing out that the so-called unnatural isomer is, in fact, widely distributed in nature, that it is probable that lactic acid is racemised in the human gut, and that a careful distinction must therefore be made between administration of racemic acid by mouth and by injection.
>
> \- *Arnold, 1966*

Why Did the D(-) Isomer Cause So Much Concern?

Since the 1920's acidified milk formulas had been advised by pediatricians for feeding infants when breast feeding was not possible or insufficient. The idea to acidify (and modify) whole cow's milk was based on the assumption that an organic acid and in particular lactic acid would improve digestibility, adsorption, and assimilation. It was also believed that some prevention of bacterial infection of the product was achieved as well as suppression of pathogenics in the alimentary canal (Droese and Stolley, 1963).

Droese and Stolley cast doubts on the advantages of acidified milk formulas because of the possible occurrence of acidosis – an abnormally high level of acidity of the blood and body tissues. Lowering of the stool pH of newborn infants by about 0.6 pH points had indeed been detected by Jacobs and Christian (Jacobs and Christian, 1957). These investigators also found, however, that babies on average grew faster on an acifified formula than on a formula composed of cow's milk and maltose-dextrin to which no lactic acid had been added.

As prematurely born infants normally are on the verge of acidosis further investigations were carried out on the effect of acidified milk formulas on prematures. Goldman et al. (Goldman et al., 1961) found that acidification indeed had a negative influence on the growth of small premature infants. [The formula applied, Pelargon of the Nestlé Co., contained about 0.4% of racemic lactic acid.]

Goldman et al. in the discussion of their findings raised the question of the difference in metabolism of the two enantiomers of lactic acid:

> The [L(+) isomer] occurs normally in human metabolism, and the [D(-) isomer] is a foreign substance that is metabolized differently. Administered [L(+)-lactate] is largely converted to glycogen, or at least promotes the formation of glycogen, while administered [D(-)-lactate] is oxidized or excreted in the urine. It is, therefore, probable that the major portion of the increased urinary lactic acid is the [D(-)isomer] and that the excretion of this isomer is largely responsible for acidosis.
>
> - *Goldman et al., 1961*

These considerations of Goldman et al. were based on earlier studies of the metabolism of the two enantiomers of lactic acid in animals (see FDA, 1975)[1].

Reactions of ILRA Members

The matter was put on the agenda of the meeting of March 9 and 10, 1967 of ILRA's Technical Committee. Commercial viewpoints clearly governed the discussion which was "long and diffuse", as Glimsholt wrote in the minutes.

Arnold reported on the actual isomeric contents of the commercial products: Boehringer 94% L(+), Schiedam 70%, Bowmans 60% and Køge 55%. [Bowmans had analyzed the competing products because of a certain interest in preparing crystalline L(+)-lactic acid.] Arnold had learned in the meantime that evidence relating to the toxicity of D(-)-lactic acid had come from the Toxicological Department of the University of Mainz. Boehringer was requested to get more information about those studies.

Some members expressed the view that a ban could be put on D(-) which "could also mean an excellent counteract from our side on a possibly coming synthetical lactic acid, but we would then have to make the pure L(+) isomer" (minutes of the meeting).

Then members' reactions to market inquiries were discussed. Only Boehringer replied to questions about L(+) content in the affirmative. Stapel of Schiedam said "that his company would reply that insistence on L(+) is nonsense". Van Velthuijsen, the new head of the just established HVA Research Department, however, expressed in his account of the meeting the view that ILRA members should now seek methods to produce lactic acid with the lowest possible D(-) content.

HVA Research Develops a Method for Making Pure L(+)-Lactic Acid

A research project to change the fermentation process to obtain the lowest possible content of D(-) was one of the first major projects of the new HVA Research Laboratory at Schiedam. Steps to investigate the possibilities of converting to L(+) production had already been taken immediately upon receiving a copy of Arnold's letter of January 3, 1966; on the 10th of the same month a visit was paid to professor Wikén of the Kluyver Laboratory of Microbiology of the Technical University of Delft to discuss the matter.

When the new HVA Laboratory was opened in the summer of 1966 the preparatory work at the Kluyver Laboratory was continued in Schiedam. Results were finally achieved in the summer of 1967: a change in the nutrient composition induced the factory culture of *Lactobacilli* [?] to form lactic acid with an L(+) content of over 90%. A long adaptation period appeared to be required before the bacterium (or combination of bacteria) achieved the productivity of the former strain. As of December 1967 Schiedamsche Melkzuur Fabriek manufactured lactic acid of the same enantiomeric quality (over 90% L(+) content) as Boehringer had been selling since the early 1960's.

The success came too late to recapture the largest client, the Nestlé company. This company had decided in the summer of 1967 to switch over from "racemic" to L(+)-lactic acid for their acidified baby milk formula. In most countries the switch was made to a formula acidified by means of an in-house fermentation process with *Streptococcus lactis*; in Holland the company continued to buy lactic acid from SMF until 1969, but now specifying a minimum L(+) content of 90%. During 1967 SMF sold Boehringer product to Nestlé Holland to keep at least this manufacturer of acidified baby milk. From early 1968 onwards Schiedam product was supplied, until 1970 when the Dutch branch of Nestlé also switched to in-house fermentation (Table 12,I).

Other important manufacturers of acidified baby milk formulas stayed on as customers, their total usage of L(+)-lactic acid was lower than that of Nestlé alone, however.

Table 12.I. SMF. Total sales of lactic acid (in t 50%) to Nestlé.

Year	t	Year	t
1964	285	1968	194
1965	335	1969	169
1966	348	1970	30
1967	171		

Source: SMF/CCA Archives.

The Fate of Natural L(+)-Lactic Acid in Baby Food

An exhaustive literature review and a number of evaluations of the health aspects of the use of lactic acid were published in subsequent years by the U.S. Food and Drug Administration (FDA, 1975, and FDA, 1978).

The general conclusion of these studies was that the problem was not of importance for the United States:

> Lactic acid has been used to acidify infant formulas. Initially, formulas made from fresh cow milk were acidified as a means of reducing curd tension but with improvements in milk processing (e.g. homogenization, evaporation), low curd tension has been achieved without acidification. While formulas acidified with lactic acid are still marketed in Europe and elsewhere, commercially available infant formulas in the United States, ..., are not acidified.
>
> *- FDA, 1978*

The best investigation discussed in the 1978 FDA report is that of Ballabriga et al. (Ballabriga et al., 1970):

Prematures fed on [high protein] milks acidified with racemic lactic acid or D(-)-lactic acid during the first two months of life develop "biochemical" metabolic acidosis, frequently accompanied by the following clinical signs: bad overall condition, loss of weight, dehydration, regurgitation and vomiting.

...

The conclusion to be drawn from this work is that when acidified milks are used in premature feeding the formulas must contain exclusively L(+)-lactic acid ....

- *Ballabriga et al., 1970*

The Joint FAO/WHO Expert Committee on Food Additives reevaluated the use of lactic acid in the diet of very young infants (15th Report, 1972), but did not find additional evidence which would "indicate a need to change the opinion expressed in the 9th Report...".

At the time CCA did not gain anything from their successful redesign of the fermentation process[2]. Doubt cast by pediatric investigations on the efficacy of acidifying such formulas at all returned this application of lactic acid also in Europe into an application of marginal interest only (See for instance Droese and Stolley, 1963).

SMF Penetrates the United Kingdom

Even before the health issue was raised by the WHO/FAO experts Stapel used the L(+) content of Schiedam's lactic acid (about 70%) to start sales in Great Britain. L(+)-lactic acid did not fall under the definition of lactic acid in the British duty regulation, and could therefore be imported tax free, instead of being charged with a 23% duty. Sales of edible lactic began in 1964.

From the date these exports (Table 12,II) started it is clear that the health issue had nothing to do with this trade. Stapel [or the HVA management?] had one very good reason to try to get access to new markets:

In 1963 Boehringer had started up fermentation again, which meant a market loss of about 600 t edible lactic acid for SMF. The company therefore was in dire need of additional markets for its expanded plant. The British market was a good target because Bowmans position was believed to have become weak due to the loss of the stearoyl lactylates market (Patterson) to Monsanto. Moreover, Schiedam's edible was (much) better in quality than Bowmans'.

Table 12.II. British imports of Dutch lactic acid.

1963	1	1967	101
1964	43	1968	147
1965	72	1969	202
1966	86	1970	174

Source: British import statistics.

Bowmans, of course, protested with the British custom officials, and in 1967 succeeded in bringing L(+)-lactic under the same tax as other qualities of lactic acid. The fact that SMF now had increased the L(+) level to a figure slightly above 90% was of no avail.

SMF did not want to give up the (small) share it had gained in the British market and decided to lower the ex-works sales price by an amount even surpassing the duty of 23%, from Fl 1.54/kg to Fl 1.15/kg.

In the fall of 1968 Bowmans reacted against the negative influence the Dutch exports had on their market volume and sales price by complaining about dumping of Dutch lactic on the U.K. market. CCA, the succeeding company of SMF, replied to an inquiry by the British Embassy in the Hague that prices for 50 t contracts were Fl 1.42/kg, and that for larger contracts substantially lower prices would be quoted. Customers requiring so much lactic acid per year did not exist in Holland, however. Hence no invoices or quotations with such low prices could be shown.

On December 23, 1970, the Financial Times reported how the issue finally was settled:

> The Department of Trade and Industry has decided not to impose an anti-dumping duty on lactic acid imported from the Netherlands. The Dutch company concerned in the DTI investigation has agreed not to export edible and technical lactic acid to the U.K. at dumped prices.

CCA prices were then raised by 10%, and the duty stayed at 23%. This lasted until April 1973 when Great Britain entered the European Common Market and as a first step the tariff was lowered to 18.4%. The episode marks the beginning of intensive export activities by SMF, which were later continued and expanded by CCA.

Lactic Acid Companies Want to Diversify

In the preceding discussion casual attention was paid to the change in ownership of Schiedamsche Melkzuurfabriek in 1969. This happened at almost the same time as the change in ownership of Bowmans.

The sale of Bowmans had much to do with the feeling that the company had become too small in the growing international market of food additives and ingredients. Arnold had tried to diversify by expanding the range of products. In 1964 a synthetic malic acid plant had been built, after a lot of teething troubles it came fully on stream in 1966. Within a few years this venture became a big commercial success (Chemical Age, June 7, 1971).

This success came too late, or was considered too small. Arnold as well as the Board of Bowmans believed that the company would be more viable when part of a larger group. Different views about the manner in which this could be achieved led to the departure of Arnold. A few months later (January 1969) Bowmans was sold to the Croda Group. It then became incorporated in Croda's Food Products Division under the name of Croda Bowmans Chemicals Limited. Other major products of the Food Products' Division were gelatin and food emulsifiers (including stearoyl lactylates). From then on lactic acid was just one of the many food ingredients sold by Croda Bowmans. No further efforts were made to enlarge the market for lactic acid.

The owner of Schiedamsche Melkzuurfabriek, HVA, also held the view that diversification was needed. In one respect the considerations were similar to those of Bowmans: making and selling one (or two) small volume products in the huge chemical market did not provide a sufficient basis for a prosperous future. The other argument was the still uncertain future of the HVA company: besides a flourishing but entirely local Ethiopian sugar enterprise, the company had only some minor activities in the trade of tropical products and as a consultant in the field of tropical agro-industry. Diversification into sucrochemistry or, more generally agrochemistry, was therefore viewed as a possible means of broadening the basis of the company. In 1968 contact with CSM was established about a joint venture with the purpose of large scale diversification in the field of sucrochemistry.

Centrale Suiker Maatschappij Becomes Interested in Lactic Acid

After the failure to produce lactic acid by alkaline degradation of molasses, the CSM Research Department at Gorinchem had worked out a process for making lactic acid by fermentation, but now the board of the company turned down the project. It did not want to become the second lactic acid producer in the small Dutch market. Instead it supported the follow-up of another research project by approving the erection of a sorbitol-mannitol pilot plant on the site of the old sugar mill at Gorinchem where the Research Department was also located (1966).

From the lactic acid studies a regular contact had evolved with HVA, both companies being interested in agrochemistry, or more specifically in sucrochemistry. HVA, before setting up a research department of its own, had contacted CSM in 1965 about the possibility of doing research together, or, preferably, of joining forces in sucrochemical production as well as in research. Differences about the evaluation of assets and good will prevented the closing of a cooperation contract at the time.

A few years later merger talks were taken up again. CSM was more interested then before because of the difficult position of their sorbitol venture (Sorbitol was becoming a commodity product and, hence, large investments were needed to reach an economical scale of production.) and the continuing high costs of the Research Department. HVA also had an additional reason for being interested: the company became apprehensive of the logistic problems which would arise when sales of SMF products would increase as planned. In particular the handling of growing amounts of bagged calcium lactate worried the Amsterdam management.

Calcium Lactate in the Treatment of Calcium Deficiency

Increasing demands for calcium lactate came quite unexpectedly. Until about 1963 sales of this product had been fluctuating around 100 to 150 t/a, a modest increase only over the 50 to 75 t/a sold in the beginning of the 1940's. Hence prospects of calcium lactate had not been an important item when taking over SMF. This situation changed dramatically in the next years when average yearly increasees from 40–50% were recorded. A new formulation of a calcium supplement for the treatment of osteoporosis and calcium deficiency in general was responsible for most of this growth.

In 1967 sales reached a level of 555 t. The shipment of this volume of a solid product which naturally had to be packed in bags already posed considerable problems. When growth would continue, as market studies indicated, obstruction of one of the (two) gates of the factory would become a regularly occurring phenomenon.

The Schiedam plant was located in a century old crowded district of the city of Schiedam locked up between two former distilleries (Figure 12,4). One factory gate was used to receive raw materials, the other for the shipment of final products. Factory personnel could hardly get in or out of the plant when a truck was loading at the main entrance.

Fig. 12,4. Aerial photograph of part of Schiedam with SMF plant. The contours of the SMF plant are marked in white. Reproduced with permission of A. Sijthoff, The Hague.

Again a crucial decision had to be taken: either to build an entirely new plant or to move (part of the) production to an existing building, for instance to the old sugar mill of CSM at Gorinchem (the subject of the 1965 negotiations).

Chemie Combinatie Amsterdam CCA is Founded

In the fall of 1968 HVA and CSM reached agreement on the merger of their sucrochemical interests. HVA brought to the venture the profitable lactic acid plant of SMF, the lactic acid plant of Sínteses, which was near the break-even point, and a research organization which cost almost as much as the total profit SMF was making. CSM's part in the venture consisted of the sorbitol/mannitol pilot production plant at Gorinchem and an also rather costly research department. The new 50/50 joint venture (combination of chemical interest) of CSM and HVA, both head-quartered in Amsterdam, was to be called N.V. Chemie Combinatie Amsterdam CCA.

CCA became operative on January 1, 1969. The management of the new company was appointed in the same year: Johan Michael Jacob Prakken (1922), until then heading the HVA operations in Ethiopia, was made responsible for finance and commercial affairs; Dr. Harm Benninga (1924), former vice director of research of a leading starch derivatives company in the Netherlands (Scholten Honig), was charged with technical affairs, research and diversification.

The new management knew nothing about lactic acid production nor sales. In the beginning they thought that prospects for lactic acid were modest, certainly not much better than those of sorbitol and mannitol. When they assumed their functions in the course of 1969 the interim management of CCA had already embarked on a large investment plan by which sorbitol/mannitol production was to be brought on what was believed to be a commercially viable level (2,000 t sorbitol and 300 t mannitol per year). At the same time new fermentation capacity was built in Gorinchem to take over calcium lactate production from Schiedam.

Within a few years Prakken and Benninga learned, however, that the scale of production of sorbitol and mannitol was still too small to be viable. Moreover, due to EEC agricultural policies, the cost of the raw material sugar was much higher than that of the starch hydrolysates which the two main European competitors used for sorbitol production. Last but not least, sorbitol made from sugar was for quality reasons not accepted as well in the market as starch based sorbitol.

More attention was then given to lactic acid and lactic derivatives. Production capacity in Gorinchem was increased to facilitate the eventual closing down of the Schiedam plant. Export activities were extended in breadth and depth. Thanks to this policy sales of lactic acid salts and esters in the first three years of CCA's existence continued to grow by 15–20% per year.

The financial performance of the company improved in line with the increase in sales: the first two years showed large deficits, much improvement was seen in 1972 and 1973. The year 1974 was the first year in which the profit reached an acceptable level.

Two Competitors Stop Manufacturing Lactic Acid

The increasing sales activities of CCA made lactic acid production for some other European manufacturers an even more marginal affair than before. The first competitor to stop production was Kemisk Vaerk Køge.

In September 1972 the ILRA members were invited to Copenhagen to learn about KVK's decision to stop production as well as sales of lactic acid and derivatives (van Velthuijsen, internal report, 1973). The lactic acid plant with a maximum output of 600 t/a was too small and had been operated in the red for most of the past 25 years. Moreover, corrosion problems necessitated the replacement of the evaporator and some other equipment, investments the company was not prepared to make in view of the market conditions[3].

The main market for KVK's lactic acid products, Scandinavia, was taken over by CCA and Croda Bowmans.

The next company to stop production was the oldest European producer C.H. Boehringer Sohn of Ingelheim. A few months after the Copenhagen meeting Boehringer approached CCA to discuss a contract for the supply of all their lactic acid requirements by CCA. The opportunity was immediately embraced by CCA as a means to enlarge the plant at Gorinchem to a more economical size and to pave the way for take-over of the Schiedam production in the near future. (The need to rationalize production and to concentrate everything in one plant was also considered a necessity because of the fact that Rhône-Poulenc was building a combined oxalic/lactic acid plant in Chalampé, Alsace, with a lactic acid capacity of 5,000 t/a. The threat of so much additional [synthetic] lactic acid production had also been one of the reasons of Boehringer for stopping production.)

In the course of 1973 CCA became the exclusive supplier of Boehringer for edible and pharmaceutical lactic acid, calcium lactate and ethyl lactate. In 1974 total sales of CCA surpassed a volume of 5,000 t of lactic acid plus derivatives, almost 70% more than the 1972 volume! CCA had now become the largest producer by far of fermentation lactic acid in the world.

Profitability remained low, however, partly because production of lactic acid was divided over two plants, partly also because the sorbitol/mannitol operation was gradually becoming a drain on the company resources. The sole managing director Benninga (Prakken left the company in January 1974 to become director of HVA.) would soon be facing two large additional, and unexpected problems:

1. In 1974 world market sugar prices rose to unprecedented levels, which

would entail a complete change in EEC agricultural policy regarding the use of world market sugar for chemical derivatization. When world market sugar prices came down again in 1976 industrial users of sugar had to pay the full or almost full EEC intervention price, which was almost as disastrous for CCA as the German sugar policy had been for Boehringer's lactic acid produciton some twenty years earlier.

2. On March 26, 1974, an explosion occurred during start-up trials in the new oxalic/lactic acid plant of Rhône-Poulenc at Chalampé, killing an electrician and wounding five others (Le Monde, April 29, 1977). For more details see this Chapter, 448). The management of Rhône-Poulenc informed CCA and Boehringer within days about the accident and expressed their doubt about the future of the lactic acid plant. (The oxalic acid plant, which had already been in operation for some months, was not affected by the explosion). Within a month CCA decided to grasp the opportunity and start construction of large additional facilities in the Gorinchem plant with the dual intention of absorbing the expected additional growth in the market as well as closing down the old SMF plant at Schiedam. This decision, taken in a hurry for preemptive reasons, was not sufficiently supported by a well-considered processing plan, however. When production and engineering staff tried to start-up the new plant in 1975 fermentation productivity fell back to 1972 levels instead of increasing! Total disaster was only prevented thanks to maximum assistance from the old SMF plant at Schiedam. It took about one and a half years to solve the problems in the fermentation department. Valuable outside help and assistance was recevied from Boehringer experts. Production volume of the Gorinchem plant reached the planned levels only in 1978, whereupon the Schiedam plant was closed down.

The European Economic Community

Before turning our attention to the almost fatal blow the EEC Commission delivered to the (lactic acid) fermentation industry in the 1970's, a short summary of the history of the EEC agricultural policy is called for.

The basic treaty in which the establishment of a European Economic Community was agreed upon by the six founding nations, was the Treaty of Rome (March 25, 1957). The principles of the treaty called for

- removal of trade barriers between the member nations,
- one [trade] policy toward non-member countries,
- promotion of free competition in the common market,
- mobility of capital, labor, etc.

Tariffs on industrial goods were cut by 40% in the following five years. In terms of trade and industrial activity the Common Market was very successful from the start.

1963 saw the merger between the EEC, the European Community for Coal and Steel, and Euratom, the agency for peaceful development of atomic energy. A joint executive, "The Commission", dealing with all the affairs of these three intereuropean agencies was set up in 1967. In 1967 a uniform sales tax system, the value added tax system, was introduced by the member nations. The customs union (for industrial goods) was completed in 1968.

The establishment of a free internal market for agricultural products proved to be a hard nut to crack. In all member nations protective measures for agricultural products and related industries had evolved during the preceding one hundred years. In earlier chapters it has been shown how such measures for instance influenced the German beet sugar industry in the second half of the nineteenth century to make Germany at the turn of the century a dominant factor on the world sugar market.

How negative agricultural policies can be for the industrial user of agricultural raw materials is attested by the fate of the oldest European lactic acid manufacturer: The autarkic measures imposed in the 1930's by the Hitler regime on agricultural commodities like potatoes and beet sugar more or less destroyed Boehringer's position as exporter of lactic acid. After World War II Boehringer had no fair chance of reestablishing itself as the leading lactic acid manufacturer, this time because of the agricultural policy of the German Federal Republic.

The Common Agricultural Policy (CAP) of the EEC

Agreement on a common agricultural policy was finally reached in 1966. The new policy brought the protective policies of the member nations under a common denominator:

The principles included protection of the income of the European farmer by guaranteeing fixed prices for most of his produce. High variable external tariffs were established as part of the system to restrict and control foreign imports of agricultural products. The variable part of the import duty on an agricultural commodity was, in principle, related to the difference in price guaranteed by the "Commission" and the world market price.

The CAP came into force in the summer of 1968. The success of the common agricultural policy was certainly as impressive as that of the common industrial policy established about ten years earlier. Agricultural production of all products falling under the intervention (= guaranteed prices) system grew in less than a decade to unprecedented levels, making the Community self-supporting for most agricultural commodities.

The regime for sugar differed on a few essential points from that for other agricultural commodities, like for instance grains and potatoes (the sources of starch). Beet sugar was the only surplus commodity due to the high financial rewards beet growers had gained for many decades in most European countries. National pricing and quota systems supported by import quota or high tariffs blocking imports of world market sugar were responsible for this situation.

The EEC only modified existing systems: it gave each sugar producing country a production quota, a certain amount for which the (yearly) fixed intervention was valid. In addition a second quota was given for which a lower intervention price was fixed. The first quota, the A quota, was more or less based on the estimated "normal" consumption, the second, the B quota, gave producers a certain leverage against yearly variations in production and consumption. Production exceeding the A plus B quotas were not bought by "Brussels", but had to be exported to the world market. This volume of sugar was called C sugar.

Understandably from the historical point of view intervention prices for sugar were set at values which ensured the North European farmer that sugar beet growing would remain the most rewarding activity. Protection against world market sugar was provided by the implementation of a variable duty, bringing the cost of world market sugar for the European consumer at a level somewhat above the intervention price. The Brussels Commission, aware of the problem of high sugar prices for the chemical industry, wrote in the preamble of Council Regulation No. 44/67/EEC of February 21, 1967, that a procedure should be implemented to provide [the chemical industry] a production restitution in the form of a total or partial exemption from the [variable import] levy. In case the national production did not provide excess [that is C] sugar for the chemical industry, that industry would be allowed to buy the required amount directly and duty free on the international market.

In Article 6, paragraph 2 of the Regulation principles for the elaboration of the production refund system for the chemical industry were given:

- a list of products had to be established for which the system would be applicable,
- a procedure for fixing the maximum amount of refund had to be laid down,
- unanimity in the council of ministers [the six governments] was required to establish the rules.

The Regulation also provided the possibility to obtain (supplementary) export refunds on the sales of manufactured products in case the production refund fell short of the intended cost reduction to world price level.

Council Regulation No. 215/67/EEC containing the worked out conditions under which the chemical industry could obtain sugar at near world market

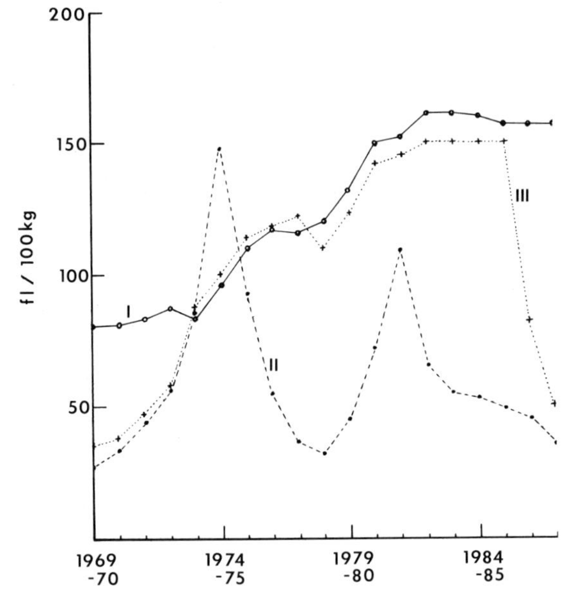

Fig. 12,5. EEC and world market sugar prices.

I EEC intervention price for sugar
II world market sugar price
III price paid by CCA

prices, was issued on June 27, 1967. A list with 41 product groups to which the refund system could be applied was attached to the Directive.

A particular clause in this Regulation specified that the level of production refund should not "discourage" the use of cereals and molasses in the manufacture of chemicals which could also be produced from these raw materials.

After intensive lobbying by the chemical industry, and, separately, by the starch industry, the rules for the protection of starch based products against identical or almost identical sugar based products was enacted in Council Regulation No. 765/68/EEC of June 18, 1968. This Regulation distinguished between products which could be made from sugar or starch (Annex I products), products for which sugar or molasses were suitable raw materials (Annex II products), and products for which sugar was the only feed stock in actual use (Annex III products).

The most important Annex I product was sorbitol, in Annex II citric acid was dominant. Annex III listed 33 product groups including lactic acid and derivatives. For Annex III products the sugar cost would stay close to world market price until the fall of 1973 when the world market price in a steep ascent surpassed the EEC intervention price.

The World Market Boom of 1974 and European Community Sugar Policy

Economic upheavals and depressions can, in hindsight, usually be very well explained by economists. The boom in commodity products of 1974 is no exception to this "rule". The commercial world, however, was taken by surprise, and, as usual, made things worse by overreacting. The chronology of events was:

1971: President Nixon decides to abolish the convertibility of the U.S. dollar into gold at a fixed price, and to devaluate the dollar. In 1973 the value of the dollar had dropped by 25% against strong European currencies like the Deutschmark and the Dutch guilder.

1972: Publication of the Forrester and Meadows report "The Limits to Growth", in which "evidence" was presented about the nearing exhaustion of mineral resources, and the coming shortage of energy.

October 1973: Egypt and Syria attack Israel in the Yom Kippur War; Arabian countries declare an oil boycott against the United States.

The oil boycott (and lowered Arabian production) led to panic on the world commodity market, grain prices doubled within months, sugar prices climbed to fivefold values. These prices far exceeded guaranteed European Community prices. As a first consequence the production refund on sugar was reduced to zero per October 1, 1973, a few days before the Yom Kippur War began.

For the next agricultural year (1974–1975) intervention prices were raised by 15%, partly to give the farmers compensation for the high cost of inflation, but also because of the prevailing view that the old days of unlimited availability of cheap energy would never return, and, consequently, renewable resources could fetch much higher prices. The agricultural lobby succeeded in the following two years to get additional large price increases despite the tumbling of world market sugar (and grain) prices to the low levels of 1970 and before. In July 1976 the sugar intervention price was fixed at a value 41% above the 1973–1974 level, more than two times the then registered world market price! (See Figure 12,5)

In September 1974, when sugar on the world market had reached the very high price of 0.76 $/kg the Commission had taken an important step by making a minor change in Regulation No. 1009/67/EEC:

- the paragraph reading "a production refund is given for the manufacture of certain products" was replaced by the formulation:
- "a production refund **can be given** for the manufacture", etc.

Reason: the scarcity of sugar on the world market and the difficult supply position in certain regions of the Community (preamble of the Regulation).

Unrelated to these events, but of great importance for the future of the European Community was the entrance of the United Kingdom (January 1, 1973). The British producer of lactic acid, Croda Bowmans Chemicals, now also became subject to the workings of the EEC agricultural policy.

The Original Production Refund System Did Not Return

Sources of information:

> The European Council of Chemical Manufacturers' Federations, CEFIC, Brussels;
> Nederlandse Suiker Raad, The Hague.

In November 1974 the world market reached its highest monthly average of 1.26 $/kg. A year later world market prices fell below the EEC intervention price, and already in 1976 sugar could again be bought on the international market for half the Common Market price!

Request from Dutch officials to the Commission to reinstitute the production refund for Annex III products initially were met with the remark that the chemical industry had drawn large benefits from the period of high world market prices when it was allowed to use much cheaper Common market sugar. These benefits could easily take care of the (supposedly) short period of low international prices.

In 1977 the answer was that EEC budget problems made the reinstitution of the sugar production refund highly unattractive. The clause in the Regulation: “a production refund can be given”, instead of “a production refund is given”, now was used for budgetary purposes, without any consideration of the impact on the sugar consuming industry. The argument that Common Market manufactured products were being replaced by similar products from outside producers was met with scepticism: the Commission would like to see import statistics proving such arguments. It could also be helpful when the industry produced estimates of the increase in consumption when sugar again was made available at near world market prices, Brussels’ bureaucrats added.

While the delaying tactics of the Commission kept the European producers of sugar based chemicals at bay imports from so-called third countries began to rise for most of the chemicals concerned[4]. Particularly illuminating are the Spanish exports of lactic acid to EEC destinations (See Table 12,III).

These exports were made by the relatively small firm of Luis Ayuso which had just built a new and modern lactic acid production unit with an initial capacity of over 3,000 t/a at Montmeló, 30 km north of Barcelona, and only 150 km from the EEC border.

Exports of the Ayuso company to Common Market countries evidently already began when world sugar prices were very high. Ayuso then paid Spanish inland prices, which differed not too much from internal EEC prices. When world market prices again dropped to low values, Ayuso benefitted from the Spanish export system, which, although not bringing the cost of sugar down to world market level, nevertheless gave the company quite a competitive edge over EEC producers of lactic acid. The Table shows the advantage the Spanish manufacturer drew from the situation.

In 1977 Ayuso further expanded production to a level of 4,000 or 5,000 t/a

Table 12.III. Spanish exports to the Common Market.

year	exports (t/a)	average sugar costs for CCA (fl/kg)	average world market price (fl/kg)
1973	0	0.73	0.72
1974	83	0.94	1.17
1975	292	1.07	1.21
1976	535	1.15	0.74
1977	627	1.19	0.46
1978	1098	1.16	0.34
1979	1298	1.17	0.38
1980	1818	1.33	0.59
1981	1247	1.43	0.91
1982	1689	1.47	0.87

Sources: Official Spanish Export Statistics; CCA Archives; CEFIC (Working Group Carbohydrates).

(trade rumors gave different figures). The total volume of exports reached 2,443 t in 1980 (official Spanish statistics), a figure which supports the estimate above.

A New Watered-down Production Refund System

Finally, in the spring of 1978, the Commission responded to the requests of the hard pressed manufacturers of sugar derivatives. The response was a less than halfhearted one, bringing hardly any relief to the industry:

As of August 1978 the industry became entitled to a production refund related to the height of the refund given to users of starch (according to the Annex I scheme of Regulation No. 765/68/EEC). The amount of the refund was made equivalent to the amount given to producers of starch based glucose. This worked out at Fl 0.116/kg, against a price differential with the world market of Fl 0.82/kg!

The reason for departing from the original refund system was the alleged interchangeability of corn and sugar for the manufacture of the chemical derivatives concerned (Preamble of Council Regulation No. 1400/78/EEC).

The starch lobby had vigorously presented this (not incorrect) view to defend the position of supplier of glucose and dextrose for industrial use it had built up during the period of high sugar prices.

The Commission, however, had fully neglected the fact that the Annex I products (for the manufacture of which starch and sugar were in 1967 considered to be interchangeable) of Regulation No. 765/68/EEC were protected against foreign imports by a special variable duty related to the difference between the EEC and the world market price of corn. Such a duty could not be imposed for example on lactic acid imports, because of the so-called GATT Agreement[5].

As a consequence of GATT, tariffs on Annex III products of Regulation No. 765/68/EEC could only be increased after GATT consultations. A discussion of a variable levy system was hardly feasible, and the whole idea of negotiating exemptions for a small number of products really conflicted with the (GATT) idea that agreements should be sought on general principles.

Lobbying to obtain sugar (and starch) for world market prices therefore went on. For years the officials of the Brussels Commission were exposed to arguments and data (like those presented in Figure 12,5 and Table 12,III) pointing at the impossibility of the chemical industry to defend its home market, the EEC, against third country competitors.

The Cost of Sugar Again Nears World Market Level

Only eight years later, in 1986, under mounting pressure from the industry, and from politicians who wanted to promote industrial activities based on "renewable resources", the Commission finally yielded. In Council Regulation No. 1010/86/EEC of March 25, 1986, the dire position of the chemical producers of sugar derivatives was not mentioned at all, however. The only reason given in the preamble for reintroducing a system allowing the chemical industry to obtain sugar at near world market prices was to promote sales of sugar [for derivatives production] to much higher levels. [CEFIC had presented the Commission forecasts on the use of renewable resources for industrial purposes showing a fivefold increase in sugar use to 500,000 t/a and a doubling of starch use to 2,600,000 t/a.]

A complication of the new system (which will not be treated here) is that the production red for sugar depends on that for corn. It still remains one of the basic principles of the EEC administration that the use of starch should not be discouraged by the sugar regime. A comprehensive discussion of these matters is given in a paper by Harris (Harris, 1987), entitled "The Provision of Carbohydrates for the European Community's Biotechnology and Chemical Industries".

The dramatic effect Regulation No. 1010/86/EEC had on sugar costs is shown in Figure 12,5. The Regulation made an end to a period in which EEC industries sold chemical derivatives of sugar and starch in their home market – the EEC – below cost price.

Rhône-Poulenc Tries to Make Synthetic Lactic Acid from Propylene

When in the 1960's thanks to the work of the ERRL (or was it the C.J. Patterson Company?) lactic acid was thought to have a good future as a chemical commodity, a new reaction, the manufacture of lactic acid by the oxidation of propylene with nitrogen peroxide, became a subject of industrial research. Two companies obtained patents on commercial manufacturing methods: BASF, German Patent 1,192,179, granted in 1965; and Rhône-Poulenc, French Patent 1,465,640, granted in 1966.

The research work of these companies was based on a reaction discovered in 1946 by Levy and Scaife, who demonstrated that lactic acid nitrates were formed in the reaction of dinitrogen tetroxide (nitrogen peroxide) with propylene (Levy and Scaife, 1946; see also Holten, 1971).

As far as has been recorded the Rhône-Poulenc company is the sole company which tried to bring the reaction to full commercialization. The original R.P. patent, demanded on December 5, 1965, followed by two additions on February 4, 1966, and May 11, 1966, gives a detailed description of the process:

Propylene gas, preferably admixed with oxygen, is passed at a temperature near the freezing point through a solution of about 16% nitrogen peroxide in concentrated nitric acid (70%). Propylene gas is absorbed quantitatively. Oxygen is blown through cocurrently with propylene to immediately oxidize lower nitrogen oxides formed in the reaction back to nitrogen peroxide.

In this addition reaction 1-nitratopropionic acid is formed. A plausible mechanism was found by Holten in Russian scientific literature of 1964 and later:

the first step is an addition of N_2O_4 to the double bond of the propylene molecule

$$CH_3\text{-}CH{=}CH_2 + \overset{+}{O}N \cdot \overset{-}{O}NO_2 \rightarrow CH_3\text{-}\underset{}{\overset{O\text{-}NO_2}{\overset{|}{C}}}H\text{-}CH_2\text{-}NO$$

the unstable nitroso compound formed rearranges its molecular structure and is then, probably under the influence of nitric acid, hydrolyzed to 1-nitratopropionic acid.

$$CH_2\text{-}\overset{O\text{-}NO_2}{\overset{|}{C}}H\text{-}CH_2\text{-}NO \rightarrow CH_3\text{-}\overset{O\text{-}NO_2}{\overset{|}{C}}H\text{-}COOH$$

The solution obtained is subjected to a vacuum distillation to remove (and recover) most of the nitrogen peroxide/nitric acid mixture. The residue of the distillation containing all of the nitratopropionic acid and small amounts of nitric acid, is neutralized and thereupon heated to 100° C to hydrolyze nitratopropionic acid to lactic acid.

$$CH_2\text{-}\overset{O\text{-}NO_2}{\overset{|}{C}}H\text{-}COOH \rightarrow CH_3\text{-}\overset{OH}{\overset{|}{C}}H\text{-}COOH$$

Purification of the crude lactic acid solution is achieved by extraction with diisopropyl ether (the Jenemann process). Other purification methods are described in later patents.

Filing of the corresponding British application took place on November 30, 1966. When the British patent finally was granted on August 14, 1968, the text contained a paragraph which did not appear in either the original French application, nor in the additions:

> It is to be noted that during the reaction of the nitrogen peroxide with the alpha-olefines leading to nitrates of alpha-hydroxycarboxylic acids, unstable products may be formed. Hence the risk exists that uncontrollable decompositions and even explosions may occur in certain cases.
>
> *- British Patent 1,123,147, 1968*

Why did Rhône-Poulenc put a warning for "uncontrollable decompositions and even explosions" in the British patent application?

Combined Oxalic/Lactic Acid Production

In the course of 1966 a pilot installation for the combined production of oxalic and lactic acid was installed at Saint Fons, a major manufacturing site of the company where also the lactic acid fermentation plant was located. Announcement was made of a commercial plant with capacities of 15,000 t/a of oxalic acid and 5,000 t/a of lactic acid (European Chemical News, April 1, 1966).

The preparation of oxalic acid starts in the same way as that of lactic acid: propylene is reacted with nitrogen peroxide to 1-nitratopropionic acid. The 1-nitratopropionic acid is not separated from nitric acid, but, with the aid of soluble iron or chromium catalysts directly hydrolyzed to oxalic acid (French Patent Application 62,973 of May 25, 1966)[6].

In June 1966 Schiedamsche Melkzuurfabriek was invited to discuss the consequences of the new synthetic lactic acid production by Rhône-Poulenc. Dirk Stapel and Kees Geitenbeek, Stapel's successor as sales director, were received at Rhône-Poulenc's headquarters in Paris by a six men delegation headed by Mr. Juin, managing director of the Société Normande. A follow-up meeting was agreed, but did not take place.

The reason for the cancelling of the follow-up meeting became only known to the other lactic acid producers in the spring of 1967 when Rhône-Poulenc informed Boehringer that an explosion had occurred in the pilot plant putting the future of the oxalic/lactic plant in jeopardy. Six people met their death in this event (Le Monde, April 29, 1977).

Rhône-Poulenc Finds an Improved Method

In 1969 Rhône-Poulenc took up contact again with the Dutch lactic acid producer CCA, which had in the meantime taken over the affairs of SMF. On November 27, 1969, a meeting took place at Schiphol airport: research on synthetic lactic acid production from propylene had been continued to the stage that the process was now believed to be fully under control. A commercial plant had been, or was about to be commissioned, and production was envisaged for mid-1972.

In early 1970 a copy of a press announcement was conveyed to CCA informing about the planned erection of a 20,000 t/a oxalic-lactic acid plant at Chalampé (Alsace). Not much later letters were sent to Boehringer and Croda Bowmans proposing these companies to cease production of lactic acid and to take future requirements from the new plant of Rhône-Poulenc (Copies of the negative responses of both companies were mailed to CCA.)

Further discussions with CCA took place on November 17, 1972: Delivery of the plant had been retarded because of extensive safety measures which had to be installed to prevent dangerous conditions and a repetition of the 1966 explosion.

When the installation neared completion in the summer of 1973 CCA was invited to visit the works at Chalampé. From Benninga's report of the visit he and van Velthuijsen paid to the plant on June 26, 1973:

> The Chalampé petrochemical complex, founded in 1956, now is in size the third largest of Rhône-Poulenc's chemical enterprises. The complex has its own nitric acid plant and already applies a nitric acid oxidation process in the manufacture of adipic acid from cyclohexanol (700 t/day).
>
> The oxalic/lactic plant consists of three separate units, which can be operated independently: in the first one propylene is oxidized with nitric acid to an intermediate [presumably 1-nitratopropionic acid] and nitric acid is recuperated. The second section is the oxalic acid production and purification plant. These two sections have been tested thoroughly and will be put into operation on June 28.
>
> In the third section [part of] the intermediate is converted into lactic acid. Separation of lactic from nitric acid is carried out by flash evaporation of the nitric. Residual amounts of foreign acids are taken out by passing the crude acid over anion exchange resins.
>
> *- Benninga, 1973*

The lactic acid section was expected to come on stream in August, about one month after the oxalic plant would be fully operative.

Advertizing was started; the October 19, 1973 issue of European Chemical News contained a full page announcement about the addition of a new lactic acid and oxalic acid plant to the "vast chemical network" of the Société des Usines Chimiques Rhône-Poulenc.

Production did not begin, however. Start up was delayed until November, and then a potential, large customer was informed that "difficulties of a chemical nature" had caused further delay. In February 1974 CCA approached Rhône-Poulenc to get an explanation of the confusion the actions of its French colleague caused in the market:

The explanation given (verbally) was that major difficulties were encountered in the separation of lactic acid from nitric acid by distillation. As this was a matter of separation technique only, and additional equipment already had been delivered, a solution to these problems was expected shortly.

A Second Explosion Hits Rhône-Poulenc

On March 26, 1974, the lactic acid section of the new plant was hit by an explosion, killing one (an electrician from a contracting firm) and injuring five. At the request of the C.F.D.T. (Confédération française démocratique du travail, a labor union) an official committee to investigate the causes of the accident was promptly set up. Although the lactic acid section was only slightly damaged, the whole installation, including the propylene oxidation and oxalic acid production sections were closed until the outcome of the investigations by the committee (Information from Rhône-Poulenc to CCA, April 2, 1974).

The result of the official inquiry was published on April 29, 1977, by the newspaper Le Monde:

> ... the explosion of March 26 [,1974,] was the result of an acceleration of the reaction of nitric acid with lactic acid, a reaction which takes place in an extremely complex chemical environment and of which the outcome is afflicted with a large margin of unpredictability. This unpredictability renders the control and mastering of the reaction uncertain.

The experts continued (Le Monde again):

> ... the complexity of the chemical reactions do not enable Rhône-Poulenc at the present state of knowledge and know-how to correct the observed anomalies in another way then by practical trials and adaptations.

Even after this verdict company officials hesitated to confirm that the synthetic lactic acid venture was abandoned. [The oxalic acid plant was restarted in 1975, or maybe even already in 1974, and has been in operation since.]

Fermentation Lactic Acid Holds Its Position

Rhône-Poulenc did not try again to make lactic acid synthetically from propylene. In 1982 the company built a new fermentation plant at Melle, for the exclusive production of D(-)-lactic acid, a building block for stereoisomeric herbicides. The old fermentation plant at Saint Fons was then closed. Ayuso and CCA supplied henceforth edible L(+)-lactic acid to Rhône-Poulenc.

From 1978 on CCA succeeded in gradually improving its position on the world market of lactic acid. The EEC sugar policy did not allow the company to make lactic acid for EEC destinations from sugar at world market cost. CCA therefore concentrated its sales efforts on countries outside the Common Market, in particular on the United States and Japan. Derivatives (salts and esters) were also promoted intensively, and better qualities of lactic acid were developed, matching the quality of synthetic lactic acid.

Table 12.IV. Production of lactic acid and lactates (calculated as 100%, in 1,000 t/a).

company	1962	1972	1982
Monsanto	–	6.4	6.7
Musashino	0.3	1.9	5.6
total synthetic	0.3	8.3	12.3
Luis Ayuso	0.6	1.0	3.0
(Croda-)Bowmans	1.8	1.8	1.6
SMF – CCA (biochem)	1.4	2.8	5.7
IQSF	–	1.2	3.3
Sheffield [8]	0.3	0.3	0.3
other producers	6.9	4.0	–
total fermentation	11.0	11.1	13.9
grand total	11.3	19.4	26.2
yearly growth (%/a)		5.5	3.0
% produced by fermentation	97	57	53

Source: Estimates by the author (1962 and 1972); the 1982 are averaged figures from Vickroy, 1985.

Table 12,IV shows how CCA, Ayuso and IQSF (the CCA affiliated company in Brazil) managed in taking over sales volumes of other fermentative producers of lactic acid between 1962 and 1972, and from then on even succeeded in taking part of the total growth of the market in lactic acid products. Since 1982 the trend has reversed: a bigger share of the market growth now goes to the fermentation industry than to the synthetic manufacturers (private information from CCA).

Around 1962 world production of lactic acid and derivatives was about 11,000 t (100%); the number of lactic acid plants in the non-communist world amounted to about 15. Average output was about 750 t.

At the time of writing (1989) the six remaining producers are responsible for a yearly production which has more than doubled to over 30,000 t; the average output of the five largest producers is about 5,000 t/a.

Three large fermentation plants of lactic acid belong to Centrale Suiker Maatschappij of the Netherlands. The story how CSM acquired ownership of these plants (CCA in Holland, Ayuso in Spain, and IQSF in Brazil) began with the purchase by the Dutch ex-colonial enterprise HVA of a very primitive lactic acid production unit in Campos, Brazil. The founding of CCA in 1969 made CSM co-owner of the lactic acid plants of HVA at Schiedam and Campos. The next thirteen years were highly turbulent for CCA, in the first years of this period the two parent companies had to absorb very substantial losses. After 1976 financial results remained poor due to high sugar costs.

In 1982 CSM became the sole owner of CCA, when HVA in a process of eliminating all of its industrial and trading activities to become an investment company only, sold its 50% share to its partner. At about the same time results of CCA began to improve, thanks to the economy of scale reached in the preceding years and the closing of the old Schiedam factory (1978).

In 1986 CSM acquired a 75% share in the Ayuso company, the remaining 25% was bought in 1988. As a consequence about 45% of the world production of lactic acid is now concentrated in the hands of one company, CSM. The size of each of its lactic acid plants is comparable to that of the two synthetic manufacturers Monsanto and Musashino[7]. The future therefore seems to be open for both ways of producing lactic acid; market approach, the introduction of new technologies and the costs of raw materials will remain as decisive factors as they have been in the past in determining the future of lactic acid making.

Notes

[1] Already in 1935 a proposal was made to market preferably dextrorotatory acid for edible use, because of "its function in animal metabolism", and the anticipated increased use (Tatum and Peterson, 1935).

[2] Many years later esters of L(+)-lactic acid became of interest for herbicide manufacturers to produce stereoisomeric herbicides instead of racemic products of which the other stereoisomer has no herbicidal properties. In this way a 50% reduction in use could be achieved, an important reduction from an environmental point of view.

[3] This was the last technical meeting of ILRA. The Association itself was dissolved in 1974.

[4] Third countries: a trade term for non-EEC countries.

[5] In 1947 a number of industrialized countries had signed "The General Agreement on Tariffs and Trade", GATT, in which principles were laid down to prevent protectionism and to promote international economic cooperation. In a reaction to the economic integration taking place in the European Common Market, and certain signs of protectionism that went along with it, President John F. Kennedy proposed in 1962 a special conference of GATT to negotiate general tariff reductions. The conference was concluded in 1967 with agreements on average tariff reductions of 35% on industrial goods, to be implemented over a five year period.

[6] In the presence of nitric acid, hydrolysis of nitratopropionic acid to lactic acid is not possible. This is the reason that the mixture of nitric acid and nitric oxides has to be distilled off before the hydrolysis of 1-nitratopropionic acid to lactic acid can be effectuated.

[7] In 1986 Monsanto sold the Texas City plant in a leveraged buyout to a group of Houston investors (Gillentine, 1987). Management and (remaining) employees of the plant were involved in the deal as well by promising them a share in future profits. The new company, Sterling Chemicals, in an effort to quickly improve its debt/equity ratio, made a drastic cut in overhead expenses. It does not seem likely that Sterling Chemicals will spend more money on market and process development of lactic acid than Monsanto did in the past.

[8] Sheffield closed its calcium lactate plant in 1987.

References

Ballabriga, A., Conde, C., and Gallart-Catala, A., **1970.** *Helv. Paediat. Acta*, **25**, 25.

Chem. Age (London), **1971**, June 7.

Droese, W., and Stolley, H., **1963**. *Paediatr. Fortbildungskurse prax.*, **2**, 329.

Eur. Chem. News, **1966**, April 1.

FDA, **1975**. *Scientific Literature Reviews on Generally Recognized as Safe (GRAS) Food Ingredients – Lactic acid*; U.S. Department of Commerce: Springfield, VA.

FDA, **1978**. *Evaluation of the Health Aspects of Lactic Acid and Calcium Lactate as Food Ingredients*; U.S. Department of Commerce: Springfield, VA.

Gillentine, C., **1987**. *Chem. Bus.*, October 1987.

Goldman, H.J., Karelitz, S., Seifter, E., Acs, H., and Schell, N.B., **1961**. *Pediatrics,* **27**, 921.

Harris, S.A., **1987**. *J. Agri. Econ. Res.*; **38**, 423.

HVA, **1975**. *Contribution of the HVA-Organization and its Head Office in Amsterdam to the Financial Results of the Sugar Industry in Ethiopia*; Internal report of HVA: Amsterdam.

Jacobs, H.M., and Christian, J.R., **1957**. *J. Lancet*, May 1957, 157.

Levy, N., and Scaife, C.W., **1946**. *J. Chem.Soc.*, **1946**, 1100.

Tatum, E.L., and Peterson, W.H., **1935**. *Ind. Eng. Chem.*, **27**, 1493.

Vickroy, T.B., **1985**. *Lactic Acid*, Ch.38 in Vol. 3 of *Comprehensive Biotechnology*; Ed. Moo-Young, M.; Pergamon: Oxford.

W.H.O., **1972**. *Evaluation of Food Additives, Fifteenth Report of the Joint FAO/WHO Expert Committee on Food Additives*; World Health Organization: Geneva.

Subject Index

(Includes chemical names, geographical and company names)

Name Index

(bold print indicates major reference)